Sophie Goldie
Rose Jewell

A LEVEL

MATHEMATICS

FIRST AID KIT

- Support for challenging topics
- Extra practice for essential skills
- Suitable for all A-level maths specifications

Acknowledgements

Every effort has been made to trace all copyright holders, but if any have been inadvertently overlooked, the Publishers will be pleased to make the necessary arrangements at the first opportunity.

Although every effort has been made to ensure that website addresses are correct at time of going to press, Hodder Education cannot be held responsible for the content of any website mentioned in this book. It is sometimes possible to find a relocated web page by typing in the address of the home page for a website in the URL window of your browser.

Hachette UK's policy is to use papers that are natural, renewable and recyclable products and made from wood grown in well-managed forests and other controlled sources. The logging and manufacturing processes are expected to conform to the environmental regulations of the country of origin.

Orders: please contact Bookpoint Ltd, 130 Park Drive, Milton Park, Abingdon, Oxon OX14 4SE. Telephone: +44 (0)1235 827827. Fax: +44 (0)1235 400401. Email education@bookpoint.co.uk Lines are open from 9 a.m. to 5 p.m., Monday to Saturday, with a 24-hour message answering service. You can also order through our website: www.hoddereducation.co.uk

ISBN: 978 1 510 482 401

First published in 2020 by:

Hodder Education,

An Hachette UK Company

Carmelite House

50 Victoria Embankment

London EC4Y 0DZ

www.hoddereducation.co.uk

Impression number 10 9 8 7 6 5 4 3 2 1

Year 2024 2023 2022 2021 2020

Illustrations by Integra Software Services Pvt. Ltd., Pondicherry, India.

Typeset in Integra Software Services Pvt. Ltd., Pondicherry, India.

Printed in India.

A catalogue record for this title is available from the British Library.

Contents

1 Algebra

Algebra review 1 2
Algebra review 2 4
Indices and surds 6
Quadratic equations 8
Inequalities 10
Completing the square 12
Algebraic fractions 14
Proof 16

2 Further algebra

Functions 18
Polynomials 20
Exponentials and logs 22
Straight lines and logs 24
Sequences and series 26
Binomial expansions 28
Partial fractions 30

3 Graphs

Using coordinates 32
Straight line graphs 34
Circles 36
Intersections 38
Transformations 40
Modulus functions 42
Vectors 44
Numerical methods 46
The trapezium rule 48

4 Trigonometry

Trigonometry review 50
Triangles without right angles 52
Working with radians 54
Trigonometric identities 56
Compound angles 58
The form $r\sin(\theta + \alpha)$ 60

5 Calculus

Differentiation 62
Stationary points 64
Integration 66
Extending the rules 68
Calculus with other functions 70
The chain rule 72
Product and quotient rules 74
Integration by substitution 76
Integration by parts 78
Further calculus 80
Parametric equations 82

6 Statistics

Sampling and displaying data 1 84
Displaying data 2 86
Averages and measures of spread 88
Cumulative frequency graphs and boxplots 90
Grouped frequency calculations 92
Probability 94
Conditional probability 96
Discrete random variables 98
Binomial distribution 100
Hypothesis testing (binomial) 102
Normal distribution 104
Hypothesis testing (normal) 106
Bivariate data 108

7 Mechanics

Working with graphs 110
suvat equations 112
Variable acceleration 114
Understanding forces 116
Forces in a line 118
Using vectors 120
Resolving forces 122
Projectiles 124
Friction 126
Moments 128

Answers 130

A Level Mathematics: First Aid Kit

Get the most from this book

Welcome to A Level Mathematics: First Aid Kit. This book will provide you with clear explanations of the basics of every A level Mathematics topic, with examples and activities to give you a step up to the A level content you may be struggling with. With 64 clearly signposted topics, targeted, practical help is available exactly where you need it.

Included in the purchase of this book is valuable online material that provides feedback on each of the multiple-choice questions included in the Skill builder exercises. Many pages also have a link to an online resource, produced in collaboration with Casio UK, that provide activities and strategies for using your graphical calculator to improve your understanding of that topic. They can be found at https://calculators.casio.co.uk/hodder/fa-resources

Features to help you succeed

The lowdown

At the start of each topic there is a set of key points that explain the basics in plain language.

Rewind and fast forward

Topics are linked to previous ones where the content is needed – you can work backwards from the problem you feel to the root cause of the problem or skip ahead to see where the topic leads to.

Example

A short, clear example of the key point in the lowdown.

Hint

Expert tips are given throughout the book to support your thinking.

Watch out!

These callouts tell you the common mistakes and misconceptions, so you know how to avoid them!

Get it right

These longer worked examples give a step-by-step guide and full worked solutions to longer examples with hints to guide your thinking. You might want to try the question on your own first before you look at the solution.

QR code for the Casio resources

Use your smartphone to go straight to the resource for the page you are working through. There are ideas and investigations to help you build your understanding with your graphical calculator. Many of the ideas are also suitable for a scientific calculator or graph drawing software.

You are the examiner ✪

This is your chance to assess whether a solution is right and to find errors and misconceptions in someone else's work. This will help you to overcome any misunderstandings you may have and to develop checking strategies.

Skill builder ✓

The short exercise provides practice questions at an introductory level for each topic. There are routine questions to consolidate understanding and build confidence. There are multiple-choice questions that have been carefully designed to highlight common mistakes and misconceptions. Some of the more difficult questions have additional hints to help you through to success.

Answers

Answers to all questions are provided in the back of this book. Worked solutions and answers to the rest of the questions are provided in the back of this book.

Algebra review 1

Year 1

THE LOWDOWN

① In **algebra**, symbols (called **variables**) are used to represent numbers.
A number on its own is a called a **constant.**
A **term** is made from numbers and variables alone or multiplied together or on their own.
An **expression** is made by adding or subtracting one or more terms. It has no =.

Example $5 - \frac{3xy}{2} + x^2$ is an expression with three terms.

Hint: You can use letters (such as n, x, y or θ) to represent:
- any number (a **variable**)
- an **unknown** number that you want to work out.

② You can **expand brackets** by multiplying them out.

Example Expand and simplify $(5x + 4)(3x - 2)$

	3x	−2
5x		
+4		

→

	3x	−2
5x	$15x^2$	$-10x$
+4	$12x$	-8

So $(5x + 4)(3x - 2) = 15x^2 + 2x - 8$

Hint: Multiply each term in the second bracket by each term in the first.

③ You can **factorise** an expression by rewriting it with brackets.

Example Factorising $8x^2y^2 - 12xy^3$ gives $4xy^2(2x - 3y)$

Hint: 8 and 12 have a **common factor** of 4
x and x^2 have a common factor of x
y^3 and y^2 have a common factor of y^2.

④ You can put one expression equal to another to make an **equation.**
An equation has an = and is usually only true for certain values.

Example Solve

$$4x + 3 = 2 - 5x$$
$+5x$ (both sides)
$$9x + 3 = 2$$
-3 (both sides)
$$9x = -1$$
$\div 9$ (both sides)
$$x = -\frac{1}{9}$$

Watch out! Keep the equation balanced by doing the **same** to both sides!

Hint: Check you are right by substituting your answer back into the equation.

⑤ An **identity** is an equation that is true for all values of x.
The symbol ≡ is used to show that an equation is an identity.

Example $3(x + 1) \equiv 3x + 3$

GET IT RIGHT

a) Solve $\frac{3}{x-2} = \frac{4}{2x}$.

b) Factorise $15x^2 - 10xy + 12x - 8y$.

Solution:

a) Clear fractions: $\frac{3}{x-2} = \frac{4}{2x}$

Expand brackets: $3(2x) = 4(x - 2)$

$6x = 4x - 8 \Rightarrow 2x = -8 \Rightarrow x = -4$

b) There is no common factor… $15x^2 - 10xy + 12x - 8y$
so factorise pairs of terms: $= 5x(3x - 2y) + 4(3x - 2y)$
Now factorise fully: $= (5x + 4)(3x - 2y)$

▶▶ See page 8 for using the **grid method to factorise**.

Hint: You can check you are right by expanding the brackets. Use a grid to help you.

YOU ARE THE EXAMINER

Sam and Lilia have both made some mistakes in their maths homework.
Which questions have they got right?
Where have they gone wrong?

SAM'S SOLUTION

1 Simplify $(5x + 1) - (x - 2)$
$(5x + 1) - (x - 2) = 5x + 1 - x - 2$
$= 4x - 1$

2 Simplify $3(ab)$
$3(ab) = 3ab$

3 Factorise $12x^2y - 8x^4y^3 + 4xy$
$12x^2y - 8x^4y^3 + 4xy = 4xy(3x - 2x^2y^2)$

4 Expand $(3x - 2)(x - 5)$
$(3x - 2)(x - 5) = 3x^2 - 2x - 15x + 10$
$= 3x^2 - 17x + 10$

5 Solve $\frac{x+3}{2} - \frac{2x+5}{3} = 4$
$6 \times \frac{x+3}{2} - 6 \times \frac{2x+5}{3} = 6 \times 4$
$3(x+3) - 2(2x+5) = 24$
$3x + 9 - 4x - 10 = 24$
$-x = 25$
$x = -25$

LILIA'S SOLUTION

1 Simplify $(5x + 1) - (x - 2)$
$(5x + 1) - (x - 2) = 5x + 1 - x + 2$
$= 4x + 3$

2 Simplify $3(ab)$
$3(ab) = 3a \times 3b$
$= 9b$

3 Factorise $12x^2y - 8x^4y^3 + 4xy$
$12x^2y - 8x^4y^3 + 4xy =$
$4xy(3x - 2x^3y^2 + 1)$

4 Expand $(3x - 2)(x - 5)$
$(3x - 2)(x - 5) = 3x^2 - 2x + 15x - 10$
$= 3x^2 + 13x - 10$

5 Solve $\frac{x+3}{2} - \frac{2x+5}{3} = 4$
$3(x+3) - 2(2x+5) = 4$
$3x + 9 - 4x - 10 = 4$
$-x = 5$
$x = -5$

SKILL BUILDER

1 Simplify these expressions.

a) $5x(4xy - 3) + 3y(2x - y)$ b) $2x(3x - 3) - (2 - 5x)$ c) $3x(1 - 2x) - 3x(4x - 2)$

Hint: Be careful when there is a negative outside the bracket.

2 Expand and simplify.

a) $(x + 5)(x - 4)$ b) $(2x + 3)(3x - 5)$ c) $(4x - 2y)(2y - 3x)$

Hint: Use the grid method to help you.

3 Factorise each expression fully.

a) $12a^2 + 8a$ b) $8b^2c - 4b$ c) $3bc - 2b + 6c^2 - 4c$

Hint: Make sure the terms inside the brackets in your answer don't have a common factor.

4 Solve these equations.

a) $5(2x - 4) = 13$ b) $4(x + 3) = 5(3 - x)$ c) $3(2x + 1) - 2(1 - x) = 2(9x + 4)$

Hint: Expand any brackets first.

5 Solve these equations.

a) $\frac{x}{4} = \frac{2x - 1}{3}$ b) $\frac{1}{4x + 3} = \frac{2}{5 - x}$ c) $\frac{x}{5} - 6 = \frac{2x + 3}{3}$

Hint: Clear any fractions first.

Algebra review 2

Year 1

THE LOWDOWN

① **Simultaneous equations** are two equations connecting two unknowns.

② To solve simultaneous equations, **substitute** one equation into the other.

Example $2x + y = 10$ and $y = 3 - x$

Substituting for y gives $2x + 3 - x = 10 \Rightarrow x = 7$

Since $y = 3 - x$ then $y = 3 - 7 = -4$

Or you can add/subtract one equation to **eliminate** one of the unknowns

Example

$$2x + 3y = 27$$
$$+ \; x - 3y = 9 \quad ①$$
$$3x = 36 \Rightarrow x = 12$$

From ① $12 - 3y = 9$ so $y = 1$

③ A **formula** is a rule connecting two or more variables.

The **subject** of a formula is the variable calculated from the rest of the formula.

The subject is a letter on its own on one side and is not on both sides of the =.

Example $v = u + at$ or $s = ut + \frac{1}{2}at^2$

④ You can **rearrange** a formula to make a different variable the subject.

Rearrange the formula in the same way you would solve an equation.

Example $v = u + at$

Swap sides: $u + at = v$

Subtract u from both sides: $at = v - u$

Divide both sides by t: $a = \frac{v - u}{t}$

Watch out! Your calculator may solve equations for you but in the exam you may lose marks if you don't show enough working!

Hint: Don't forget to find the values of both unknowns.

Check your answer works for both equations.

GET IT RIGHT

a) Solve $3a + 2b = -5$ and $4a + 3b = -4.5$.

b) Make x the subject of $y = \frac{2x + b}{x - 3c}$.

Solution:

a) To eliminate b, multiply each equation to get $6b$ in both.

Then subtract one equation from the other.

$3a + 2b = -5$ ① ① × by 3 $9a + 6b = -15$

$4a + 3b = -4.5$ ② ② × by 2 $-\; 8a + 6b = -9$

$a = -6$

Use equation ① to find b: $3 \times -6 + 2b = -5$

$2b = 13 \Rightarrow b = 6.5$

b) Clear the fraction: $y(x - 3c) = 2x + b$

Expand the brackets: $xy - 3cy = 2x + b$

Gather x terms on one side: $xy - 2x = b + 3cy$

Factorise: $x(y - 2) = b + 3cy$

Divide by $(y - 2)$: $x = \frac{b + 3cy}{y - 2}$

▶▶ See page 18 for **inverse functions** and page 38 for **more on simultaneous equations**.

Hint: You can **eliminate** an unknown if both equations have the **same** amount of that unknown.

Look at the signs of that unknown:

Different signs: ADD Same signs: SUBTRACT

YOU ARE THE EXAMINER

Mo and Lilia have both missed out some work in their maths homework. Complete their working.

MO'S SOLUTION

Make a the subject of $s = ut + \frac{1}{2}at^2$

Swap sides: $ut + \frac{1}{2}at^2 = s$

Subtract ut from both sides:

$\frac{1}{2}at^2 = \square$

Multiply both sides by $\square$:

$at^2 = \square$

Divide both sides by $\square$:

$a = \square$

LILIA'S SOLUTION

Solve $3x + 4y = -1$ ①

$6x - 5y = -8.5$ ②

Multiply equation ① by 2: $\square$ ③

Use ③ and ① to eliminate x:

$\square$

$\square \quad 6x - 5y = -8.5$

$\square = \square$

So $y = \square$

Using equation ①:

$3x + 4 \times \square = -1$

which gives $x = \square$

SKILL BUILDER

1 Which one of the following is the correct x-value for the linear simultaneous equations $5x - 3y = 1$ and $3x - 4y = 4$?

A $x = \frac{16}{29}$ **B** $x = \frac{16}{11}$ **C** $x = -\frac{8}{11}$ **D** $x = -\frac{3}{11}$ **E** $x = \frac{5}{29}$

2 Which one of the following is the correct y-value for the linear simultaneous equations $5x - 2y = 3$ and $y = 1 - 2x$?

A $y = -9$ **B** $y = \frac{5}{9}$ **C** $y = \frac{7}{9}$ **D** $y = -\frac{1}{9}$ **E** $y = -\frac{7}{3}$

3 Solve **a)** $2a + b = 14$, $a = 2b - 3$ **b)** $2c + 3d = 14$, $4c + 3d = 10$ **c)** $5e - 3f = 11$, $4e + 2f = 18$

4 Decide whether each statement is true (T) or false (F).

a) $y = x + \frac{5}{2} \Rightarrow 2y = x + 5$

Hint: Substitute in values for x to check your answers.

b) $y = 9x^2 \Rightarrow y = (3x)^2$

c) $\frac{2}{3}(x - 4) = 6 \Rightarrow 2(3x - 12) = 18$

5 Make y the subject of $ax + by = c$.

6 Make x the subject of $y = \frac{ax + 1}{x + by}$.

Hint: Look at the 'Get it right' box.

7 Make x the subject of $y = \frac{5}{\sqrt{a^2 - x^2}}$.

Hint: Follow these steps:

1 Multiply both sides by $\sqrt{a^2 - x^2}$.
2 Divide both sides by y.
3 Square both sides to remove the square root.
4 Make x^2 the subject and then square root.

Indices and surds

GCSE Review/Year 1

THE LOWDOWN

① 3^4 means $\underbrace{3 \times 3 \times 3 \times 3}_{4 \text{ of them}}$; 4 is the **index** or **power** and 3 is the base.

② **Laws of indices**

$a^n = \underbrace{a \times a \times \ldots \times a}_{n \text{ of them}}$ $\qquad a^0 = 1$ $\qquad a^1 = a$

$a^m \times a^n = a^{m+n}$ $\qquad \dfrac{a^m}{a^n} = a^{m-n}$ $\qquad (a^m)^n = a^{m \times n}$

③ **Fractional and negative indices**

$a^{-n} = \dfrac{1}{a^n}$ $\qquad a^{\frac{1}{n}} = \sqrt[n]{a}$ $\qquad a^{\frac{m}{n}} = \sqrt[n]{a^m} = \left(\sqrt[n]{a}\right)^m$

Remember $\sqrt{\ }$ means the positive square root only.

④ You can use **roots** to solve some equations involving powers.

Example $\quad x^4 = 20 \Rightarrow x = \pm\sqrt[4]{20} = \pm 2.11$ (to 3 s.f.)

⑤ $\sqrt{2} = 1.414213\ldots$ the decimal carries on forever and never repeats.
$\sqrt{2}$ is a **surd**; a surd is a number that is left in square root form.
When you are asked for an **exact answer** then you often need to use surds.

⑥ To **simplify** a surd use square factors of the number under the root.

Example $\quad \sqrt{72} = \sqrt{36 \times 2} = 6\sqrt{2}$

⑦ **Useful rules:**

$\sqrt{xy} = \sqrt{x}\sqrt{y}$ $\qquad \sqrt{x} \times \sqrt{x} = x$ $\qquad (x + y)(x - y) = x^2 - y^2$

⑧ A fraction isn't in its simplest form when there is a surd in the bottom line.
To simplify it you need to **rationalise the denominator**.

Example $\quad \dfrac{3}{\sqrt{5}} = \dfrac{3 \times \sqrt{5}}{\sqrt{5} \times \sqrt{5}} = \dfrac{3\sqrt{5}}{5}$

Hint: You might like this memory aid: $a^{\frac{m}{n}} = \sqrt[n]{a^m}$.
A fractional index is like a flower, the bottom's the root and the top's the power!

Watch out!
When the power is even then there are two roots: one positive and one negative.
When the power is odd, there is only one root.

▶▶ See page 8 for the **difference of two squares**.

GET IT RIGHT

Rationalise the denominator of each fraction: **a)** $\dfrac{14}{3 + \sqrt{2}}$ **b)** $\dfrac{1}{5 - 2\sqrt{3}}$

Solution:

Step 1 a) $\dfrac{14}{(3 + \sqrt{2})}$

$= \dfrac{14(3 - \sqrt{2})}{(3 + \sqrt{2})(3 - \sqrt{2})}$

Step 2 $= \dfrac{14(3 - \sqrt{2})}{3^2 - (\sqrt{2})^2}$

Step 3 $= \dfrac{14(3 - \sqrt{2})}{9 - 2}$

$= \dfrac{14(3 - \sqrt{2})}{7}$

$= 2(3 - \sqrt{2})$

b) $\dfrac{1}{(5 - 2\sqrt{3})}$

$= \dfrac{1(5 + 2\sqrt{3})}{(5 - 2\sqrt{3})(5 + 2\sqrt{3})}$

$= \dfrac{(5 + 2\sqrt{3})}{5^2 - (2\sqrt{3})^2}$

$= \dfrac{(5 + 2\sqrt{3})}{25 - 4 \times 3}$

$= \dfrac{5 + 2\sqrt{3}}{13}$

Hint: Step 1: Multiply 'top' and 'bottom' lines by the 'bottom line with the opposite sign'. Don't forget brackets!
Step 2: Multiply out the brackets. Remember $(x + y)(x - y) = x^2 - y^2$
Step 3: Simplify.

YOU ARE THE EXAMINER

Sam and Nasreen have both made some mistakes in their maths homework. Which questions have they got right? Where have they gone wrong?

SAM'S SOLUTION

1 Solve $x^3 + 1 = 28$

$x^3 = 27$

Cube root: $x = \pm 3$

2 Simplify $\frac{4ab^4}{(2a^2b)^3}$

$= \frac{4ab^4}{8a^6b^3} = \frac{b}{2a^5}$

3 Simplify $\sqrt{147} - \sqrt{27}$

$\sqrt{147} - \sqrt{27} = \sqrt{120} = \sqrt{4 \times 30} = 2\sqrt{30}$

4 Evaluate $64^{-\frac{3}{2}}$

$64^{-\frac{3}{2}} = \frac{1}{64^{\frac{3}{2}}} = \frac{1}{(\sqrt{64})^3} = \frac{1}{8^3} = \frac{1}{512}$

NASREEN'S SOLUTION

1 Solve $x^3 + 1 = 28$

$x^3 = 27$

Cube root: $x = 3$

2 Simplify $\frac{4ab^4}{(2a^2b)^3}$

$= \frac{4ab^4}{2a^6b^3} = 2a^5b$

3 Simplify $\sqrt{147} - \sqrt{27}$

$\sqrt{147} - \sqrt{27} = \sqrt{49 \times 3} - \sqrt{9 \times 3}$

$= 7\sqrt{3} - 3\sqrt{3} = 4\sqrt{3}$

4 Evaluate $64^{-\frac{3}{2}} = 64^{\frac{2}{3}} = (\sqrt[3]{64})^2 = 4^2 = 16$

SKILL BUILDER

*Don't use your calculator; only use it to **check** your answers!*

1 Find the value of $\left(\frac{1}{3}\right)^{-2}$

A -9 **B** $\frac{1}{9}$ **C** 9 **D** $-\frac{1}{9}$ **E** $-\frac{2}{3}$

2 Find the value of $\frac{(2x^4y^2)^3}{10\left(x^3\sqrt{y^5}\right)^2}$, giving the answer in its simplest form.

A $\frac{1}{5}x^6y$ **B** $\frac{2}{25}x^6y$ **C** $\frac{4x^6}{5y^4}$ **D** $\frac{4}{5}x^6y$ **E** $\frac{8x^{12}y^6}{10x^6y^5}$

3 Simplify $(2 - 2\sqrt{3})^2$, giving your answer in factorised form.

A 16 **B** $8(2 - \sqrt{3})$ **C** $-8(1 + \sqrt{3})$ **D** $16 - 8\sqrt{3}$ **E** $4(4 - \sqrt{3})$

4 Decide whether each statement is true or false.
Write down the correct statement, where possible, when the statement is false.

a) $-4^2 = 16$ b) $\sqrt{9} = \pm 3$ c) $\sqrt{a} + \sqrt{b} \equiv \sqrt{a+b}$

d) $\sqrt{16a^2b^4} \equiv 4ab^2$ e) $\sqrt{a^2 + b^2} \equiv a + b$ f) $(\sqrt{a} + \sqrt{b})(\sqrt{a} - \sqrt{b}) \equiv a - b$

5 Rationalise the denominator:

a) $\frac{4}{\sqrt{2}}$ b) $\frac{4}{3 + \sqrt{5}}$ c) $\frac{2}{6 - 3\sqrt{2}}$

6 Solve these equations.

a) $x^2 - 49 = 15$ b) $5x^3 - 27 = 13$ c) $10x(x - 4) = 4(1 - 10x)$

Quadratic equations

Year 1

THE LOWDOWN

① A **quadratic equation** can be written in the form $ax^2 + bx + c = 0$.

② You can solve some quadratic equations by factorising.

Example Solve $x^2 - 10x + 16 = 0$

$ac = 1 \times 16 = 16$ and $b = -10$, so $-10x$ splits to $-2x - 8x$

x^2	$-2x$
$-8x$	16

➡

	x^2	$-2x$
x		
-8	$-8x$	16

➡

	x	-2
x	x^2	$-2x$
-8	$-8x$	16

x is the HCF of x^2 and $-8x$

-8 is the HCF of $-8x$ and 16

From the grid: $(x-2)(x-8) = 0$

So $x = 2$ or $x = 8$

Hint: Use a grid and split the middle term. Look for two numbers that multiply to give ac and add to give b. Take out the common factors.

③ Make sure you can recognise these special forms:

- The difference of two squares: $x^2 - a^2 = (x + a)(x - a)$.
- Perfect squares: $x^2 + 2ax + a^2 = (x + a)^2$.

④ Watch out for quadratic equations in disguise!

Rewrite them as a quadratic equation first.

Example Solve $2^{2y} - 10 \times 2^y + 16 = 0$

Let $x = 2^y$, so $x^2 - 10x + 16 = 0$

Since $x = 2$ or $x = 8$ then $2^y = 2$ or $2^y = 8$

So $y = 1$ or $y = 3$

Hint: You can write 2^{2y} as $(2^y)^2$.

▶▶ See page 12 for **completing the square** and page 20 for **factorising cubic equations**.

GET IT RIGHT

Show that there is only one solution to $2y + 5\sqrt{y} - 3 = 0$.

Solution:

Let $x = \sqrt{y}$, so the equation is $2x^2 + 5x - 3 = 0$

$ac = 2 \times -3 = -6$ and $b = 5$, so $5x$ splits to $-x + 6x$

$2x^2$	$6x$
$-x$	-3

➡

	$2x^2$	$6x$
$2x$		
-1	$-x$	-3

➡

	x	$+3$
$2x$	$2x^2$	$6x$
-1	$-x$	-3

From the grid: $(x+3)(2x-1) = 0$

So $x = -3$ or $x = \frac{1}{2}$

So $\sqrt{y} = -3$ which is impossible,

or $\sqrt{y} = \frac{1}{2} \Rightarrow y = \frac{1}{4}$, so there is only one solution.

Hint: Since $y = (\sqrt{y})^2$,

$2(\sqrt{y})^2 + 5\sqrt{y} - 3 = 0$

$2x^2 \quad + 5x \quad - 3 = 0$

Watch out: Make sure you find the solutions for y and not just for x.

Watch out: $\sqrt{\ }$ means the **positive square root**, so it is never negative.

YOU ARE THE EXAMINER

Which one of these solutions is correct? Where are the errors in the other solution?

Solve $5x^4 = 10x^2$

LILIA'S SOLUTION

$5x^4 = 10x^2$

Divide by $5x^2$: $x^2 = 2$

Take the square root: $x = \sqrt{2}$

PETER'S SOLUTION

Let $y = x^2$, so $5y^2 = 10y$

Rearrange: $5y^2 - 10y = 0$

Factorise: $5y(y - 2) = 0$

Solve: $y = 0$ or $y = 2$

So $x^2 = 0$ or $x^2 = 2$

Take the square root: $x = 0$ or $x = \pm\sqrt{2}$

SKILL BUILDER

1 Which of the following is the solution of the quadratic equation $x^2 - 5x - 6 = 0$?

A $x = -6$ or $x = 1$ **B** $x = -2$ or $x = 3$ **C** $x = 2$ or $x = 3$

D $x = -3$ or $x = 2$ **E** $x = -1$ or $x = 6$

2 Factorise $6x^2 + 19x - 20$

A $(x + 4)(6x - 5)$ **B** $(3x + 10)(2x - 2)$ **C** $(x + 20)(6x - 1)$

D $(3x + 4)(2x - 5)$ **E** $(6x + 10)(x - 2)$

3 Which of the following is the solution of the quadratic equation $2x^2 - 9x - 18 = 0$?

A $x = \frac{3}{2}$ or $x = -6$ **B** $x = \frac{9}{2}$ or $x = -2$ **C** $x = 6$ or $x = 3$

D $x = -\frac{9}{2}$ or $x = 2$ **E** $x = -\frac{3}{2}$ or $x = 6$

4 Factorise.

a) $x^2 + 7x + 12$ b) $x^2 - 2x - 15$ c) $x^2 + 6x + 9$

d) $x^2 - 12x + 36$ e) $x^2 - 49$ f) $4x^2 - 100$

g) $2x^2 + 7x + 3$ h) $6x^2 + x - 15$ i) $9x^2 - 12x + 4$

5 Solve these equations by factorising.

a) $x^2 + 3x - 10 = 0$ b) $2x^2 - x - 1 = 0$ c) $2x^2 = 11x - 12$

d) $4x^2 = 5x$ e) $4x^2 + 25 = 20x$ f) $9x^2 = 25$

6 Solve these equations by factorising.

a) i) $x^2 - 8x + 12 = 0$ ii) $y^4 - 8y^2 + 12 = 0$ iii) $z - 8\sqrt{z} + 12 = 0$

b) i) $x^2 - 10x + 9 = 0$ ii) $3^{2y} - 10 \times 3^y + 9 = 0$ iii) $z - 10\sqrt{z} + 9 = 0$

c) i) $4x^2 + 3x - 1 = 0$ ii) $4y^4 + 3y^2 - 1 = 0$ iii) $4z + 3\sqrt{z} - 1 = 0$

7 Daisy factorises $f(x) = 5x^2 + 25x + 20$ by dividing through by 5 to give $f(x) = x^2 + 5x + 4$.

a) Explain why Daisy is wrong.

b) Ahmed says $f(x) = 5x^2 + 25x + 20$ and $g(x) = x^2 + 5x + 4$ have the same roots. Is Ahmed right?

Inequalities

Year 1

THE LOWDOWN

1. An **inequality** states that two expressions are not equal.
   You can solve an inequality in a similar way to solving an equation.
   Remember:
   - keep the inequality sign instead of =
   - when you multiply or divide by a negative number, **reverse** the inequality
   - the solution to an inequality is a **range of values**.

   Example $10 - 2x < 4$
   Subtract 10 from both sides: $-2x < -6$
   Divide both sides by -2: $x > 3$

2. To solve a quadratic inequality:
   - replace the inequality sign with = solve to find the critical values a and b
   - the solution is either between the critical values: $a < x < b$.

   OR at the extremes: $x < a$ or $x > b$.
   To decide which solution is correct you can:
   - **test** a value between a and b see if it satisfies the inequality
   - **look** at a sketch of the graph of the quadratic.

   Example Solve $2x^2 - 4 \geqslant 14$
   $2x^2 - 4 = 14$
   $x^2 = 9$
   $\Rightarrow x = -3$ or $x = 3$
   Since $x^2 \geqslant 9$
   then $x \leqslant -3$ or $x \geqslant 3$

3. You can use **set notation** to write the solutions to inequalities
   - $x < a$ or $x > b$ is written as $\{x : x < a\} \cup \{x : x > b\}$
   - $a \leqslant x \leqslant b$ is written as $\{x : a \leqslant x\} \cap \{x : x \leqslant b\}$

Watch out! Two regions need two inequalities to describe them!

Watch out! Note the word 'or'; x can't be both less than -3 and greater than 3!

Hint: $\cup$ means **union.**
$A \cup B$ means anything in set A or set B (or both).
$\cap$ means **intersection.**
$A \cap B$ means anything that is in set A and in set B.

GET IT RIGHT

a) Draw the graphs of $y = x^2 - 3$ and $y = 2x$ on the same axes.
b) Solve $x^2 - 3 = 2x$.
c) Solve the inequality $x^2 - 3 \leqslant 2x$.

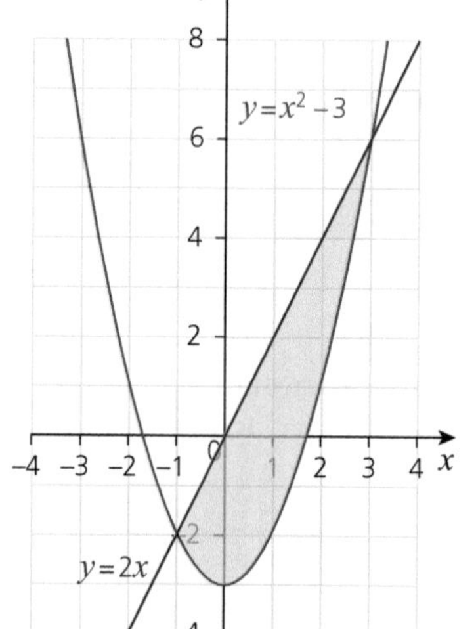

Solution:

a) $y = x^2 - 3$ is the same shape as $y = x^2$ but translated 3 squares down.
$y = 2x$ is a straight line through the origin, with gradient 2.

b) $x^2 - 3 = 2x$
Subtract 2x: $x^2 - 2x - 3 = 0$
Factorise: $(x + 1)(x - 3) = 0$
Solve: $x = -1$ and $x = 3$

c) So critical values are $x = -1$ and $x = 3$.
You want $y = x^2 - 3$ to be below $y = 2x$
so from the graph you can see the solution is $-1 \leqslant x \leqslant 3$.

Hint: Use a solid line, ——, to show the line **is** included.
Use a dashed line, - - -, to show the line **is not** included.

▶▶ See page 12 for the **discriminant** and section 3 for **graphs**.
▶▶ See pages 94–97 for **Venn diagrams**.

YOU ARE THE EXAMINER

Which one of these solutions is correct? Where are the errors in the other solution?

Solve $2x^2 + 5x \geqslant 12$

SAM'S SOLUTION

Rearrange $2x^2 + 5x - 12 \geqslant 0$

Find critical values: $2x^2 + 5x - 12 = 0$

$(2x - 3)(x + 4) = 0$

Critical values: $x = -4$ or $x = \frac{3}{2}$

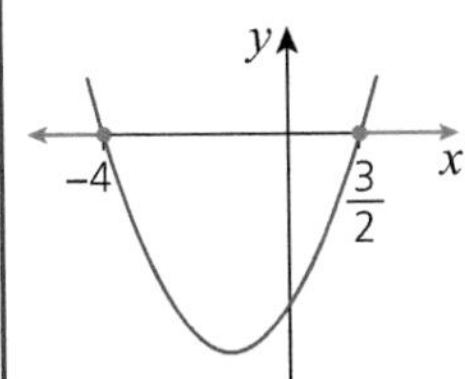

So the solution is $x \leqslant -4$ or $x \geqslant \frac{3}{2}$

LILIA'S SOLUTION

Find critical values: $2x^2 + 5x = 12$

$x(2x + 5) = 12$

$2 \times 6 = 12$, so $x = 2$ or $2x + 5 = 6$

So critical values are $x = 2$ or $x = \frac{1}{2}$

Test a value between $x = \frac{1}{2}$ and $x = 2$:

Try $x = 1.8$

$2 \times 1.8^2 + 5 \times 1.8 = 15.48 \geqslant 12$

So the solution is $\frac{1}{2} \leqslant x \leqslant 2$

SKILL BUILDER

1 Solve $x + 7 < 3x - 5$.

A $x > 6$ **B** $x < 1$ **C** $x > 1$ **D** $x < 6$ **E** $x > 4$

2 Solve $\frac{2(2x + 1)}{3} \geq 6$.

A $x \geqslant 5$ **B** $x \geqslant \frac{3}{2}$ **C** $x > 4$ **D** $x \geqslant 4$ **E** $x > 5$

3 The diagram shows the lines $y = 3x - 3$ and $y = -x + 5$.

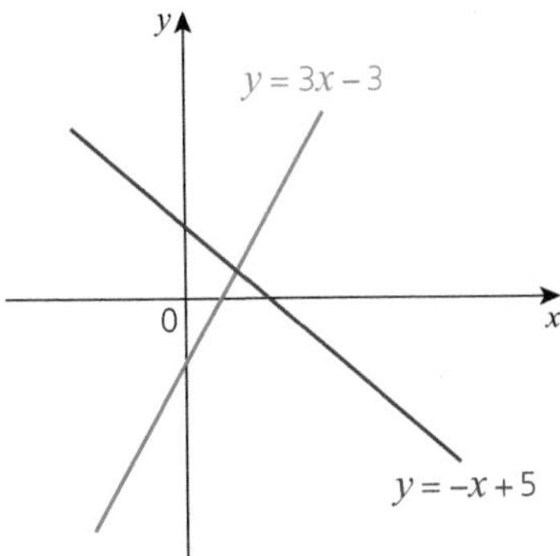

For what values of x is the line $y = 3x - 3$ above the line $y = -x + 5$?

A $x < 4$ **B** $x < 2$ **C** $x > \frac{1}{2}$ **D** $x > 2$ **E** $x > 4$

4 Solve the inequality $x^2 + 2x - 15 \leqslant 0$.

A $-5 \leqslant x \leqslant 3$ **B** $-3 \leqslant x \leqslant 5$ **C** $3 \leqslant x \leqslant 5$

D $x = -5$ or $x = 3$ **E** $x \leqslant -5$ or $x \geqslant 3$

5 Solve the inequality $6x - 6 < x^2 - 1$.

A $x < -1$ or $x > 7$ **B** $1 < x < 5$ **C** $x < 1$ or $x > 7$

D $x < -5$ or $x > -1$ **E** $x < 1$ or $x > 5$

6 Use set notation to write the solutions to the inequalities in **questions 4** and **5**.

7 A ball is thrown up in the air.

The height h metres of the ball at time t seconds is given by $h = 2 + 15t - 5t^2$.

a) Find the times when the ball reaches a height of 12 metres.

b) Find when the height of the ball is:

i) above 12 metres ii) below 12 metres.

Completing the square

Year 1

THE LOWDOWN

① You can write a quadratic expression in the form $a(x + p)^2 + q$.
This is called **completing the square**.
Follow the steps to complete the square for a quadratic in the form $ax^2 + bx + c$.

Example		$2x^2 - 12x + 7$
Step 1	Factor a from first 2 terms:	$2(x^2 - 6x) + 7$
Step 2	Take the coefficient of x:	-6
Step 3	Halve it:	-3
Step 4	Square the result:	$+9$
Step 5	Add and subtract this square:	$2\left[x^2 - 6x + 9 - 9\right] + 7$ (this is a perfect square)
Step 6	Factorise:	$= 2[(x - 3)^2 - 9] + 7$
Step 7	Simplify:	$= 2(x - 3)^2 - 11$

② The **turning point** of the graph of $y = a(x + p)^2 + q$ is at $(-p, q)$.
It is symmetrical about the line $x = -p$.

Example $y = 2x^2 - 12x + 7$ has a turning point at $(3, -11)$.

③ You can use the **quadratic formula** to solve a quadratic equation in the form $ax^2 + bx + c = 0$.

$$x = \frac{-b \pm \sqrt{b^2 - 4ac}}{2a}$$

The square root of a negative number isn't real.
So when $b^2 - 4ac$ is negative the quadratic equation has no real roots.
$b^2 - 4ac$ is called the **discriminant**, it tells you how many roots to expect.

When the discriminant is…	Positive $b^2 - 4ac > 0$	Zero $b^2 - 4ac = 0$	Negative $b^2 - 4ac < 0$
The number of real roots is…	2	1 (repeated)	0
When a is positive graph is: Note when a is negative the curve is 'upside down' (Positive a, Negative a)	Two real roots	One real root	No real roots

Hint: The graph of a quadratic equation $y = ax^2 + bx + c$ is a curve called a **parabola.**
The curve is

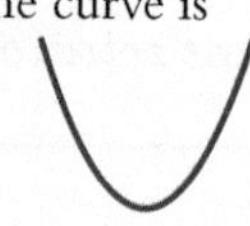

Positive shaped when a is +ve and

Negative shaped when a is −ve.

Hint: When a question asks for **exact solutions** then you should leave your answer as a surd (with the $\sqrt{\ }$).

▶▶ See page 38 for **simultaneous equations.**
◀◀ See page 8 for **quadratic equations.**
◀◀ See page 10 for **inequalities.**

GET IT RIGHT

The equation $16x^2 + kx + 9 = 0$ has no real roots.
Work out the possible values of k.

Solution:

Use the discriminant: $b^2 - 4ac < 0$

$a = 16$, $b = k$ and $c = 9$: $k^2 - 4 \times 16 \times 9 < 0$

$k^2 - 576 < 0 \Rightarrow k^2 < 576$

Find the critical values: $k^2 = 576 \Rightarrow k = \pm 24$

So $-24 < k < 24$

Watch out! Use the **discriminant** when a question asks about the number of roots.

Hint: When $k = \pm 24$ then $b^2 - ac = 0$ so the equation has one repeated root.

YOU ARE THE EXAMINER

Mo and Lilia have both missed out some work in their maths homework. Complete their working.

MO'S SOLUTION

a) Write $y = 5x^2 + 10x + 8$ in the form

$y = a(x + b)^2 + c$

$y = 5x^2 + 10x + 8$

$= 5(x^2 + \square x) + 8$

$= 5\left(x^2 + \square x + \square - \square\right) + 8$

$= 5\left[\left(x + \square\right)^2 - \square\right] + 8$

$= 5\left(x + \square\right)^2 + \square$

b) $y = 5x^2 + 10x + 8$ has $\square$ real roots because $b^2 - 4ac \square 0$.

LILIA'S SOLUTION

a) Find the exact solutions of

$3(x - 4)^2 - 15 = 0$

$3(x - 4)^2 = \square$

$(x - 4)^2 = \square$

$x - 4 = \pm\sqrt{\square}$

$x = \square \pm \sqrt{\square}$

b) $y = 3(x - 4)^2 - 15$

The coordinates of the turning point are

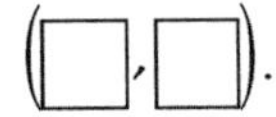

The equation of the line of symmetry is $x = \square$.

SKILL BUILDER

1 Which of the following is the exact solution of the quadratic equation $2x^2 - 3x - 4 = 0$?

A $x = \dfrac{-3 \pm \sqrt{41}}{4}$ **B** $x = \dfrac{3 \pm \sqrt{41}}{4}$ **C** $x = \dfrac{3 \pm \sqrt{23}}{4}$ **D** $x = \dfrac{-3 \pm \sqrt{23}}{4}$

E There are no real solutions.

Hint: Use the quadratic formula and use your calculator to check.

2 Write $x^2 - 12x + 3$ in completed square form.

A $(x - 6)^2 + 3$

B $(x - 12)^2 - 141$

C $(x + 6)^2 - 33$

D $(x - 6 - \sqrt{33})(x - 6 + \sqrt{33})$

E $(x - 6)^2 - 33$

3 The curve $y = -2(x - 5)^2 + 3$ meets the y-axis at A and has a maximum point at B. What are the coordinates of A and B?

A A(0, 3) and B(5, 3)

B A(0, −47) and B(5, 3)

C A(0, −47) and B(5, −6)

D A(0, 3) and B(−10, −47)

E A(0, −47) and B(−10, 3)

4 Four of the following statements are true and one is false. Which one is false?

A $3x^2 - 2x + 1 = 0$ has no real roots.

B $2x^2 - 5x + 1 = 0$ has two distinct real roots.

C $9x^2 - 6x + 1 = 0$ has one repeated real root.

D $x^2 + 2x - 5 = 0$ has two distinct real roots.

E $4x^2 - 9 = 0$ has one repeated real root.

5 Find the exact solutions of $3(x + 5)^2 - 9 = 0$.

Hint: You don't need to expand the bracket!

6 The equation $3x^2 + 7x + k = 0$ has two real roots. Work out the possible values of k.

Hint: Look at the 'Get it right' box.

7 The equation $5x^2 + kx + 5 = 0$ has no real roots. Work out the possible values of k.

Algebraic fractions

Year 1

THE LOWDOWN

① An algebraic fraction is a fraction with a letter symbol in the denominator. You can **simplify** an algebraic fraction by **cancelling common factors**. Factorise the expressions on the top and bottom of the fraction first.

Example $\dfrac{x^2-1}{x^2-2x-3} = \dfrac{(x+1)(x-1)}{(x+1)(x-3)} = \dfrac{x-1}{x-3}$

You can combine algebraic fractions in the same way as you would ordinary fractions.

② **Addition and subtraction** – find a common denominator.

Example
$$\frac{2}{x+4}+\frac{3}{x} = \frac{2\times x}{(x+4)\times x}+\frac{3\times(x+4)}{x\times(x+4)}$$
$$= \frac{2x}{x(x+4)}+\frac{3(x+4)}{x(x+4)}$$
$$= \frac{2x+3(x+4)}{x(x+4)} = \frac{5x+12}{x(x+4)}$$

Sometimes you need to rewrite a term as a fraction.

Example $\dfrac{5}{x^2}-2x = \dfrac{5}{x^2}-\dfrac{2x}{1} = \dfrac{5}{x^2}-\dfrac{2x\times x^2}{1\times x^2} = \dfrac{5-2x^3}{x^2}$

③ **Multiplication** – multiply numerators (tops) and multiply denominators (bottoms).

Example $\dfrac{4}{3-2x}\times\dfrac{5}{x+2} = \dfrac{20}{(3-2x)(x+2)}$

④ **Division** – flip the second fraction and multiply.

Example
$$\frac{4x}{x-5}\div\frac{x}{2x+1} = \frac{4x}{x-5}\times\frac{2x+1}{x}$$
$$= \frac{4x(2x+1)}{x(x-5)} = \frac{4(2x+1)}{x-5}$$

Watch out! You can only cancel common factors. For example, $\dfrac{x}{x+4}$ doesn't simplify to $\dfrac{1}{4}$ as x isn't a factor of the bottom line.

Hint To rewrite with a common denominator, multiply the top **and** bottom of each fraction by the bottom of the other fraction.

Hint When you 'flip' a fraction you are finding the **reciprocal**. The reciprocal of $\dfrac{2}{x+1}$ is $\dfrac{x+1}{2}$. The reciprocal of x is $\dfrac{1}{x}$.

GET IT RIGHT

Simplify $\dfrac{\left(\dfrac{x^2-25}{6x}\right)}{\left(\dfrac{x-5}{2x^3}\right)}$.

Solution:

Step 1 Rewrite as a division: $\dfrac{\left(\dfrac{x^2-25}{6x}\right)}{\left(\dfrac{x-5}{2x^3}\right)} = \dfrac{x^2-25}{6x}\div\dfrac{x-5}{2x^3}$

Step 2 Flip the second fraction and multiply: $= \dfrac{x^2-25}{6x}\times\dfrac{2x^3}{x-5}$

Step 3 Simplify: $= \dfrac{2x^3(x^2-25)}{6x(x-5)}$

$$= \frac{2x^3(x-5)(x+5)}{6x(x-5)} = \frac{x^2(x+5)}{3}$$

◀◀ **Factorising quadratics** on page 8 and **simplifying expressions** on page 2.
▶▶ **Partial fractions** on page 30.

YOU ARE THE EXAMINER

Which one of these solutions is correct? Where is the error in the other solution?

Simplify $\dfrac{x+1}{x-\dfrac{1}{x}}$

SAM'S SOLUTION

$$\frac{x+1}{\frac{x^2-1}{x}} = \frac{(x+1)(x^2-1)}{x}$$

$$= \frac{(x+1)(x+1)(x-1)}{x}$$

$$= \frac{(x+1)^2(x-1)}{x}$$

LILIA'S SOLUTION

$$\frac{x+1}{\frac{x^2-1}{x}} = (x+1) \div \frac{x^2-1}{x}$$

$$= (x+1) \times \frac{x}{x^2-1}$$

$$= \frac{x(x+1)}{x^2-1}$$

$$= \frac{x(x+1)}{(x+1)(x-1)}$$

$$= \frac{x}{x-1}$$

SKILL BUILDER

1 Simplify the expression $\dfrac{x^2-9}{x^2-x-12}$ as far as possible.

A $\dfrac{9}{x-12}$ **B** $\dfrac{3}{4}$ **C** $\dfrac{x+3}{x-4}$ **D** $\dfrac{x-3}{x-4}$ **E** $\dfrac{x+3}{x+4}$

2 Simplify $\dfrac{6x^3}{(x+1)^2} \times \dfrac{3x+3}{2x}$ as far as possible.

A $9x$ **B** $\dfrac{6x^3(x+3)}{(x+1)^2}$ **C** $\dfrac{3x^2(3x+3)}{(x+1)^2}$ **D** $\dfrac{12x^2}{x+1}$ **E** $\dfrac{9x^2}{x+1}$

3 Decide whether each of the following statements is true or false.

a) $\dfrac{1}{x+4} \equiv \dfrac{1}{x} + \dfrac{1}{4}$ b) $\dfrac{2x}{x+y} \equiv \dfrac{2}{y}$ c) $\dfrac{1}{xy} \equiv \dfrac{1}{x} \times \dfrac{1}{y}$

d) $\dfrac{1}{4} \div x \equiv \dfrac{1}{4x}$ e) $\dfrac{x^2}{y^2} \equiv \dfrac{x}{y}$ f) $\dfrac{x+4}{y} \equiv \dfrac{x}{y} + \dfrac{4}{y}$

Hint: Substitute in values of x and y to check that you get the same answer for both sides.

4 Write as a single fraction.

a) $\dfrac{x+3}{2x^2} - \dfrac{5}{x}$ b) $\dfrac{1}{x} + \dfrac{1}{y}$ c) $\dfrac{x}{2} - \dfrac{x-3}{x}$

Hint: Start by finding a common denominator.

5 Write as a single fraction.

a) $\dfrac{x^2+y}{2x} \div \dfrac{x}{y^2}$ b) $\dfrac{x^2-25}{2x} \times \dfrac{x^2}{x^2+6x+5}$ c) $\dfrac{x^2}{3+\dfrac{2}{x}}$

Proof

Year 1 and 2

THE LOWDOWN

① When you are asked to **prove** or show a statement or **conjecture** is true you must **show all your working.**

You can use:

a) **Proof by direction argument (deduction)**

You start from a known result and then use logical argument – you usually need to use algebra.

b) **Proof by exhaustion**

You test every possible case.

c) **Proof by contradiction**

Start by assuming the conjecture is false and then use logical argument to show this leads to a contradiction.

d) **Disproof by counter example**

You only need one counter example to disprove a conjecture.

② Here are some useful starting points for questions on proof.

An **integer** is a whole number.

A **rational number** can be written as a fraction.

Example $9 = \frac{9}{1}$, $0.75 = \frac{3}{4}$ and $0.\dot{4}2857\dot{1} = \frac{3}{7}$

An **irrational number** can't be written as a fraction – its decimal part carries on for ever and never repeats.

Example $\pi = 3.14159....$, $e = 2.71828...$ and $\sqrt{2} = 1.41421...$

Consecutive numbers are numbers that follow one after another other in order.

Examples: 3 consecutive integers: $n, n + 1, n + 2$ or $n - 1, n, n + 1$
3 consecutive even numbers: $2n, 2n + 2, 2n + 4$
3 consecutive odd numbers: $2n + 1, 2n + 3, 2n + 5$
3 consecutive square numbers: $n^2, (n + 1)^2, (n + 2)^2$

▶▶ **Proofs in trigonometry** on pages 56 and 58.

Hint: Often proofs use the following symbols:

⇒ implies or 'leads to'

A shape is a square ⇒ the shape is a quadrilateral.

⇐ implied by or 'follows from'

The shape is a polygon ⇐ a shape has 5 sides.

⇔ implies and is implied by or 'leads to and follows from'

A shape has 3 straight sides ⇔ the shape is a triangle.

GET IT RIGHT

a) Use proof by exhaustion to show that, with the exceptions of 2 and 3, every prime number is either 1 more or 1 less than a multiple of 6.

b) Find a counter example to prove that the following statement is not true.
$n^2 > n$ for all values of n

c) Prove that the sum of two consecutive odd numbers is a multiple of 4.

Solution:

a) A multiple of 6 is not prime as 1, 2, 3 and 6 are factors.
1 more than a multiple of 6 is odd, so could be prime.
2 more than a multiple of 6 is even so is not prime.
3 more than a multiple of 6 is a multiple of 3 so is not prime.
4 more than a multiple of 6 is even so is not prime.
5 more than a multiple of 6 is odd, so could be prime.
5 more than a multiple of 6 is the same as 1 less than a multiple of 6.
So all prime numbers are either 1 more or 1 less than a multiple of 6.

b) When $n = 0.5$, $0.5^2 = 0.25 < 0.5$, so $n^2 < n$ when $0 < n < 1$

c) $2n + 1 + 2n + 3 = 4n + 4 = 4(n + 1)$
Since 4 is a factor then the sum of two consecutive odd numbers is a multiple of 4.

Hint: When you are asked to find a counter example it is often a good idea to test negative numbers and numbers less than 1.

Watch out! You must write a conclusion.

Hint: To show a number is a multiple of 4, you need to show 4 is a factor.

YOU ARE THE EXAMINER

Sam's proof is muddled up and Lilia's proof is incomplete.
Correct both proofs.

SAM'S SOLUTION

Proof that there are an infinite number of primes:

Assume that there are a finite number of primes and list all the primes.

A: Either way there is a new prime.

B: None of the primes on the list is a factor of Q (there is always a remainder of 1).

C: So Q is either prime, or it has a prime factor not on the list of all primes.

D: Multiply all the primes together, call the result P.

E: Let $Q = P + 1$

This contradicts my assumption that there is a finite list of primes.

LILIA'S SOLUTION

Proof that $\sqrt{2}$ is irrational:

Assume $\sqrt{2}$ is rational, so it can be written as $\sqrt{2} = \frac{a}{b}$ where $\frac{a}{b}$ is a fraction in its lowest terms.

Squaring both sides gives $\square = \frac{a^2}{b^2}$

Rearranging gives $2b^2 = \square$

a^2 is ____ since $2\times$ any number is ____.

So a must be ____.

Let $a = 2n$, so $2b^2 = (2n)^2 = \square$

Dividing by 2 gives $b^2 = \square$

So b^2 is ____ and so b is ____.

But if a and b are both ____ then $\frac{a}{b}$ can't be in its ____ terms.

This contradicts my original assumption so $\sqrt{2}$ must be ____.

SKILL BUILDER

1 'For all values of n greater than or equal to 1, $n^2 + 3n + 1$ is a prime number.'
Which value of n gives a counter-example that disproves this conjecture?

A $n = 7$ **B** $n = 2$ **C** $n = 8$ **D** $n = 6$

2 Below is a proof that appears to show that $2 = 0$.
The proof must contain an error.
At which line does the error occur?

Let $a = b = 1$

[Line 1] $\Rightarrow a^2 = b^2$

[Line 2] $\Rightarrow a^2 - b^2 = 0$

[Line 3] $\Rightarrow (a + b)(a - b) = 0$

[Line 4] $\Rightarrow a + b = 0$

[Line 5] $\Rightarrow 2 = 0$

A line 1 **B** line 2 **C** line 3 **D** line 4 **E** line 5

3 Look at the following statements about non-zero numbers, x and y. Which of them are true?

(1) $xy = 1 \Rightarrow x = 1, y = 1$

(2) $xy = 1 \Leftarrow x = \frac{1}{y}$

(3) $xy = 1 \Leftarrow x = \frac{1}{y}$ or $y = \frac{1}{x}$

A (1) only **B** (2) only **C** (3) only **D** (1) and (3) **E** (2) and (3)

Functions

THE LOWDOWN

① A **function** is a rule. It maps a number in one set to a number in another set. A function is usually written using letters such as f, g or h.

Example $f(x) = x^2$ and $g(x) = 4x - 1$ are functions.
$f(3) = 3^2 = 9$ and $g(2) = 4 \times 2 - 1 = 7$

② The **set of inputs** (x-values) to a function is the **domain**.
The set of outputs from the function is the range.
Every input has **exactly one** output, but some inputs may map to the same output. You say a function is **one-to-one** or **many-to-one**.

Example $f(x) = x^2$, $-3 \leqslant x \leqslant 3$ has a domain of $-3 \leqslant x \leqslant 3$ and a range of $0 \leqslant f(x) \leqslant 9$ since $f(0) = 0$ and $f(3) = f(-3) = 9$

③ For a composite function gf(x) you apply g to the output of f.
Make sure you apply the right function first – start with x and work outwards.

Example Given $f(x) = x^2$ and $g(x) = 4x - 1$, find i) $g(f(x))$ ii) $fg(x)$.

i) $x \xrightarrow{f} x^2 \xrightarrow{g} 4x^2 - 1$ so $gf(x) = 4x^2 - 1$

ii) $x \xrightarrow{g} 4x - 1 \xrightarrow{f} (4x - 1)^2$

④ The inverse function f^{-1} reverses the effect of the function.
It maps the range back onto the domain.
Note: $ff^{-1}(x) = f^{-1}f(x) = x$

Example $f(x) = x^3$ so

$2 \xrightarrow{f(2)} 8 \xrightarrow{f^{-1}(8)} 2$

The graphs of a function and its inverse function are reflections of each other in the line $y = x$.

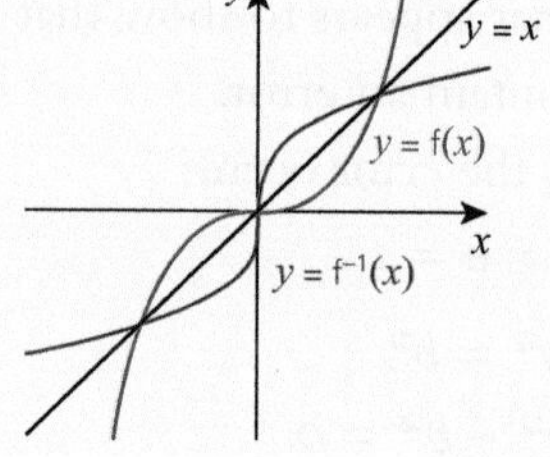

Hint: $f(x) = x^2$ means square the input number.
$g(x) = 4x - 1$ means multiply the input by 4 and subtract 1.

Watch out! Draw a graph to help you find the range.

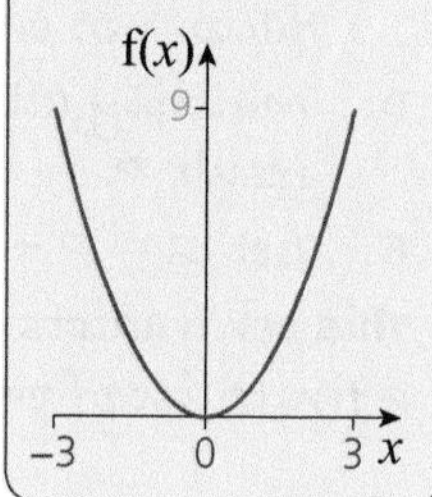

Hint:
$3 \xrightarrow{f} 9 \xrightarrow{g} 35$
so $gf(3) = 35$

Watch out! In general, $gf(x)$ is not the same as $fg(x)$.

Watch out! A function can only have an inverse function if it is a one-to-one function.

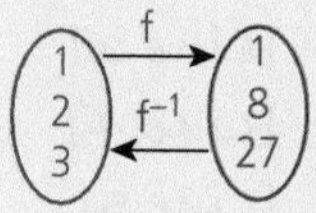

GET IT RIGHT

You are given $f(x) = 3x - 2$, $x \in \mathbb{R}$ and $g(x) = \sqrt{x - 3}$, $x \geqslant 3, x \in \mathbb{R}$

a) Write down the range of each function.

b) Find fg(12)and gf(x).

c) Find $f^{-1}(x)$.

Hint: $\mathbb{R}$ is the set of real numbers – this means all numbers, positive negative, rational, irrational and 0.

Solution:

a) $f(x) \in \mathbb{R}$ and $g(x) \geqslant 0$, $g(x) \in \mathbb{R}$

b) $12 \xrightarrow{g} 3 \xrightarrow{f} 7$ so $fg(12) = 7$

$x \xrightarrow{f} 3x - 2 \xrightarrow{g} \sqrt{3x - 2 - 3}$ so $gf(x) = \sqrt{3x - 5}$

c) Step 1 Replace f(x) with y: $y = 3x - 2$

Step 2 Interchange y and x: $x = 3y - 2$

Step 3 Rearrange to make y the subject: $\frac{x + 2}{3} = y$

Step 4 Replace y with $f^{-1}(x)$: $f^{-1}(x) = \frac{x + 2}{3}$

Watch out! The square root of a negative number isn't real, so g(x) is only valid when $x \geqslant 3$. This is called **restricting the domain.**

YOU ARE THE EXAMINER

Parts of each of these solutions are wrong. Where are the mistakes?

You are given $f(x) = \frac{1}{2x+3}$.

a) What value of x must be excluded from the domain of $f(x)$?

b) Solve $f^{-1}(x) = 1$.

SAM'S SOLUTION

a) $x \neq -\frac{3}{2}$

b) $f(x) = \frac{1}{2x+3}$

$\Rightarrow f^{-1}(x) = 2x+3$

Solve $f^{-1}(x) = 1$

$\Rightarrow \quad 2x+3 = 1$

$\Rightarrow \quad 2x = -2$

$\Rightarrow \quad x = -1$

LILIA'S SOLUTION

a) $x \neq 0$

b) $\frac{1}{2x+3} = y$

Swap x and y: $\frac{1}{2y+3} = x$

Make y the subject: $1 = 2xy + 3x$

$\Rightarrow 2xy = 1 - 3x \Rightarrow y = \frac{1-3x}{2x}$

Solve $f^{-1}(x) = \frac{1-3x}{2x} = 1$

$1 - 3x = 2x \Rightarrow 5x = 1 \Rightarrow x = \frac{1}{5}$

SKILL BUILDER

1 The function f is defined as f: $x \rightarrow \sqrt{2x-3}$. Write down a suitable domain for f.

A $x \in \mathbb{R}$ **B** $x \geqslant 0$ **C** $x \geqslant \frac{3}{2}$ **D** $x \geqslant 3$

2 The function g is defined as $g(x) = x^2 - 2x - 1$ for $-2 \leqslant x \leqslant 2$. What is the range of the function?

A $-1 \leqslant g(x) \leqslant 7$ **B** $g(x) \leqslant 7$ **C** $g(x) \geqslant -2$ **D** $-2 \leqslant g(x) \leqslant 7$

Hint: Sketch the graph.

3 The functions f and g are defined for all real numbers x as $f(x) = x^2 - 2$ and $g(x) = 3 - 2x$. Find an expression for the function fg(x) (for all real numbers x).

A $fg(x) = 7 - 2x^2$ **B** $fg(x) = -2x^3 + 3x^2 + 4x - 6$ **C** $fg(x) = 4x^2 - 12x + 7$

D $fg(x) = 7 + 4x^2$ **E** $fg(x) = 1 - 2x^2$

4 The function f is defined by $f(x) = 2x^3 - 1$. Find the inverse function $f^{-1}(x)$.

A $f^{-1}(x) = \frac{\sqrt[3]{x+1}}{2}$ **B** $f^{-1}(x) = \frac{1}{2x^3 - 1}$ **C** $f^{-1}(x) = \sqrt[3]{\frac{x+1}{2}}$ **D** $f^{-1}(x) = \frac{\sqrt[3]{x}+1}{2}$

5 The diagram shows the graph of $y = g(x)$.

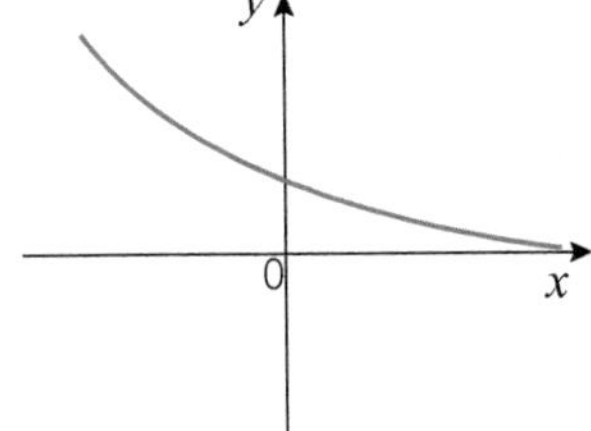

Which of the diagrams below shows the graph of $y = g^{-1}(x)$?

A

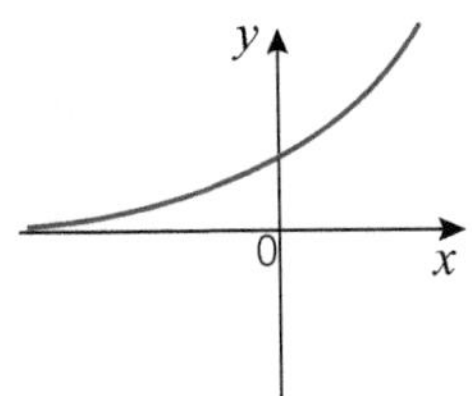

B

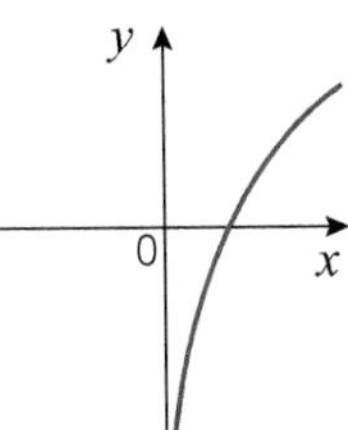

C

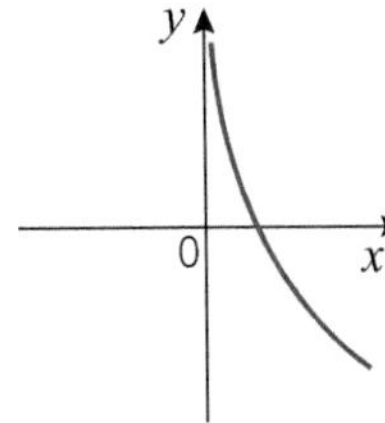

D

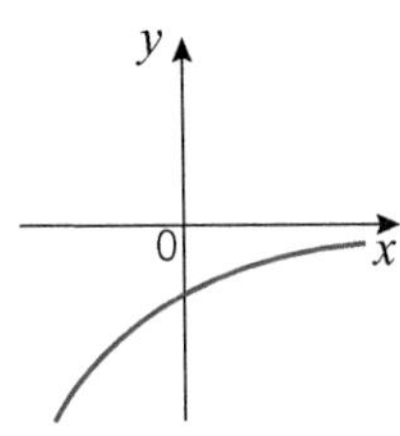

6 The functions p and q are defined as $p(x) = \frac{1}{x}, x \neq 0$ and $q(x) = 2x - 1, x \in \mathbb{R}$.

a) Find an expression for the function pq(x). b) Solve pq(x) = x.

Polynomials

Year 1

THE LOWDOWN

1. A **polynomial** equation only has positive integer powers of x.
   The order of a polynomial is the highest power of x.

Example $f(x) = 2x^3 + 3x^2 - 18x + 8$ has order 3 (a cubic).

Hint: Any 'whole number' roots will be factors of the constant term. So you can spot roots by checking factors of +8.

2. The **factor theorem** says that if $(x - a)$ is a factor of $f(x)$ then $f(a) = 0$ and $x = a$ is a root of the equation $f(x) = 0$.

Example $f(2) = 2 \times 2^3 + 3 \times 2^2 - 18 \times 2 + 8 = 16 + 12 - 36 + 8 = 0$
So $(x - 2)$ is a factor of $f(x) = 2x^3 + 3x^2 - 18x + 8$

Watch out! The quadratic factor often factorises.

3. Use the factor theorem and the grid method to factorise a polynomial.

Example $2x^3 + 3x^2 - 18x + 8 = (x - 2) \times$ (some quadratic)

Step 1 Draw a grid
Write x and -2 on one side and fill in $2x^3$ and 8.

×			
x	$2x^3$		
-2			8

Step 2 What do you multiply x by to get $2x^3$? Answer: $x \times 2x^2 = 2x^3$.
Now complete the 1st and last columns.

×	$2x^2$		-4
x	$2x^3$		$-4x$
-2	$-4x^2$		8

See page 8 for **factorising quadratics** and page 4 for **solving simultaneous equations**.

Step 3 You have $-4x^2$ and the equation says $+3x^2$ so you need $+7x^2$.

	$2x^2$		-4
x	$2x^3$	$+7x^2$	$-4x$
-2	$-4x^2$		8

Step 4 What do you multiply x by to get $+7x^2$? Answer: $x \times 7x = 7x^2$.
Now you can complete the grid.

	$2x^2$	$+7x$	-4
x	$2x^3$	$+7x^2$	$-4x$
-2	$-4x^2$	$-14x^2$	$+8$

So $2x^3 + 3x^2 - 18x + 8 = (x - 2)(2x^2 + 7x - 4)$
Factorising the 2nd bracket gives $(x - 2)(2x - 1)(x + 4)$

5. Now you can sketch the graph of $y = f(x)$.

Example $y = 2x^3 + 3x^2 - 18x + 8$
When $x = 0$: $y = 8$
Since $y = (x - 2)(2x - 1)(x + 4)$
when $y = 0$: $x = -2$, $x = \frac{1}{2}$, $x = -4$

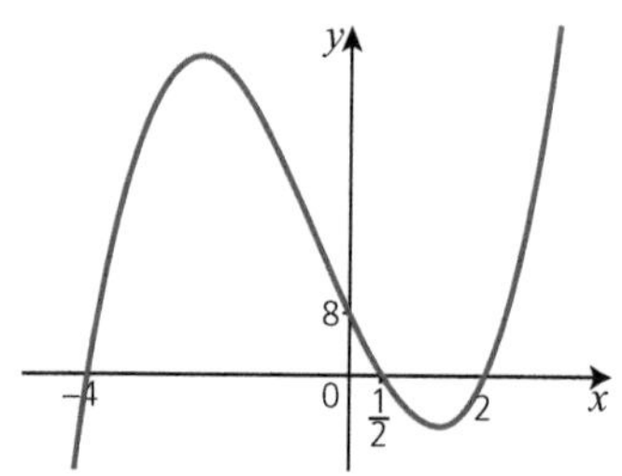

Watch out! A cubic has up to 2 turning points.
If the sign of x^3 is negative the curve is 'upside down' like this:

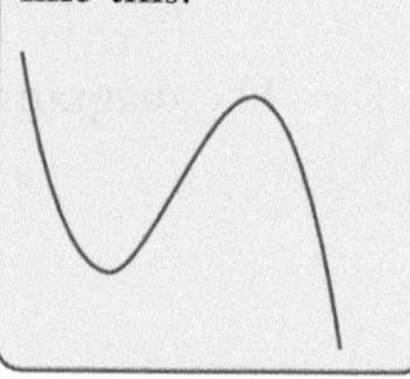

GET IT RIGHT

$f(x) = 2x^3 + ax^2 + bx - 3$ has a factor of $(x + 3)$ and a root $x = 1$.
Find the values of a and of b.

Solution:

$(x + 3)$ is a factor $\Rightarrow f(-3) = 0 \Rightarrow 2 \times (-3)^3 + a \times (-3)^2 + b \times (-3) - 3 = 0$
$\Rightarrow -54 + 9a - 3b - 3 = 0 \Rightarrow 9a - 3b = 57$
$x = 1$ is a root $\Rightarrow f(1) = 0 \Rightarrow 2 \times 1^3 + a \times 1^2 + b \times 1 - 3 = 0$
$\Rightarrow 2 + a + b - 3 = 0 \Rightarrow a + b = 1$
Using your calculator to solve gives $a = 5$ and $b = -4$

Hint: There are 2 unknowns so you need to form 2 equations and solve them simultaneously.

YOU ARE THE EXAMINER

Which one of these solutions is correct? Where are the errors in the other solution?

a) Show that $(3x - 2)$ is a factor of $f(x) = 3x^3 - 5x^2 - 4x + 4$.

b) Use the factor theorem to find two more factors of $f(x)$.

SAM'S SOLUTION

a) $f\left(\frac{2}{3}\right) = 3 \times \left(\frac{2}{3}\right)^3 - 5 \times \left(\frac{2}{3}\right)^2 - 4 \times \frac{2}{3} + 4$

$= \frac{24}{27} - \frac{20}{9} - \frac{8}{3} + 4 = 0$

So $(3x - 2)$ is a factor.

b) Check factors of 4: $\pm1, \pm2, \pm4$

$f(1) = 3 \times 1 - 5 \times 1 - 4 \times 1 + 4 = -2$

$f(2) = 3 \times 2^3 - 5 \times 2^2 - 4 \times 2 + 4 = 0$

$f(4) = 3 \times 4^3 - 5 \times 4^2 - 4 \times 4 + 4 = 100$

$f(-1) = 3(-1)^3 - 5(-1)^2 - 4(-1) + 4 = 0$

Factors are $(x - 2)$ and $(x + 1)$

LILIA'S SOLUTION

a) $f\left(\frac{2}{3}\right) = 0$

b) Check factors of 4:

$1, 2, 4, -1, -2, -4$

$f(1) = -2$

$f(2) = 0$

$f(4) = 100$

Check negative factors of 4:

$f(-1) = 0$

So the factors are $x = 2$ and $x = -1$

SKILL BUILDER

1 You are given that $f(x) = 4x^3 - x + 3$ and $g(x) = 2x^2 - 3x + 4$.
Which of the following polynomials is $f(x) + g(x)$?

A $6x^3 - 4x + 7$ **B** $4x^3 + 2x^2 - 2x + 7$ **C** $6x^2 - 4x + 7$

D $4x^3 + 2x^2 + 2x + 7$ **E** $4x^3 + 2x^2 - 4x + 7$

2 You are given $f(x) = 2x^3 - 3$ and $g(x) = 3x^2 + x - 2$. Which of the following is $f(x) \times g(x)$?

A $5x^5 + 2x^4 - 4x^3 - 9x^2 - 3x + 6$ **B** $6x^5 + x^4 - 2x^3 - 9x^2 - 3x + 6$

C $6x^5 + 2x^4 - 4x^3 + 9x^2 + 3x - 6$ **D** $6x^5 + 2x^4 - 4x^3 - 9x^2 - 3x + 6$

E $6x^5 + 2x^4 - 4x^3 - 9x^2 - x - 2$

Hint: Use a grid to help you.

3 Which of the following graphs represents the shape of the curve $y = -x^4 + 5x^3 - 5x^2 - 5x + 6$?

A

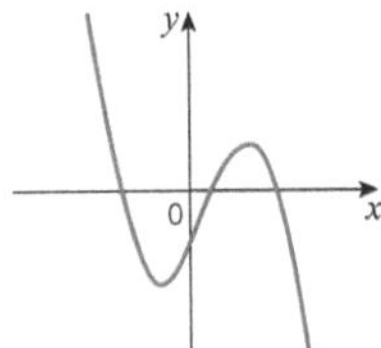

B

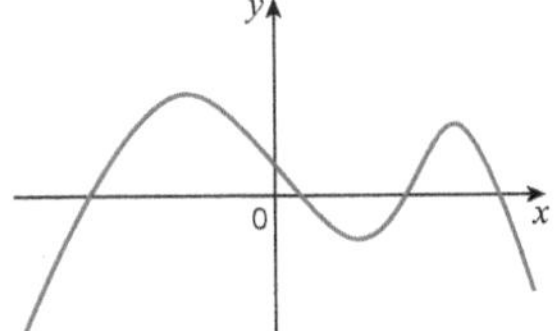

C

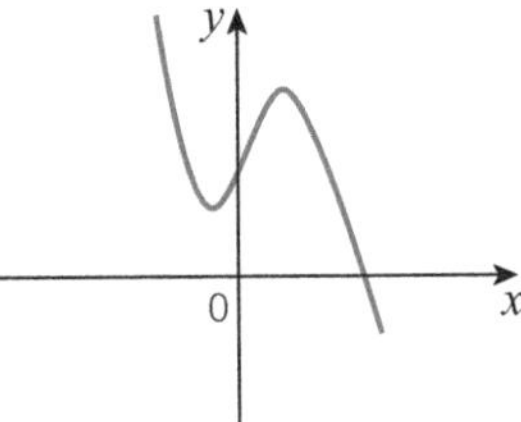

D

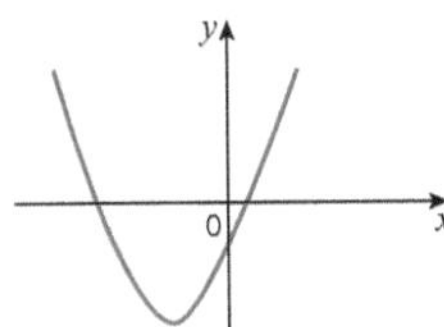

E

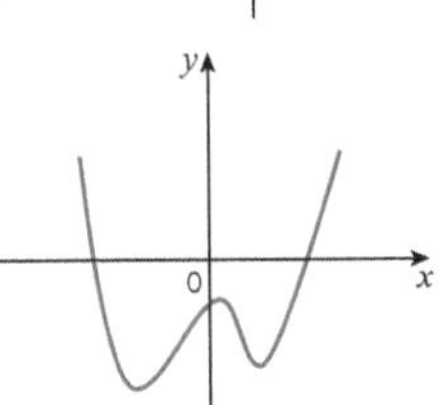

Hint: A polynomial of order n has at most $n - 1$ turning points.

4 You are given that $2x^3 + x^2 - 7x - 6 = (x - 2)(2x^2 + bx + 3)$. The value of b is:

A -3 **B** 5 **C** 3 **D** -5 **E** $-\frac{1}{4}$

5 A polynomial is given by $f(x) = x^3 - 3x^2 - 7x - 15$. Which one of these is a factor of $f(x)$?

A $(x - 1)$ **B** $(x - 3)$ **C** $(x - 5)$ **D** $(x + 1)$ **E** $(x + 3)$

6 a) Given $(2x - 1)$ is a factor of $f(x) = 2x^3 - 3x^2 + kx + 6$, find the value of k.

b) Factorise $f(x)$ completely.

c) Sketch the graph of $y = 2x^3 - 3x^2 + kx + 6$.

Exponentials and logs

Year 1

THE LOWDOWN

① An **exponential function** has the variable as the power, e.g. $y = 10^x$.
The exponential function is $y = e^x$ where e = 2.718 281…
The **base** a can be any positive number; so $y = a^x$ is also always positive.

② A **logarithm** is the **inverse** of an exponential function.

Example $10^2 = 100$ so $\log_{10} 100 = 2$, say 'log to base 10 of 100 is 2'
$10^3 = 1000$ so $\log_{10} 1000 = 3$

③ You can have other bases, but base 10 and base e are the most common.
$\log_{10} x$ is often written as $\log x$ or $\lg x$.
$\log_e x$ is called the natural logarithm and written as $\ln x$.

④ Make sure you know these useful rules:

- $\log x + \log y = \log xy$ Example $\log 4 + \log 5 = \log(4 \times 5) = \log 20$
- $\log x - \log y = \log \frac{x}{y}$ Example $\log 12 - \log 2 = \log \frac{12}{2} = \log 6$
- $\log x^n = n \log x$ Example $\log 5^2 = 2 \log 5$
- $a^1 = a$ so $\log_a a = 1$ Example $\ln e = 1$ and $\log_{10} 10 = 1$
- $a^0 = 1$ so $\log_a 1 = 0$ Example $\log_{10} 1 = 0$ and $\ln 1 = 0$

⑤ The graph of $y = a^x$

- passes through (0, 1) and (1, a)
- never touches the x-axis.

The graph of $y = \log_a x$

- is a reflection of $y = a^x$ in the line $y = x$
- passes through (1, 0) and (a, 1)
- never touches the y-axis
- is not valid for negative x.

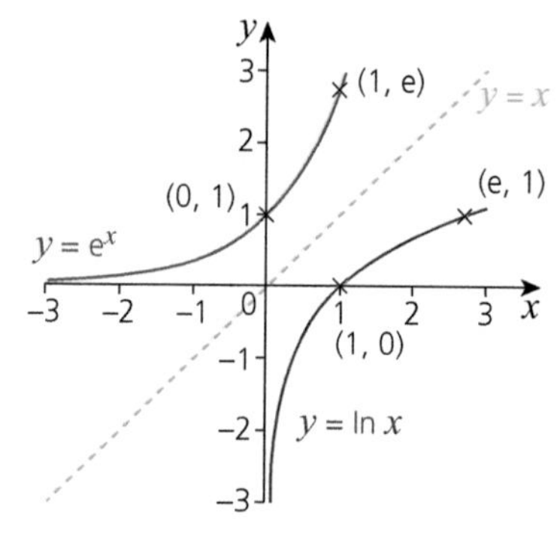

◀◀ **Laws of indices** on page 6.
▶▶ **Straight lines and logs** on page 24.

Hint: So the $\log_{10}$ of any number between 100 and 1000 is between 2 and 3. You say 'log to base 10 of 100 is 2'

Use your calculator to check these

Hint: In these examples 10 and e are the bases.

Hint: For examples... $2^4 = 16$ so $\log_2 16 = 4$

Hint: Exponential functions are used to model exponential growth and decay.

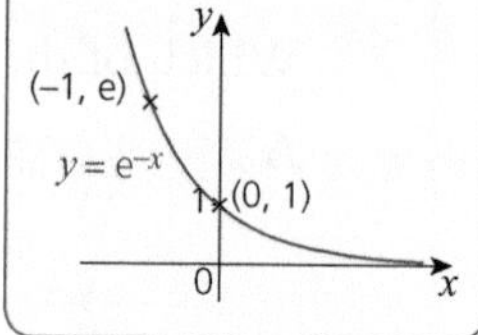

GET IT RIGHT

Solve

a) $4^x = 100$

b) $\ln \frac{1}{x} + 2 \ln x + \ln 3 = 2$

Solution:

a) Take logs of both sides: $\log(4^x) = \log 100$
Use $\log x^n = n \log x$: $x \log 4 = \log 100$
Divide by log 4: $x = \frac{\log 100}{\log 4} = 3.32$ (to 3 s.f.)

b) Use $\log x^n = n \log x$: $\ln \frac{1}{x} + \ln x^2 + \ln 3 = 2$
Use $\log a + \log b = \log ab$: $\ln\left(\frac{1}{x} \times x^2 \times 3\right) = \ln 3x = 2$
Raise both sides as a power of e: $3x = e^2 \Rightarrow x = \frac{1}{3}e^2$

Hint: You can solve in one step if your calculator will evaluate $\log_4 100$.

Hint: 'Undoing the ln' is the same as 'finding e to the power…' They are inverse functions.

YOU ARE THE EXAMINER

Which one of these solutions is correct? Where are the errors in the other solution?

The number of fish, P, in a lake is modelled by the formula $P = 2000e^{-0.05t}$ where t is the time, in years. After how many years is the population of fish half the initial population?

PETER'S SOLUTION

At $t = 0, P = 2000 \times e^{-0.05\times 0}$

$= 2000 \times 1 = 2000$

Population is halved:

$$2000e^{-0.05t} = 1000$$

$$e^{-0.05t} = \frac{1000}{2000} = \frac{1}{2}$$

Take ln of both sides:

$$\ln e^{-0.05t} = \ln\left(\frac{1}{2}\right)$$

$$-0.05t = \ln\left(\frac{1}{2}\right)$$

Divide by -0.05:

$$t = \frac{\ln\left(\frac{1}{2}\right)}{-0.05} = 13.86\ldots$$

So after 13.9 years (to 3 s.f.)

LILIA'S SOLUTION

At $t = 0, P = 2000 \times e^{-0.05\times 0} = 2000$

Population is halved:

$$2000e^{-0.05t} = 1000$$

$$e^{-0.05t} = \frac{1000}{2000} = \frac{1}{2}$$

Take ln of both sides:

$$\ln e^{-0.05t} = \ln\left(\frac{1}{2}\right)$$

Using rules of logs:

$$-0.05t\ln e = \ln\left(\frac{1}{2}\right)$$

So

$$-0.05t = \frac{\ln\left(\frac{1}{2}\right)}{\ln e} = \ln\frac{1}{2} - \ln e$$

Divide by -0.05:

$$t = \frac{\ln\frac{1}{2} - \ln e}{-0.05} = 33.86\ldots$$

So after 33.9 years (to 3 s.f.)

SKILL BUILDER

1 Write $\log 12 - 3\log 2 + 2\log 3$ as a single logarithm.

A $\log 12$ **B** $\log 13.5$ **C** $\log\frac{1}{6}$ **D** $\log 10\frac{2}{3}$

2 Use logarithms to the base 10 to solve the equation $2.5^x = 1000$ to 2 decimal places.

A 2.90 **B** 3.00 **C** 7.54 **D** 7.538 82

3 Simplify $\frac{1}{2}\log 64 - 2\log 2$ writing your answer in the form $\log x$.

A $\log 8$ **B** $\log 2$ **C** $\log 32$ **D** 0.301

4 Express $\log\sqrt{x} + \log x^{\frac{7}{2}} - 2\log x$ as a single simplified logarithm.

A $2\log x$ **B** $\log x^2$ **C** $\log(x^{-\frac{3}{2}} + x^{\frac{3}{2}})$ **D** $\log(\frac{x^{\frac{1}{2}} \times x^{\frac{7}{2}}}{x^2})$

5 The value, £V, of an investment varies according to the formula $V = Ae^{0.1t}$, where t is the time in years. The investment is predicted to be worth £10 000 after 5 years.
Find the value of A to the nearest £.

A £16 487 **B** £6065 **C** £7358 **D** £6065.31

6 Make x the subject of $\ln(3x + 2) = 5t$.

A $x = \frac{1}{3}e^{(5t-2)}$ **B** $x = \frac{5}{3}(t - \ln 2)$ **C** $x = \frac{1}{3}(e^{5t} + 2)$

D $x = \frac{1}{3}e^{5t} - 2$ **E** $x = \frac{1}{3}(e^{5t} - 2)$

7 Solve these equations. Leave your answers in exact form.

a) $\log_3 x - \log_3(2x - 1) = \log_3 4$ **b)** $\ln x + \ln x^2 = 6$ **c)** $\log_{10} x^3 - \log_{10}\sqrt{x} = 5$

Hint: Use the rules of logs to write the left-hand side of the equation as a single log.

8 **a)** Solve $2x^2 + 5x - 3 = 0$ **b)** Hence solve $2 \times 3^{2y} + 5 \times 3^y - 3 = 0$ to 3 s.f.

Hint: Replace 3^y with x and remember $(3^y)^2 = 3^{2y}$

Straight lines and logs

Year 1

THE LOWDOWN

① You can use logs to rewrite $y = kx^n$ so that the graph is a straight line.

Example Rewrite $y = kx^n$ to produce a straight line graph.

Take logs of both sides: $\log y = \log(kx^n)$

Use $\log ab = \log a + \log b$: $\log y = \log k + \log x^n$

Use $\log x^n = n\log x$: $\log y = \log k + n\log x$

$\downarrow \quad \downarrow \quad \downarrow \quad \downarrow$

Compare with $y = mx + c$: $y = c + m\ x$

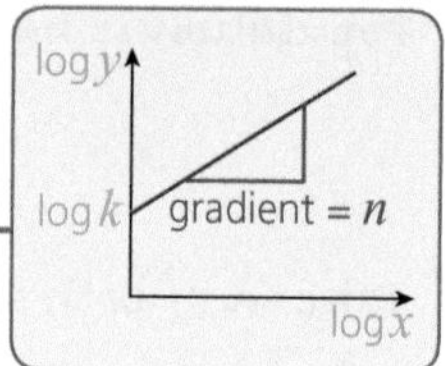

So plotting $\log y$ against $\log x$ gives a straight line with **gradient** n and y-intercept $\log k$.

② You can also use logs to rewrite $y = ka^x$ so that the graph is a straight line.

Hint: It doesn't matter which base you use, but usually ln or $\log_{10}$ are used.

Example Rewrite $y = ka^x$ to produce a straight line graph.

$y = ka^x$

Take logs of both sides: $\log y = \log(ka^x)$

Use $\log ab = \log a + \log b$: $\log y = \log k + \log a^x$

Use $\log a^x = x\log a$: $\log y = \log k + x\log a$

$\downarrow \quad \downarrow \quad \downarrow \quad \downarrow$

Compare with $y = mx + c$: $y = c + x\ m$

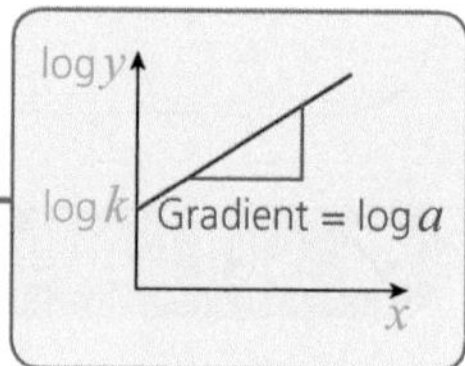

So plotting $\log y$ against x gives a straight line with **gradient** $\log a$ and y-intercept $\log k$.

GET IT RIGHT

The relationship between y and x is $y = kx^n$ where k and n are constants.

x	2	3	4
y	4	13.5	32

a) Using the table of values, plot the graph of $\log_{10} y$ against $\log_{10} x$.

b) Find the equation for y in terms of x.

Solution:

a) Make a table of values and draw a graph.

$\log_{10} x$	0.301	0.477	0.602
$\log_{10} y$	0.602	1.13	1.51

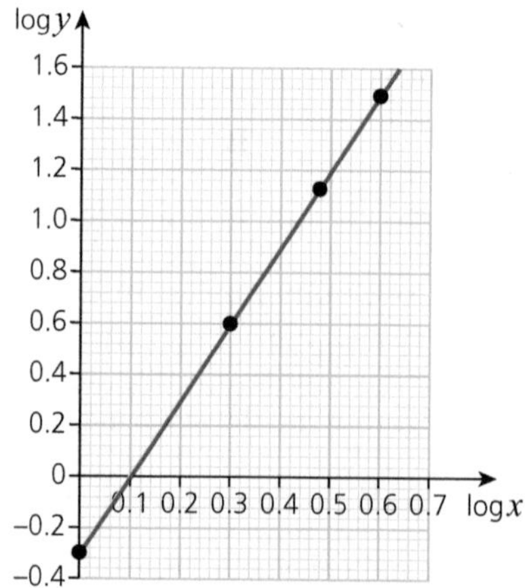

b) $y = kx^n \Rightarrow \log y = \log k + n\log x$

Read from graph

y-intercept $= \log k = -0.3$

So $k = 10^{-0.3} = 0.5$

The gradient, $n = \dfrac{1.51 - (-0.3)}{0.602 - 0}$

$= 3$ (to 1 s.f.)

So $y = 0.5 \times x^3$

YOU ARE THE EXAMINER

Which one of these solutions is correct? Where are the errors in the other solution?

The diagram shows the graph of $\ln y$ against x.

Find the equation for y in terms of x.

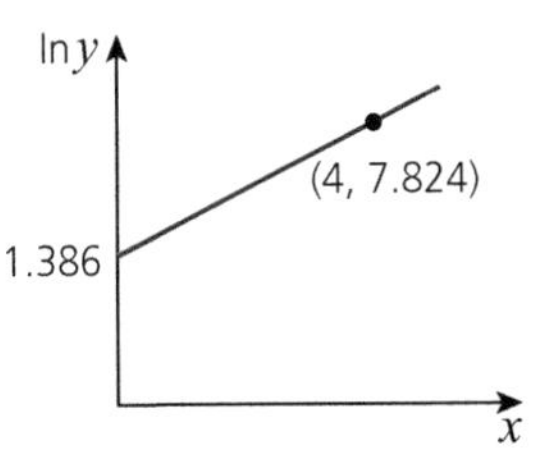

SAM'S SOLUTION

Gradient $= \dfrac{7.824 - 1.386}{4 - 0}$

$= 1.6095$

y-intercept $= 1.386$

Equation of line: $\ln y = 1.6095x + 1.386$

So $y = e^{1.6095x + 1.386}$

$y = e^{1.6095x} \times e^{1.386}$

$y = (e^{1.6095})^x \times 4 = 4 \times 5^x$

LILIA'S SOLUTION

Gradient $= \dfrac{7.824 - 1.386}{4 - 0} = 1.6095$

y-intercept $= 1.386$

Equation of line is $y = 1.6095x + 1.386$

SKILL BUILDER

1 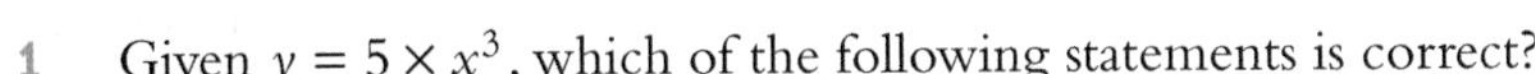
Given $y = 5 \times x^3$, which of the following statements is correct?

A $\ln y = 3\ln x + \ln 5$ **B** $\ln y = 5\ln x + \ln 3$

C $\ln y = x\ln 5 + \ln 3$ **D** $\ln y = x\ln 3 + \ln 5$

2 Given $y = 2 \times 7^x$, which of the following statements is correct?

A $\log y = 2\log x + \log 7$ **B** $\log y = 7\log x + \log 2$

C $\log y = x\log 7 + \log 2$ **D** $\log y = x\log 2 + \log 7$

3 The graph shows the result of plotting $\log_{10} y$ against x.

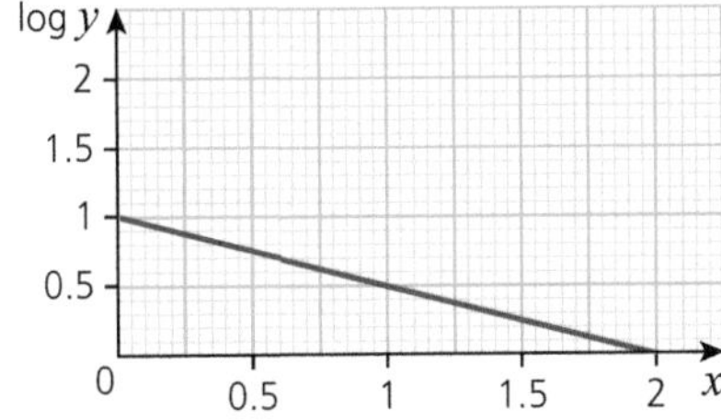

The relationship between x and y is of the form $y = k \times T^x$. The values of T and k, to 2 d.p. are:

A $T = 0.32$ and $k = 10$ **B** $T = 0.32$ and $k = 1$ **C** $T = -\frac{1}{2}$ and $k = 10$ **D** $T = -\frac{1}{2}$ and $k = 1$.

4 In an experiment, a variable, y, is measured at different times, t. The graph shows $\log_{10} y$ against $\log_{10} t$.

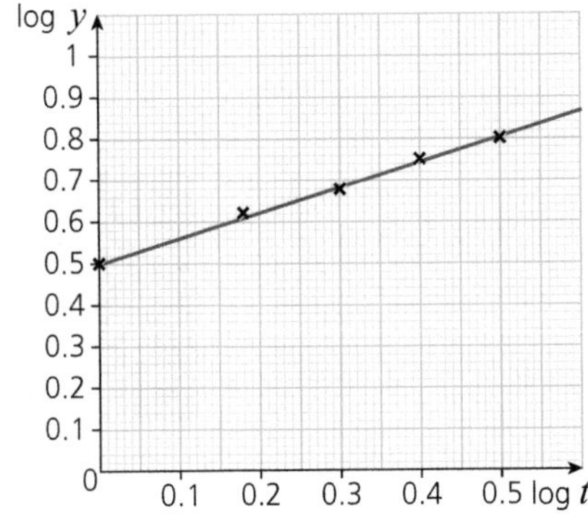

The relationship between y and t is:

A $y = 3.16t^{1.67}$ **B** $y = 3.16t^{0.6}$ **C** $y = 0.5t^{0.6}$ **D** $3.16t^{-0.22}$

Sequences and series

Year 2

THE LOWDOWN

① A **sequence** is an ordered set of numbers that follow a rule.
Each number in a sequence is called a **term**.
An **infinite sequence** continues forever and has no last term.

② A **series** is found by adding the terms in the sequence.

Example 4, 7, 10, 13, 16, ... is an infinite sequence.
4 + 7 + 10 + 13 + 16 + ... + 61 is a finite series.

③ In an **arithmetic sequence** the difference between one term and the next is always the same:

- the first term is a and the common difference is d
- the nth term is $a_n = a + (n-1)d$
- the sum of the first n terms is $S_n = \frac{n}{2}[2a + (n-1)d]$.

④ In a **geometric sequence** the ratio, r, between adjacent terms is the same:

- to find the common ratio, r, divide one term by the term before it
- the nth term is $a_n = ar^{n-1}$
- the sum of the first n terms is $S_n = \frac{a(1-r^n)}{(1-r)}$.

⑤ A **convergent geometric series** has a common ratio between -1 and 1.
The sum to infinity is $S = \frac{a}{1-r}$.

Example $27 + 9 + 3 + 1 + \frac{1}{3} + \frac{1}{9} + \ldots$ has a sum to infinity because
$r = \frac{9}{27} = \frac{1}{3}$ and so $-1 < r < 1$
The sum to infinity is $S = \frac{27}{1 - \frac{1}{3}} = 40.5$

⑥ Sometimes the capital Greek letter sigma Σ is used for a series.

Example $\sum_{r=1}^{5}(12 - 2r)$ means substitute $r = 1, 2 \ldots 5$ and add.

$$\sum_{r=1}^{5}(12-2r) = (12-2) + (12-4) + (12-6) + (12-8) + (12-10) = 30$$

⑦ Some sequences have a rule which connects each term to the one before.

Example A sequence is defined by $a_1 = 3$ and $a_{k+1} = 2a_k + 1$.
The first four terms are 3, 7, 15, 31.

Watch out!
Usually a or u is used for the terms of a sequence. So a_5 means the 5th term.
In this sequence $a_1 = 4$ and $a_5 = 16$.

$a_n = a + (n-1)d$ and $a_n = ar^{n-1}$ are called **position-to-term rules.**

Convergent means the series get closer and closer to a particular number – the **limit of the series.**
Each term is smaller than the one before so you keep on adding on a smaller and smaller number and the series **converges.**

Hint: This is called a recurrence relation.
$a_1 = 3$
$a_2 = 2 \times 3 + 1 = 7$
$a_3 = 2 \times 7 + 1 = 15$
$a_4 = 2 \times 15 + 1 = 31$

GET IT RIGHT

The 2nd term of a geometric series is -1 and the 5th term is 8.
Find the first term and the sum of the first nine terms.

Solution:

$a_2 = a \times r^1 = -1$ and $a_5 = a \times r^4 = 8$

Divide to find r: $\frac{a_5}{a_2} = \frac{a \times r^4}{a \times r} = \frac{8}{-1} \Rightarrow r^3 = -8 \Rightarrow r = -2$ (taking cube root)

$a_2 = -1 \Rightarrow a_1 = \frac{-1}{-2} = \frac{1}{2}$ so $S_n = \frac{a(1-r^n)}{(1-r)} \Rightarrow S_9 = \frac{\frac{1}{2}(1-(-2)^9)}{1-(-2)} = 85.5$

Watch out! Use brackets when you enter this into your calculator.

YOU ARE THE EXAMINER

Which parts of these solutions is correct? Where are the errors in the other solution?

An arithmetic series has first term 100 and common difference −3.

a) Write down the first 5 terms and find the *n*th term of the series.

b) The sum of the first *n* terms is *n*. Find the value of *n*.

SAM'S SOLUTION

a) $a = 100, d = -3:$

$100, 97, 94, 91, 88$

$a_n = a + (n-1)d$

$a_n = 100 + (n-1) \times (-3)$

$= 100 - 3n - 3$

$= 97 - 3n$

b) $S_n = \frac{n}{2}[2a + (n-1)d]$

$n = \frac{n}{2}[200 - 3(n-1)]$

$2n = 200n - 3n^2 + 3n$

$3n^2 - 201n = 0$

$3n(n - 67) = 0$

$n = 67$

LILIA'S SOLUTION

a) $100, 97, 94, 91, 88$

$a_n = 100 - 3(n-1)$

$= 100 - 3n + 3$

$= 103 - 3n$

b) $S_n = \frac{n}{2}[2a + (n-1)d]$

$\frac{n}{2}[200 + (n-1) \times (-3)] = n$

$100n - 3n^2 + 3n = n$

$3n^2 + 103n = n$

$3n^2 - 102n = 0$

$3n(n - 34) = 0$

$n = 34$

SKILL BUILDER

1 For each of these arithmetic sequences find:

i) the *n*th term ii) the 50th term iii) the sum of the first 100 terms.

a) $2, 5, 8, 11, \ldots$ b) $40, 37, 34, 31, \ldots$ c) $-\frac{7}{4}, -\frac{3}{2}, -\frac{5}{4}, -1, \ldots$

2 For each of these geometric series find:

i) the *n*th term ii) the 10th term iii) the sum of the first 10 terms iv) the sum to infinity (if the series converges).

a) $2 + -6 + 18 + -54 + \ldots$ b) $256 + 64 + 16 + 4 + \ldots$ c) $1 + \frac{1}{2} + \frac{1}{4} + \frac{1}{8} + \ldots$

Hint: There is only a sum to infinity if r is between −1 and 1.

3 Work out the value of:

a) $\sum_{r=1}^{5} r^2$ b) $\sum_{k=1}^{3} a_k$ given $a_k = 2k + 3$ c) $\sum_{r=1}^{20}(3r-7) - \sum_{r=1}^{18}(3r-7)$.

Hint: Look for terms which cancel.

4 The numbers $p, 4, q$ form a geometric sequence.
Which of the following values of p and q are possible?

A $p = 0, q = 8$ **B** $p = 0, q = 16$ **C** $p = 1, q = 16$ **D** $p = 6, q = 2$

5 The first three terms of an arithmetic sequence are −3, 2 and 7.
What is the sum of the first 12 terms?

A 52 **B** 294 **C** 324 **D** 648

6 The 2nd term of an arithmetic sequence is 7 and the 6th term is −5.
Three of the following statements are false and one is true. Which one is true?

A The common difference is 3. **B** The first term is 13. **C** $a_k < -10$ if $k \geq 8$. **D** $\sum_{k=2}^{8} a_k = -4$.

Binomial expansions

Year 1 and 2

THE LOWDOWN

① A **binomial expression** has two terms, such as $(2x + y)$ and $(x - a)$.

② You can raise a binomial expression to a power and expand the brackets to make a **binomial expansion.**

Example:

$$(a+b)^1 = 1a + 1b$$
$$(a+b)^2 = 1a^2 + 2ab + 1b^2$$
$$(a+b)^3 = 1a^3 + 3a^2b + 3ab^2 + 1b^3$$
$$(a+b)^4 = 1a^4 + 4a^3b + 6a^2b^2 + 4a^3b + 1b^4$$

The red numbers are binomial coefficients and are from Pascal's triangle. Each number in the triangle is found by **adding** the **two numbers above it**. You can use the ${}^nC_r$ button on your calculator to work out the red numbers.

Example ${}^4C_0 = 1$, ${}^4C_1 = 4$, ${}^4C_2 = 6$, ${}^4C_3 = 4$ and ${}^4C_3 = 1$

You can use this pattern to expand expressions such as $(2x - 3)^5$.
Writing out the whole expansion gives:

${}^5C_0 = 1$	${}^5C_1 = 5$	${}^5C_2 = 10$	${}^5C_3 = 10$	${}^5C_4 = 5$	${}^5C_5 = 1$
$(2x)^5$	$(2x)^4$	$(2x)^3$	$(2x)^2$	$(2x)^1$	$(2x)^0$
$(-3)^0$	$(-3)^1$	$(-3)^2$	$(-3)^3$	$(-3)^4$	$(-3)^5$
$1 \times 32x^5 \times 1$	$5 \times 16x^4 \times (-3)$	$10 \times 8x^3 \times 9$	$10 \times 4x^2 \times (-27)$	$5 \times 2x^1 \times 81$	$1 \times 1 \times (-243)$

$$(2x - 3)^5 = 32x^5 - 240x^4 + 720x^3 - 1080x^2 + 810x - 243$$

③ The **Binomial theorem** says when n is a positive whole number:

$$(a+b)^n = {}^nC_0a^n + {}^nC_1a^{n-1}b + {}^nC_2a^{n-2}b^2 + \ldots + {}^nC_nb^n$$

④ The **Binomial theorem** says when n is negative or a fraction you can expand expressions in the form $(1 + x)^n$ using:

$$(1+x)^n = 1 + nx + \frac{n(n-1)}{1\times2}x^2 + \frac{n(n-1)(n-2)}{1\times2\times3}x^3 + \ldots \text{ valid for } -1 < x < 1$$

◀◀ **Expanding brackets** on page 8.
▶▶ **Binomial distribution** on page 100 and **Hypothesis testing** on page 102.

A **coefficient** is the number in front of a **variable**.

Row 1: Work out the binomial coefficients.

Rows 2 and 3: Use brackets! The powers of $2x$ and -3 in each column add to give 5.

Row 4: Multiply the 3 terms above.

Watch out! You must add the terms to get the answer.

Watch out! This form of the binomial theorem is only valid when x is between -1 and 1.

GET IT RIGHT

a) Find the first four terms in the expansion of $(1 + x)^{-3}$.

b) Find the first three terms in the expansion of $\dfrac{1}{\sqrt{1-2x}}$.

Solution:

a) $(1+x)^{-3} = 1 + (-3)x + \dfrac{(-3)(-3-1)}{1\times2}x^2 + \dfrac{(-3)(-3-1)(-3-2)}{1\times2\times3}x^3 + \ldots$

$= 1 - 3x + 6x^2 - 10x^3 + \ldots$ valid for $-1 < x < 1$

b) $\dfrac{1}{\sqrt{1-2x}} = (1-2x)^{-\frac{1}{2}} = 1 + \left(-\dfrac{1}{2}\right)(-2x) + \dfrac{\left(-\frac{1}{2}\right)\left[\left(-\frac{1}{2}\right)-1\right]}{1\times2}(-2x)^2 + \ldots$

$= 1 + x + \dfrac{\left(-\frac{1}{2}\right)\left(-\frac{3}{2}\right)}{2}4x^2 + \ldots$

$= 1 + x + \dfrac{3}{2}x^2 + \ldots$ valid for $-1 < 2x < 1$ so $-\dfrac{1}{2} < x < \dfrac{1}{2}$

Watch out! This series continues forever so you will only be asked to give the first few terms.

Watch out! When you expand $(1 + ax)^n$ the expansion is valid when $-1 < ax < 1$

So $-\dfrac{1}{a} < x < \dfrac{1}{a}$.

YOU ARE THE EXAMINER

Which one of these solutions is correct? Where are the errors in the other solution?

Find the term independent of x in the expansion of $\left(2x - \frac{1}{x}\right)^{12}$.

Hint: Independent of x means there is no x in that term.

MO'S SOLUTION

The x's cancel so powers of x are equal.

The powers always add to give 12, so we need the terms in $(2x)^6$ and $\left(-\frac{1}{x}\right)^6$.

...	${}^{12}C_6 = 924$	...
...	$(2x)^6$	...
...	$\left(-\frac{1}{x}\right)^6$	...
...	$924 \times 64x^6 \times \frac{1}{x^6}$	...

Term is 59136.

NASREEN'S SOLUTION

The x's cancel so powers of x are equal.

The powers always add to give 12, so we need the terms in $2x^6$ and $-\frac{1}{x^6}$.

...	${}^{12}C_6 = 924$	...
...	$2x^6$	...
...	$-\frac{1}{x^6}$	...
...	$924 \times 2x^6 \times \left(-\frac{1}{x^6}\right)$	...

Term is −1848.

SKILL BUILDER

1 Expand these expressions.

a) $(1 + x)^4$ b) $(t - 2)^5$ c) $(1 + 2y)^3$ d) $(5 - 2p)^3$

2 Write out the binomial expansion of $(1 - 3x)^4$.

A $1+12x+54x^2+108x^3+81x^4$ **B** $1-12x+54x^2-108x^3+81x^4$

C $1-12x-18x^2-12x^3-3x^4$ **D** $1-12x+108x^2-684x^3+1944x^4$

E $1-3x+9x^2-27x^3+81x^4$

3 Simplify $(x - 1)^3 + (x + 1)^3$.

A $2x^3 + 6x$ **B** $2x^3$ **C** $2x^3 + 2x$ **D** $2x^3 + 6x^2 + 6x + 2$ **E** $8x^3$

4 Find the coefficient of x^5 in the expansion of $(2 - x)^8$.

A $-{}^8C_5$ **B** $-{}^8C_5 \times 2^5$ **C** ${}^8C_5 \times 2^3$ **D** $-{}^8C_5 \times 2^3$ **E** $-{}^8C_3 \times 2^5$

5 Find the first four terms in the binomial expansion of $(1 - 5x)^{-2}$.

A $1+10x+75x^2+500x^3$ **B** $1-10x+75x^2-500x^3$

C $1-2x+3x^2-4x^3$ **D** $1+10x+15x^2+20x^3$

6 Use the first three terms in the expansion of $\sqrt{1-x}$ to find an approximation for $\sqrt{0.95}$.

A 1.024 687 5 **B** 0.974 687 5 **C** 0.974 679 687 5 **D** 0.975 312 5

Hint: Solve $1 - x = 0.95$.

7 Find a, b and c such that:

$$\frac{1}{(1+3x)^3} \approx 1 + ax + bx^2 + cx^3.$$

A $a = 9, b = 27$ and $c = 27$ **B** $a = -3, b = 6$ and $c = -10$

C $a = -9, b = 18$ and $c = -30$ **D** $a = -9, b = 54$ and $c = -270$

Partial fractions

Year 2

THE LOWDOWN

① You learnt how to add together algebraic fractions on page 14.

Example $\frac{2}{x}+\frac{3}{x+2}=\frac{2(x+2)}{x(x+2)}+\frac{3x}{x(x+2)}=\frac{2x+4+3x}{x(x+2)}=\frac{5x+4}{x(x+2)}$

In this topic, you learn how to go **from** $\frac{5x+4}{x(x+2)}$ **back to** $\frac{2}{x}+\frac{3}{x+2}$.

This is called expressing $\frac{5x+4}{x(x+2)}$ in **partial fractions**.

② There are two forms you need to learn:

TYPE 1: Linear factors on the bottom line:

$$\frac{ex+f}{(ax+b)(cx+d)}\equiv\frac{A}{(ax+b)}+\frac{B}{(cx+d)}$$

Example Express $\frac{x+7}{(x-2)(x+1)}$ in partial fractions.

Step 1 $\frac{x+7}{(x-2)(x+1)}\equiv\frac{A}{(x-2)}+\frac{B}{(x+1)}$

Step 2 Multiply by $(x-2)(x+1)$: $x+7\equiv A(x+1)+B(x-2)$

Step 3 Choose $x=2$: $2+7=A(2+1)\Rightarrow 9=3A$ so $A=3$

Choose $x=-1$: $-1+7=B(-1-2)\Rightarrow 6=-3B$ so $B=-2$

Step 4 Write down the partial fractions: $\frac{x+7}{(x-2)(x+1)}=\frac{3}{(x-2)}-\frac{2}{(x+1)}$

TYPE 2: A **squared factor** on the bottom line:

$$\frac{ex^2+fx+g}{(ax+b)(cx+d)^2}\equiv\frac{A}{(ax+b)}+\frac{B}{(cx+d)}+\frac{C}{(cx+d)^2}$$

◀◀ **Algebraic fractions** on page 14.
▶▶ **Integration** on page 76.

Hint: The identity symbol (≡) means the equation is true for any value of x, so you can just substitute in any value of x.
$x=0$ is often a good one to use!

Hint: There are **two factors** in the bottom line so you need **two partial fractions**.

Hint: Choose values of x that make the brackets zero.

GET IT RIGHT

Express $\frac{x^2-3x-4}{x(x-2)^2}$ in partial fractions.

Solution:

Step 1 Choose right form: $\frac{x^2-3x-4}{x(x-2)^2}\equiv\frac{A}{x}+\frac{B}{(x-2)}+\frac{C}{(x-2)^2}$

Step 2 Multiply both sides by the bottom of the left-hand side (LHS)

$x^2-3x-4\equiv A(x-2)^2+Bx(x-2)+Cx$

Step 3 Choose values of x to find A and C.

Let $x=0$: $-4=A(0-2)^2\Rightarrow -4=4A$ so $A=-1$

Let $x=2$: $4-6-4=C\times 2\Rightarrow -6=2C\Rightarrow C=-3$

Look at x^2: On the left hand side there's $1x^2$.

On the right hand side there's $Ax^2+\ldots+Bx^2+\ldots$

Equate coefficients of x^2: $1=A+B$ but since $A=-1$ then $B=2$.

Step 4 Write down the partial fractions.

$$\frac{x^2-3x-4}{x(x-2)^2}\equiv-\frac{1}{x}+\frac{2}{(x-2)}-\frac{3}{(x-2)^2}$$

Hint: There are **three factors** in the bottom line so you need **three partial fractions**, the ones for B and C are to deal with the square.

Hint: If you expand the RHS you would get
$Ax^2-4Ax+4A$
$+Bx^2-2Bx+Cx$
but you only need the x^2 terms.

YOU ARE THE EXAMINER

Which one of these solutions is correct? Where are the errors in the other solution?

Express $\frac{x-1}{(2x+1)(x-2)}$ in partial fractions.

MO'S SOLUTION

$\frac{x-1}{(2x+1)(x-2)} \equiv \frac{A}{(2x+1)} + \frac{B}{(x-2)}$

So $x - 1 \equiv A(2x+1) + B(x-2)$

Let $x = 2$: $2 - 1 = A(2 \times 2 + 1)$

So $1 = 5A \Rightarrow A = 0.2$

Let $x = -\frac{1}{2}$: $-\frac{1}{2} - 1 = B\left(-\frac{1}{2} - 2\right)$

So $-1.5 = -2.5B \Rightarrow B = 0.6$

So $\frac{x-1}{(2x+1)(x-2)} \equiv \frac{0.2}{(2x+1)} + \frac{0.6}{(x-2)}$

LILIA'S SOLUTION

$\frac{x-1}{(2x+1)(x-2)} \equiv \frac{A}{(2x+1)} + \frac{B}{(x-2)}$

So $x - 1 \equiv A(x-2) + B(2x+1)$

Let $x = -\frac{1}{2}$: $-\frac{3}{2} = A\left(-\frac{1}{2} - 2\right)$

So $-\frac{3}{2} = -\frac{5}{2}A \Rightarrow A = \frac{3}{5}$

Let $x = 2 : 1 = 5B \Rightarrow B = \frac{1}{5}$

So $\frac{x-1}{(2x+1)(x-2)} \equiv \frac{3}{5(2x+1)} + \frac{1}{5(x-2)}$

SKILL BUILDER

1 Simplify the following.

a) $\frac{A}{(x-3)} \times (x+2)(x-3) + \frac{B}{(x+2)} \times (x+2)(x-3)$

b) $\frac{A}{(2x-1)} \times (2x-1)(x+5) + \frac{B}{(x+5)} \times (2x-1)(x+5)$

2 Work out the values of A and B.

a) $2x \equiv A(x-2) + B(x+3)$

b) $4x + 1 \equiv A(x-3) + B(2x-1)$

3 Write $\frac{3x+1}{(x-3)(x+2)}$ in the form $\frac{A}{(x-3)} + \frac{B}{(x+2)}$.

4 Write $\frac{2x-7}{(x+1)(x-2)}$ in the form $\frac{A}{(x+1)} + \frac{B}{(x-2)}$.

5 Write $\frac{4x^2+9x-1}{(x-1)(x+1)(x+2)}$ in the form $\frac{A}{(x-1)} + \frac{B}{(x+1)} + \frac{C}{(x+2)}$.

6 Write $\frac{4x^2-4x+3}{x(x-1)^2}$ in the form $\frac{A}{x} + \frac{B}{(x-1)} + \frac{C}{(x-1)^2}$.

7 a) Factorise $2x^2 - x - 6$.

b) Express $\frac{x-9}{2x^2-x-6}$ in partial fractions.

8 Which of the following is the correct form of partial fractions for the expression $\frac{x}{x^2-3x-4}$?

A $\frac{A}{x+4} + \frac{B}{x-1}$

B $\frac{Ax}{x+1} + \frac{Bx}{x-4}$

C $\frac{Ax+B}{x^2-3x-4}$

D $\frac{A}{x+1} + \frac{B}{x-4}$

9 Express $\frac{4+6x-x^2}{(x-1)(x+2)^2}$ as the sum of partial fractions.

A $\frac{1}{x-1} + \frac{4}{(x+2)^2}$

B $\frac{1}{x-1} - \frac{2}{x+2} + \frac{4}{(x+2)^2}$

C $\frac{1}{x-1} - \frac{2}{x+2} + \frac{4}{3(x+2)^2}$

D $\frac{6}{x+1} + \frac{4}{(x+2)^2} - \frac{1}{x-1}$

Using coordinates

THE LOWDOWN

① The position of a point on a grid is given by its **coordinates** (x, y).

Example A(−3, −2) and B(3, 1)
C(−1, 4) and D(2, −2)

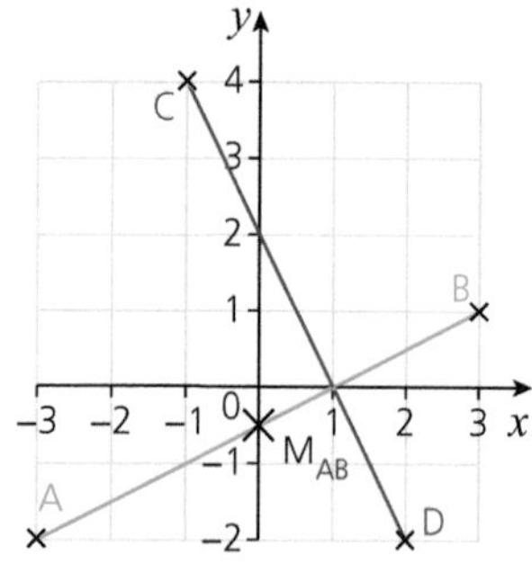

- The **midpoint** of two points $A(x_1, y_1)$ and $B(x_2, y_2)$ is halfway between them.
- Midpoint $= \left(\frac{x_1 + x_2}{2}, \frac{y_1 + y_2}{2}\right)$

Example $M_{AB} = \left(\frac{-3+3}{2}, \frac{-2+1}{2}\right) = \left(0, -\frac{1}{2}\right)$

- Use Pythagoras' theorem to find the distance between two points.

Distance AB $= \sqrt{(x_2 - x_1)^2 + (y_2 - y_1)^2}$

Example $AB = \sqrt{(3-(-3))^2 + (1-(-2))^2} = \sqrt{6^2 + 3^2} = \sqrt{45} = 3\sqrt{5}$

$CD = \sqrt{(2-(-1))^2 + (-2-4)^2} = \sqrt{3^2 + (-6)^2} = \sqrt{45} = 3\sqrt{5}$

Note: It doesn't matter which point you use as (x_1, y_1).

② The gradient (steepness) of a line is the amount y increases for every 1 that x increases.

Gradient, $m = \frac{\text{change in } y\text{-coordinates}}{\text{change in } x\text{-coordinates}} = \frac{y_2 - y_1}{x_2 - x_1}$

Example: $m_{AB} = \frac{1-(-2)}{3-(-3)} = \frac{3}{6} = \frac{1}{2}$

and

$m_{CD} = \frac{-2-4}{2-(-1)} = \frac{-6}{3} = -2$

③ **Parallel lines** have the **same** gradient:
$m_1 = m_2$

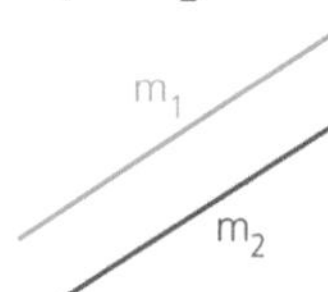

④ **Perpendicular lines** are at right angles (90°) to each other.

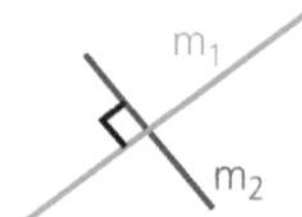

Perpendicular lines have gradients that **multiply to give −1.**

$m_1 \times m_2 = -1$ so $m_1 = -\frac{1}{m_2}$

Note: the gradients are **negative reciprocals** of each other.

Example $m_{AB} \times m_{CD} = \frac{1}{2} \times (-2) = -1$, so AB and CD are perpendicular to each other.

GET IT RIGHT

The point A has coordinates (−3, 2) and the midpoint of AB is at (−1, −1).

a) Find the coordinates of B.

b) Find the gradient of a line perpendicular to AB.

Solution:

a) Midpoint (−1, −1) is 2 right and 3 down from A(−3, 2).
So B is 2 right and 3 down from (−1, −1), so B has coordinates (1, −4).

b) A(−3, 2) and B(1, −4): $m_{AB} = \frac{-4-2}{1-(-3)} = \frac{-6}{4} = -\frac{3}{2}$

Negative reciprocal: $m_{perp} = \frac{2}{3}$

Hint: It helps to draw a diagram.

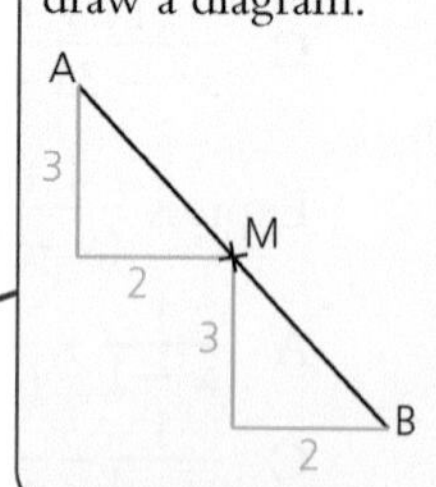

Hint: Turn the fraction 'upside down' and change the sign.

YOU ARE THE EXAMINER

Which one of these solutions is correct? Where are the errors in the other solution?

A quadrilateral ABCD has vertices at A(−5, 4), B(3, 5), C(7, −2) and D(−1, −3).

a) Find the lengths of the sides AB and CD.

b) Show that the diagonals are perpendicular to each other.

MO'S SOLUTION

a) $AB = \sqrt{(3-(-5))^2 + (5-4)^2}$

$= \sqrt{8^2 + 1^2} = \sqrt{65}$

$CD = \sqrt{(-1-7)^2 + (-3-(-2))^2}$

$= \sqrt{8^2 + (-1)^2} = \sqrt{65}$

b) $m_{AC} = \dfrac{-2-4}{7-(-5)} = -\dfrac{6}{12} = -\dfrac{1}{2}$

$m_{BD} = \dfrac{-3-5}{-1-3} = \dfrac{-8}{-4} = 2$

$m_{AC} \times m_{BD} = -\dfrac{1}{2} \times 2 = -1$

so perpendicular.

PETER'S SOLUTION

a) $AB = \sqrt{(5-4)^2 + (3-5)^2}$

$= \sqrt{-1^2 + 8^2} = \sqrt{63}$

$CD = \sqrt{(7-1)^2 - (-2-3)^2}$

$= \sqrt{36-25} = \sqrt{11}$

b) $m_{AC} = \dfrac{-2-4}{7--5} = -\dfrac{6}{2} = -3$

Negative reciprocal of −3 is $\dfrac{1}{3}$

So $m_{BD} = \dfrac{1}{3}$

SKILL BUILDER

1 The points A and B have coordinates (−1, 4) and (3, −2). What is the gradient of the line AB?

A −3 **B** $-\dfrac{2}{3}$ **C** −1 **D** −1.5 **E** 1.5

2 The midpoint of the line AB is (−2, 1). The coordinates of point B are (1, −1). What are the coordinates of A?

A (−5, 3) **B** (−3, 3) **C** (−0.5, 0) **D** (4, −3) **E** (−5, 1)

3 The points A and B have coordinates (−3, 1) and (2, 5). Find the length of the line AB.

A 3 **B** $\sqrt{41}$ **C** $\sqrt{17}$ **D** 41 **E** 5

4 The trapezium PQRS has vertices P(0, 1), Q(6, 4), R(4, z) and S(0, 5).

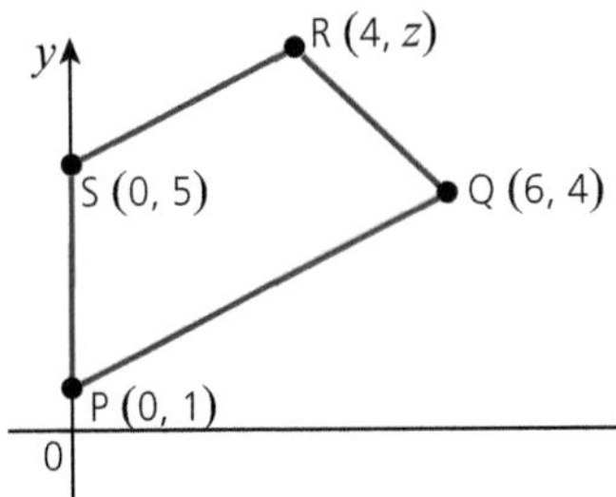

Hint: A trapezium has one pair of parallel sides.

What is the value of z?

A 13 **B** 7 **C** 8 **D** $4\frac{1}{3}$ **E** 5

5 The points A, B, C and D have coordinates (1, 5), (3, −1), (1, 2) and (x, y). CD is perpendicular to AB. Which of the following could be the coordinates of D?

A (7, 4) **B** (2, −1) **C** (2, 5) **D** (3, 8) **E** (7, 0)

6 The midpoint of the line segment PQ is $\left(-\frac{1}{2}, 1\right)$ and Q is the point (1, 3).

a) Find the coordinates of P.

PQ is perpendicular to PR and R is the point (2, y).

b) Find the coordinates of R.

c) Show that triangle PQR is isosceles.

d) What is the area of triangle PQR?

Straight line graphs

THE LOWDOWN

① The equation of a straight line can be written in the form $y = mx + c$ where m is the **gradient** and c is the y-intercept at $(0, c)$.

When m or c are not whole numbers the equation of a line is often written in the form $ax + by + c = 0$.

Example The line $3x - 2y - 6 = 0$ can be written as

$2y = 3x - 6$

$\Rightarrow y = \frac{3}{2}x - 3$ so the gradient is $\frac{3}{2}$ and y-intercept at $(0, -3)$.

② You can use $y - y_1 = m(x - x_1)$ to find the equation of a line if you are given the **gradient**, m, and the coordinates of one point on the line (x_1, y_1).

Example The equation of the line with gradient 3 through $(-2, 5)$ is

$y - 5 = 3(x - (-2))$

$\Rightarrow y - 5 = 3(x + 2)$

$\Rightarrow y - 5 = 3x + 6$

$\Rightarrow \quad y = 3x + 11$

③

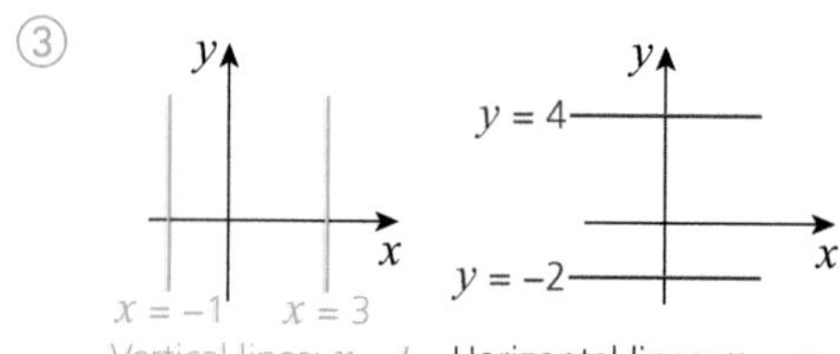

④ You can find where two lines **intersect** (meet) by solving their equations simultaneously.

Example $y = 2x + 1$ and $2y = 7 - x$ intersect at P.

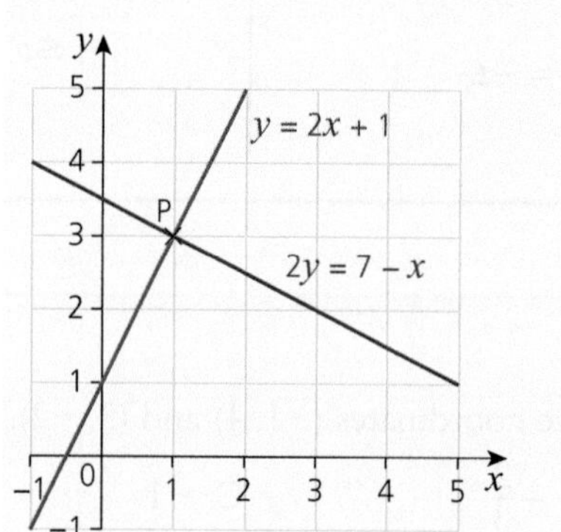

Solving gives: $2(2x + 1) = 7 - x$

$4x + 2 = 7 - x$

$\Rightarrow \quad 5x = 5$

$\Rightarrow \quad x = 1$

When $x = 1$, $2y = 7 - 1 \Rightarrow y = 3$

So the lines intersect at $(1, 3)$.

Watch out! The line has to be in the form $y = mx + c$ first!

Watch out! Take care with your signs!

Uphill gradients are positive.

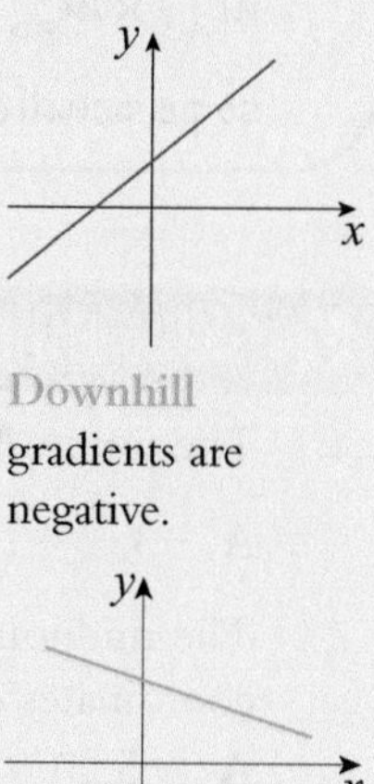

Downhill gradients are negative.

Watch out: Make sure you give your answer as coordinates.

GET IT RIGHT

The line L is perpendicular to $2x + 3y - 5 = 0$ and passes through $(2, -1)$. Find the equation of L, giving your answer in the form $ax + by + c = 0$.

Solution:

Step 1 Rearrange line into form $y = m_1x + c$: $3y = -2x + 5 \Rightarrow y = -\frac{2}{3}x + \frac{5}{3}$

Step 2 Write down gradient: $m_1 = -\frac{2}{3}$

Step 3 Find negative reciprocal: $m_2 = \frac{3}{2}$

Step 4 Use $y - y_1 = m_2(x - x_1)$ where $x_1 = 2, y_1 = -1$: $y - (-1) = \frac{3}{2}(x - 2)$

Step 5 Multiply by 2 to clear fraction: $2(y + 1) = 3(x - 2)$

Expand brackets: $2y + 2 = 3x - 6$

Write in form $ax + by + c = 0$: $3x - 2y - 8 = 0$

◀◀ **Solving simultaneous equation** on page 4 and **using coordinates** on page 32.

▶▶ **Solving non-linear simultaneous equations** on page 38 and **equations of tangents and normals** on page 62.

Watch out! Give your answer in the correct form!

★ YOU ARE THE EXAMINER

Which one of these solutions is correct? Where is the error in the other solution?

Find the equation of the line that passes through the points (2, −5) and (−3, 4).

MO'S SOLUTION

$$m = \frac{-5-4}{2-(-3)} = -\frac{9}{5}$$

$$y - 5 = -\frac{9}{5}(x-2)$$

$$\Rightarrow 5(y-5) = -9(x-2)$$

$$\Rightarrow 5y - 25 = -9x - 18$$

$$\Rightarrow 9x + 5y - 7 = 0$$

PETER'S SOLUTION

$$m = \frac{4-(-5)}{-3-2} = -\frac{9}{5}$$

$$y - 4 = -\frac{9}{5}(x-(-3))$$

$$\Rightarrow 5(y-4) = -9(x+3)$$

$$\Rightarrow 5y - 20 = -9x - 27$$

$$\Rightarrow 9x + 5y + 7 = 0$$

✓ SKILL BUILDER

1 Find the equation of each of these lines.

a) Passing through (0, 3), gradient 2

b) Passing through (1, 3), gradient 3

c) Passing through (0, −7), parallel to $y = 5x - 2$

d) Passing through (−2, 2), parallel to $y = \frac{1}{2}x - 2$

e) Passing through (1, −2), parallel to $3y - 2x = 1$

2 Find the equation of each of these lines.

a) Passing through (0, −1), perpendicular to $y = -3x - 2$

b) Passing through (−3, −2), perpendicular to $y = \frac{1}{2}x - 2$

c) Passing through (−1, 4), perpendicular to $2y + 3x + 7 = 0$

3 Find the equation of the line through (−1, 3) and (2, −3).

A $y = -6x - 3$ **B** $y = -2x + 5$ **C** $y = -2x + 1$ **D** $y = 2x + 5$ **E** $y = -6x + 17$

Hint: Work out the gradient first.

4 A line has equation $5x - 7y + 2 = 0$. Find its gradient.

A $-\frac{5}{7}$ **B** 5 **C** $\frac{5}{7}$ **D** $-\frac{7}{5}$ **E** $\frac{7}{5}$

5 Which of these is the equation of the line in the diagram?

A $x + 2y = 6$ **B** $y = -2x + 3$ **C** $2x + y = 6$

D $x + 2y = 3$ **E** $2y - x = 6$

6 Find the equation of the line perpendicular to $y = -4x + 1$ and passing through (2, 1).

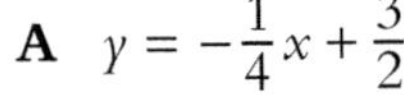

A $y = -\frac{1}{4}x + \frac{3}{2}$ **B** $y = \frac{1}{4}x + \frac{3}{2}$ **C** $y = \frac{1}{4}x + \frac{1}{2}$

D $y = 4x - 7$ **E** $y = \frac{1}{4}x + \frac{7}{4}$

7 The line L is parallel to $y = 3x - 2$ and passes through the point (−2, −1). Find the coordinates of the point of intersection with the x-axis.

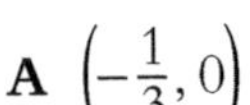

A $\left(-\frac{1}{3}, 0\right)$ **B** $\left(-\frac{5}{3}, 0\right)$ **C** (0, 5) **D** $\left(\frac{7}{3}, 0\right)$ **E** $\left(\frac{2}{3}, 0\right)$

8 Find the coordinates of the point where the lines $2x + 3y = 12$ and $3x - y = 7$ intersect.

A $\left(\frac{9}{7}, \frac{22}{7}\right)$ **B** (3, 2) **C** (9, 2) **D** (3, 6) **E** $\left(\frac{33}{8}, \frac{5}{4}\right)$

Circles

THE LOWDOWN

① Make sure you know these **circle theorems** from GCSE maths.

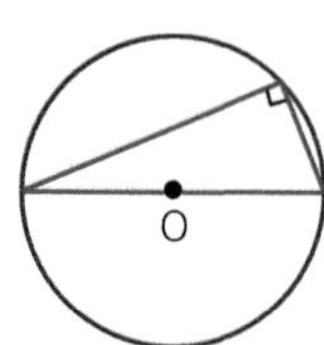

The angle in a semi-circle is 90°

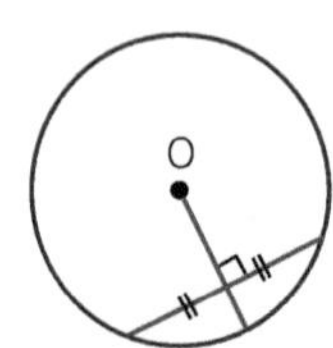

A radius bisects a chord at 90°

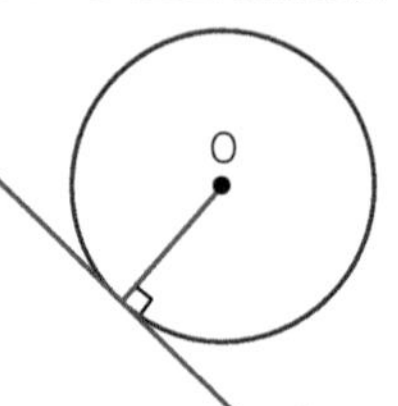

A tangent and radius meet at right-angles

A tangent is a line that touches a curve exactly once.

② The equation of a circle with radius and centre (0, 0) is $x^2 + y^2 = r^2$.

Example The radius of a circle is 4 and its centre is at (0,0). The equation of the circle is $x^2 + y^2 = 16$.

③ The equation of a circle with radius and centre (a, b) is $(x - a)^2 + (y - b)^2 = r^2$

Hint: You can work out where a circle cuts the axis by substituting $x = 0$ and $y = 0$. There is no need to expand brackets!

Example The equation of this circle with centre $(-1, 2)$ is $(x + 1)^2 + (y - 2)^2 = 9$.

The radius is 3.

It crosses the y-axis when $x = 0$:

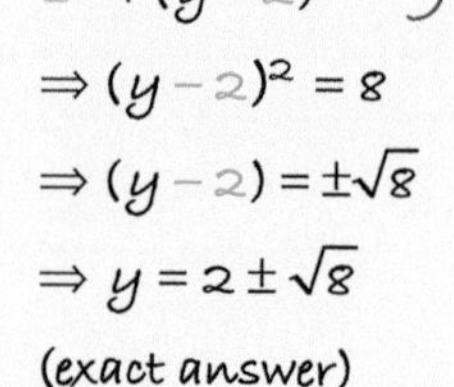

$1^2 + (y - 2)^2 = 9$

$\Rightarrow (y - 2)^2 = 8$

$\Rightarrow (y - 2) = \pm\sqrt{8}$

$\Rightarrow y = 2 \pm \sqrt{8}$

(exact answer)

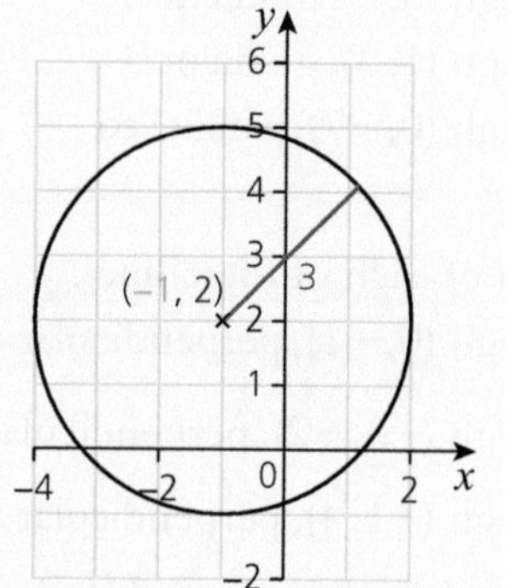

④ Sometimes the equation of a circle is written without brackets:

Example $x^2 + y^2 + 2x - 4y - 4 = 0$

Gather x and y terms: $x^2 + 2x + y^2 - 4y - 4 = 0$

Complete the square: $(x + 1)^2 - 1 + (y - 2)^2 - 4 - 4 = 0$

Write in circle form: $(x + 1)^2 + (y - 2)^2 = 9$

GET IT RIGHT

The equation of a circle is $(x - 2)^2 + (y - 3)^2 = 41$.

The point P(6, 8) lies on the circle. Find the equation of the tangent to the circle at P.

Solution:

The centre of the circle is (2, 3).

The radius and the tangent meet at right angles.

So gradient of tangent $= \dfrac{-1}{\text{gradient of radius}}$

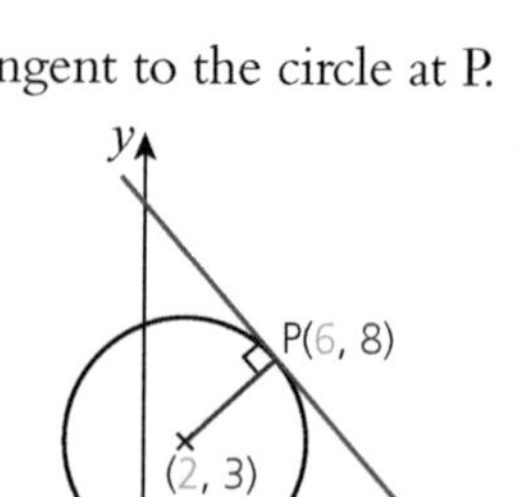

$m_{radius} = \dfrac{8 - 3}{6 - 2} = \dfrac{5}{4}$ so $m_{tangent} = -\dfrac{4}{5}$

Equation of line through (6, 8) with gradient $-\frac{4}{5}$ is $y - 8 = -\frac{4}{5}(x - 6)$

Multiply by 5, expand and simplify: $5y - 40 = -4x + 24 \Rightarrow 4x + 5y = 64$

Hint: You can show a point lies on a circle by substituting the coordinates into the equation.

Show (6, 8) lies on the circle

$(x - 2)^2 + (y - 3)^2 = 41$

Substituting gives

$(6 - 2)^2 + (8 - 3)^2 = 41$

$\Rightarrow 4^2 + 5^2 = 41$ ✓

YOU ARE THE EXAMINER

Which one of these solutions is correct? Where are the errors in the other solution?

AB is the diameter of a circle.

Given A(−1, 1) and B(5, −7), work out the equation of the circle.

MO'S SOLUTION

Centre of circle is midpoint of AB.

So C is at $\left(\frac{-1+5}{2}, \frac{1+(-7)}{2}\right) = (2, -3)$

Radius of circle is the distance between (−1, 1) and (2, −3).

So radius $= \sqrt{(-1-2)^2 + (1-(-3))^2}$

$= \sqrt{(-3)^2 + (4)^2} = \sqrt{9+16} = 5$

So the equation of the circle is

$(x-2)^2 + (y+3)^2 = 25$

PETER'S SOLUTION

Centre of circle is midpoint of AB.

So C is at $\left(\frac{-1-5}{2}, \frac{1-(-7)}{2}\right) = (-3, 4)$

Radius of circle is half the diameter.

So diameter$^2 = (-1-5)^2 + (1-(-7))^2$

$= (-6)^2 + 8^2 = 100$

So radius$^2 = 50$

So the equation of the circle is

$(x+3)^2 + (y-4)^2 = 50$

SKILL BUILDER

1 Write down the equation of each of these circles.

a) Centre (0, 0), radius 3

b) Centre (2, 5), radius 4

c) Centre (−1, −3), radius 6

d) Centre (−4, 0), radius 1

2 AB is the diameter of a circle.

Given A(−2, 4) and B(4, 0), work out the equation of the circle.

3 The equation of a circle is $x^2 + y^2 = 10$.

The point P(1, 3) lies on the circle. Find the equation of the tangent to the circle at P.

4 What is the equation of the circle with centre (1, −3) and radius 5?

A $(x+1)^2 + (y-3)^2 = 25$

B $(x-1)^2 + (y+3)^2 = 25$

C $(x+1)^2 + (y-3)^2 = 5$

D $(x-1)^2 + (y+3)^2 = 5$

E $x^2 + y^2 = 25$

5 Give the centre and the radius of the circle with equation $x^2 + y^2 + 6x - 4y - 36 = 0$.

A Centre (−3, 2), radius 7

B Centre (−6, 4), radius 6

C Centre (−3, 2), radius 6

D Centre (3, −2), radius 7

E Centre (3, −2), radius 6

Hint: Gather x and y terms and use completing the square.

6 Find where the circle $(x-1)^2 + (y+2)^2 = 16$ crosses the positive x-axis.

A $x = 1 + 4\sqrt{3}$ **B** $x = 3$ **C** $x = 1 + 2\sqrt{3}$ **D** $x = 5$ **E** $x = \sqrt{15} - 2$

Hint: Substitute $y = 0$ into the equation.

7 Find the equation of the tangent to the circle $(x-1)^2 + (y+1)^2 = 34$ at the point (6, 2).

A $y = -7x + 44$ **B** $y = -\frac{5}{3}x + 12$ **C** $y = -\frac{5}{3}x + 9\frac{1}{3}$

D $y = \frac{3}{5}x - 1\frac{3}{5}$ **E** $y = -5x + 32$

Hint: Look at the 'Get it right' box.

8 A(2, −1) and B(4, 3) are two points on a circle with centre (1, 2). What is the distance of the chord from the centre of the circle?

A $2\sqrt{5}$ **B** 5 **C** $\sqrt{10}$ **D** $2\sqrt{2}$ **E** $\sqrt{5}$

Hint: Draw a diagram and use one of the circle theorems from Point 1 in the Lowdown.

Intersections

THE LOWDOWN

① There are **three possibilities** when you draw the graphs of a straight line and a curve:

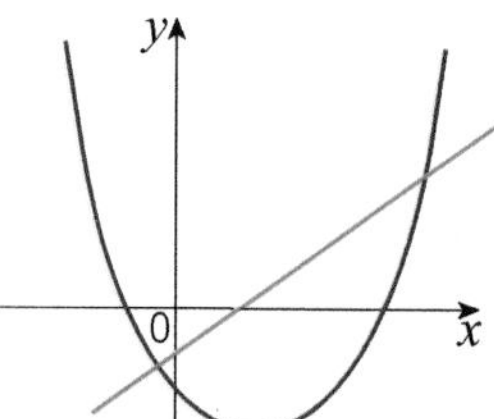

Two (or more) points of intersection

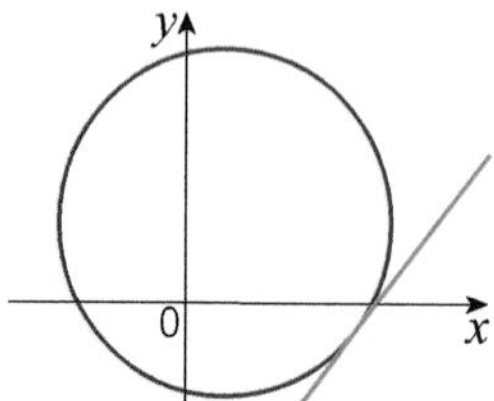

One point of intersection (a tangent)

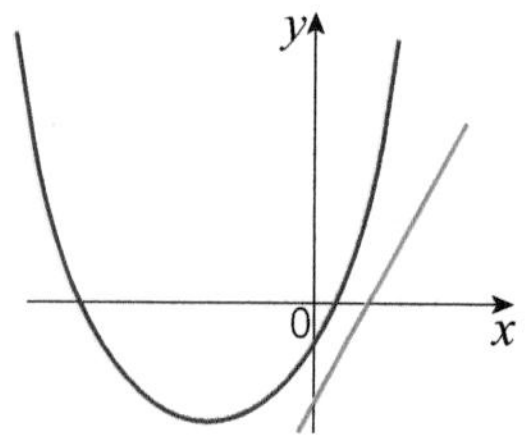

No points of intersection

② To find the coordinates of the point(s) of intersection you need to solve the equations of the line and the curve simultaneously.

Example To find where $y = x^2 + 2x$ and $y = 3x + 2$ intersect

Step 1 Equate '$y =$ line' and '$y =$ curve':

$$x^2 + 2x = 3x + 2$$

Step 2 Rearrange to $= 0$: $x^2 - x - 2 = 0$

Step 3 Solve: $(x + 1)(x - 2) = 0 \Rightarrow x = -1, x = 2$

Step 4 Find y: When $x = -1, y = 3 \times (-1) + 2 = -1$

When $x = 2, y = 3 \times 2 + 2 = 8$

Step 5 Write down the coordinates: $(-1, -1)$ and $(2, 8)$

③ **Using the discriminant** $b^2 - 4ac$ to determine if a line intersects a curve. Carry out **Steps 1** and **2** above.

Step 3 Work out $b^2 - 4ac$:

If $b^2 - 4ac = 0$ there is a repeated root $\Rightarrow$ line is a tangent

If $b^2 - 4ac < 0$ (negative) there are no real roots $\Rightarrow$ line misses the curve.

Watch out! The line could form a tangent in one place and cross the curve at another point!

Hint: Form a quadratic equation in just x (or y).

Watch out! Substitute the x-values into the **equation of the line** to find y, don't use the curve equation.

If $b^2 - 4ac$ is positive then the line meets the curve twice.

GET IT RIGHT

The line $y = 2x + k$ is a tangent to the curve $y = x^2 - 3$ at the point P. Find the value of k and the coordinates of P.

Solution:

Step 1 Both equations have '$y =$' so write them equal to each other.

$y = 2x + k$ and $y = x^2 - 3 \Rightarrow x^2 - 3 = 2x + k$

Step 2 Rearrange to give $x^2 - 2x - 3 - k = 0$

Step 3 Use $b^2 - 4ac$ where $a = 1$, $b = -2$ and $c = -3 - k$

So $b^2 - 4ac = (-2)^2 - 4 \times 1 \times (-3 - k)$

$= 4 - 4(-3 - k)$

$= 4 + 12 + 4k = 16 + 4k$

At a tangent $b^2 - 4ac = 0$ so $16 + 4k = 0 \Rightarrow k = -4$

Step 4 Solve your quadratic, using k to find the coordinates of P.

When $k = -4$ then $x^2 - 3 = 2x - 4 \Rightarrow x^2 - 2x + 1 = 0$

$\Rightarrow \quad (x - 1)^2 = 0$

So $x = 1$ (repeated root as line $y = 2x - 4$ is a tangent)

When $x = 1$ then $y = -2$ so P is at $(1, -2)$.

Watch out! Take extra care with your signs. Remember 'c' is the whole of the constant term including signs.

◀◀ **Solving quadratic equations** on page 8 and **simultaneous equations** on page 4.

▶▶ **Tangents and normals** on page 62.

YOU ARE THE EXAMINER

Which one of these solutions is correct? Where are the errors in the other solution?

Find the coordinates of the points where the line $y = x + 1$ intersects the circle $(x - 3)^2 + (y - 1)^2 = 5$

MO'S SOLUTION

Expand brackets

$x^2 - 6x - 9 + y^2 - 2y - 1 = 5$

Simplify $x^2 - 6x + y^2 - 2y - 15 = 0$

Substituting $y = x + 1$ gives

$x^2 - 6x + (x + 1)^2 - 2(x + 1) - 15 = 0$

Simplify

$x^2 - 6x + x^2 + 2x + 1 - 2x - 2 - 15 = 0$

$$2x^2 - 6x - 16 = 0$$

Solve $x = \dfrac{6 \pm \sqrt{(-6)^2 - 4 \times 2 \times (-16)}}{4}$

$$x = \frac{6 \pm \sqrt{164}}{4}$$

So $x = \dfrac{3 \pm \sqrt{41}}{2}$

PETER'S SOLUTION

Substituting $y = x + 1$ into

$(x - 3)^2 + (y - 1)^2 = 5$ gives

$$(x - 3)^2 + (x + 1 - 1)^2 = 5$$

$$\Rightarrow \quad (x - 3)^2 + x^2 = 5$$

Expand brackets: $x^2 - 6x + 9 + x^2 = 5$

Rearrange: $2x^2 - 6x + 4 = 0$

Divide by 2: $x^2 - 3x + 2 = 0$

Factorise: $(x - 1)(x - 2) = 0$

So $x = 1$ or $x = 2$

Using equation of line to find y:

When $x = 1$ then $y = 1 + 1 = 2$

When $x = 2$ then $y = 1 + 1 = 3$

So the line meets the circle at (1, 2) and (2, 3).

SKILL BUILDER

1 Find the coordinates of the points where each line intersects the curve.

a) $y = x^2$ and $y = x + 6$ b) $x^2 + y^2 = 9$ and $y = x + 3$

2 Simon is solving the simultaneous equations $y(1 - x) = 1$ and $2x + y = 3$. Simon's working is shown below.

Line X Rearrange second equation: $y = 2x - 3$

Line Y Substitute into first equation: $(2x - 3)(1 - x) = 1$

$$-2x^2 + 5x - 3 = 1$$

$$2x^2 - 5x + 4 = 0$$

Line Z Discriminant $= 5^2 - 4 \times 2 \times 4 = -25 - 32 = -57$

There are no real solutions.

Simon knows that he must have made at least one mistake, as his teacher has told him that the equations do have real solutions. In which line(s) of the working has Simon made a mistake?

A Line X only **B** Line Y only **C** Lines Y and Z **D** Line Z only **E** Lines X and Z

3 Find the coordinates of the points where the line $y = 2x + 3$ intersects the curve $y = x^2 + 3x + 1$.

A $(-2, 5)$ and $(1, -1)$ **B** $(-4, -5)$ and $(-1, 1)$ **C** $(-2, -9)$ and $(1, 5)$

D $(-1, 1)$ and $(2, 7)$ **E** $(-2, -1)$ and $(1, 5)$

4 Look at the simultaneous equations $3x^2 + 2y^2 = 5$ and $y - 2x = 1$.

Which one of the following is the correct pair of x-values for the solution of these equations?

A $x = \frac{3}{11}$ or $x = -1$ **B** $x = \pm\sqrt{\frac{3}{11}}$ **C** $x = \frac{3}{7}$ or $x = -1$

D $x = -\frac{3}{11}$ or $x = 1$ **E** $x = -\frac{3}{7}$ or $x = 1$

Hint: Start by rearranging the equation of the line into the form $y = \ldots$

5 The line $y = x - k$ forms a tangent to the curve $y = x^2 + 3x - 4$ at the point P.

Find the value of k and the coordinates of P.

Hint: Look at the 'Get it right' box.

Transformations

GCSE Review/Year 1/2

THE LOWDOWN

① **Transformations** are used to **translate** (move), **reflect** or **stretch** a graph.

② Translation of $y = \mathrm{f}(x)$

- $y = \mathrm{f}(x) \rightarrow y = \mathrm{f}(x - a)$ is a translation by vector $\begin{pmatrix} a \\ 0 \end{pmatrix}$ (translate a units to the right)
- $y = \mathrm{f}(x) \rightarrow y = \mathrm{f}(x) + a$ is a translation by vector $\begin{pmatrix} 0 \\ a \end{pmatrix}$ (translate a units up).

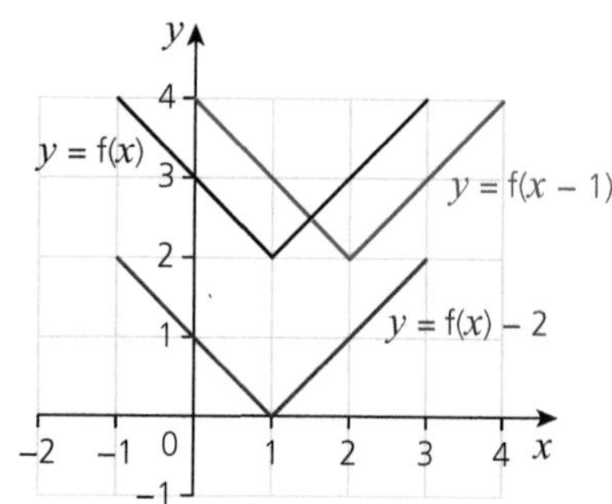

③ One-way stretch of $y = \mathrm{f}(x)$

- $y = \mathrm{f}(x) \rightarrow y = \mathrm{f}(ax)$ is a one-way stretch, scale factor $\frac{1}{a}$, parallel to x-axis
- $y = \mathrm{f}(x) \rightarrow y = a\mathrm{f}(x)$ is a one-way stretch, scale factor a, parallel to y-axis.

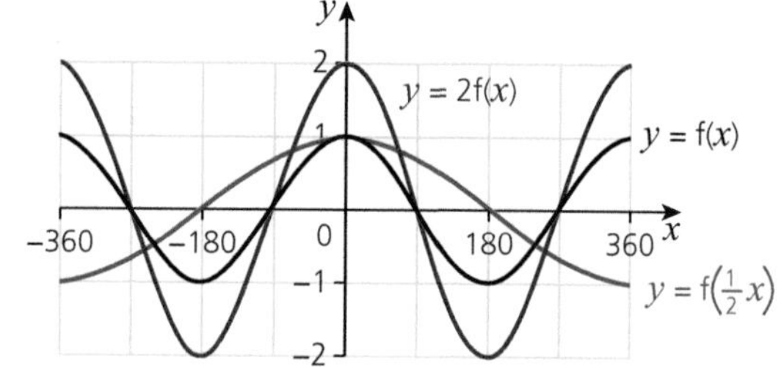

④ Reflection of $y = \mathrm{f}(x)$

- $y = \mathrm{f}(x) \rightarrow y = \mathrm{f}(-x)$ is a reflection in y-axis
- $y = \mathrm{f}(x) \rightarrow y = -\mathrm{f}(x)$ is a reflection in x-axis.

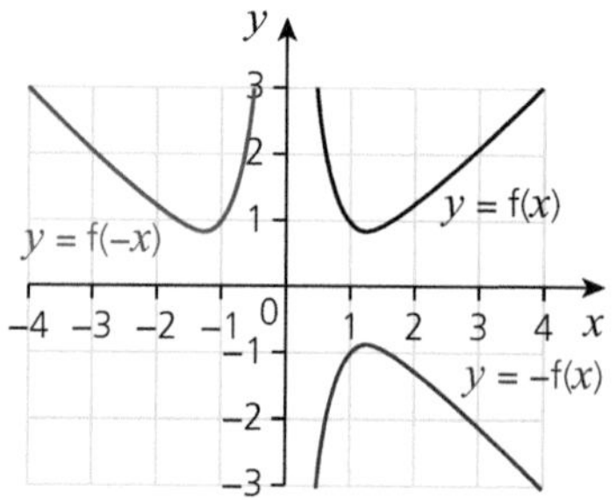

GET IT RIGHT

The diagram shows the graph of $y = \mathrm{f}(x)$. There are turning points at A(−2, 5) and B(1, −3).

Write down the coordinates of the turning points of the curve.

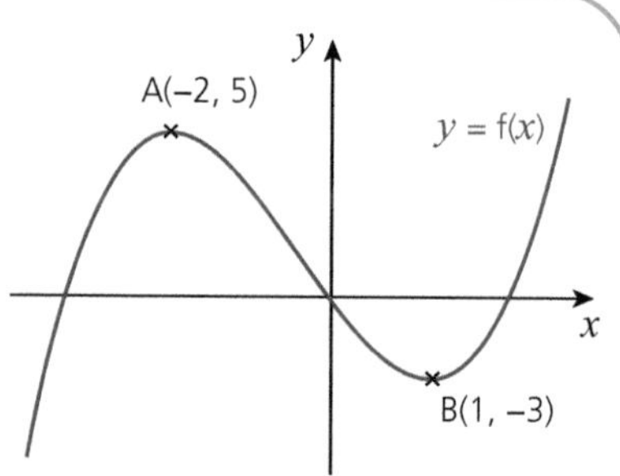

◀◀ **Functions** on page 18.

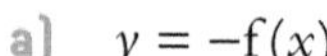

a) $y = -\mathrm{f}(x)$

b) $y = \mathrm{f}(2x)$

c) $y = \mathrm{f}(x + 2) + 3$

Solution:

a) $y = -\mathrm{f}(x)$ is a reflection in the x-axis: A(−2, 5) → A′(−2, −5)
B(1, −3) → B′(1, 3)

Hint: Change the sign of the y-coordinates.

b) $y = \mathrm{f}(2x)$ is a one-way stretch, s.f. $\frac{1}{2}$, parallel to the x-axis.

A(−2, 5) → A′(−1, 5) and B(1, −3) → B′$\left(\frac{1}{2}, -3\right)$

Hint: Halve the x-coordinates.

c) $y = \mathrm{f}(x + 2) + 3$ is a translation by the vector $\begin{pmatrix} -2 \\ 3 \end{pmatrix}$

A(−2, 5) → A′(−4, 8) and B(1, −3) → B′(−1, 0)

Hint: Each point moves 2 left and 3 up.

YOU ARE THE EXAMINER

Which one of these solutions is correct? Where are the errors in the other solution?

a) Describe the transformation that maps the curve $y = e^x$ onto the curve $y = 1 + e^{-x}$.

b) The curve $y = x^2 + 3x$ is translated by the vector $\begin{pmatrix} 1 \\ 2 \end{pmatrix}$.
Find the equation of the new curve.

SAM'S SOLUTION

a) Translation by the vector $\begin{pmatrix} 0 \\ 1 \end{pmatrix}$ and a reflection in the y-axis carried out in either order.

b) $y = (x - 1)^2 + 3(x - 1) + 2$

Expand brackets:

$y = x^2 - 2x + 1 + 3x - 3 + 2$

Simplify: $y = x^2 + x$

PETER'S SOLUTION

a) Translation by the vector $\begin{pmatrix} 0 \\ 1 \end{pmatrix}$ followed by a reflection in the x-axis.

b) $y = (x + 1)^2 + 3(x + 1) + 2$

Expand brackets:

$y = x^2 + 2x + 1 + 3x + 3 + 2$

Simplify: $y = x^2 + 5x + 6$

SKILL BUILDER

For questions 1 and 2 you will need to use this diagram.

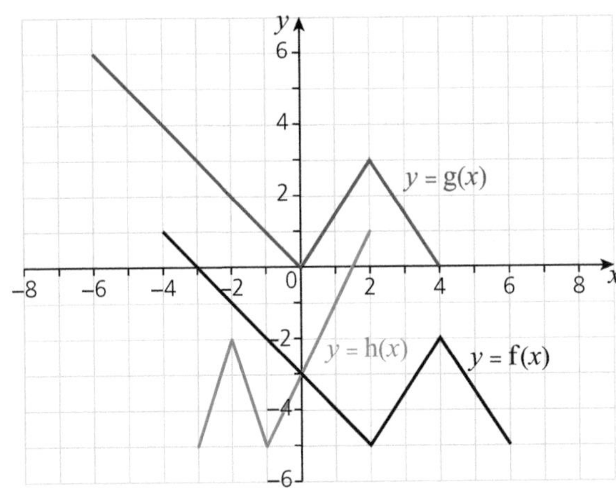

1 The diagram shows the graph of $y = f(x)$ and its image $y = g(x)$ after a transformation. What is the equation of the image?

A $y = f(x + 2) - 5$
B $y = f(x - 2) + 5$
C $y = f(x - 2) - 5$
D $y = f(x + 2) + 5$
E $y = f(x + 5) + 2$

2 The diagram shows the graph of $y = f(x)$ and its image $y = h(x)$ after a transformation. What is the equation of the image?

A $y = f(2x)$
B $y = f(\frac{1}{2}x)$
C $y = f(-2x)$
D $y = f(-\frac{1}{2}x)$
E $y = -\frac{1}{2}f(x)$

3 The curve $y = x^2 - 4x$ is translated and the equation of the new curve is $y = (x - 1)^2 - 4(x - 1) + 2$. What are the coordinates of the vertex of the new curve?

A $(1, 2)$
B $(1, -2)$
C $(3, -2)$
D $(1, 5)$
E $(1, -6)$

4 The curve $y = x^2 - 2x + 3$ is translated through $\begin{pmatrix} 2 \\ -4 \end{pmatrix}$. What is the equation of the new curve?

A $y = x^2 + 2x - 1$
B $y = x^2 - 6x + 11$
C $y = x^2 - 6x + 15$
D $y = 4x^2 - 4x - 1$
E $y = x^2 - 6x + 7$

5 The curve $y = x^2$ is first translated 2 units to the right and 1 unit vertically upwards, and then reflected in the y-axis. Which of the following is the equation of the new graph?

A $y = x^2 + 4x + 5$
B $y = -x^2 + 4x - 5$
C $y = x^2 - 4x + 5$
D $y = -x^2 - 4x - 5$

6 The graph of $y = f(x)$ has a maximum point at $(-3, 2)$. Which of the following is the maximum point of the graph of $y = 2 + 3f(x)$?

A $(-3, 12)$
B $(-7, 8)$
C $(-1, 4)$
D $(-3, 8)$

Modulus functions

Year 2

THE LOWDOWN

① $|x|$ is the **modulus of x** and it means the size of x.

Example $|3| = 3$ and $|-3| = 3$ so $|x| = 3$ means x is either 3 or -3.

② $|2x + 1| = 3$ is an equation – it has **two solutions**.

Example To solve $|2x + 1| = 3$

Step 1 Rewrite the equation as 2 separate equations without the modulus signs

$|2x+1| = 3$ → $2x + 1 = 3$ and $2x + 1 = -3$

Step 2 Solve each equation separately $2x + 1 = 3 \Rightarrow x = 1$

$2x + 1 = -3 \Rightarrow x = -2$

Step 3 Check your answers! $|2 \times 1 + 1| = |3| = 3$ ✓

$|2 \times (-2) + 1| = |-3| = 3$ ✓

◀◀ **Solving inequalities** on page 10 and **graphs of straight lines** on page 34.
▶▶ **Integration** on page 70.

③ The graph of the modulus function is **V** shaped.
Here is the graph of $y = |x|$.
So when $x = 2, y = |2| = 2$ and when $x = -2, y = |-2| = 2$.

④ Follow these steps to draw the graph of $y = |ax + b|$:

- Draw the unmodulated graph.
- Reflect the negative part of the graph in the x-axis.

You can use graphs to help you solve inequalities.

The graph of $y = |ax + b|$ has two branches.
The left hand branch, \, is $y = -(ax + b)$ and the right hand branch, /, is $y = ax + b$.

Example Solve $|2x + 1| < 3$

True when x is between -2 and 1

So $-2 < x < 1$ is the solution.

Hint: To solve $|2x + 1| < 3$, look for where the red graph is below 3.

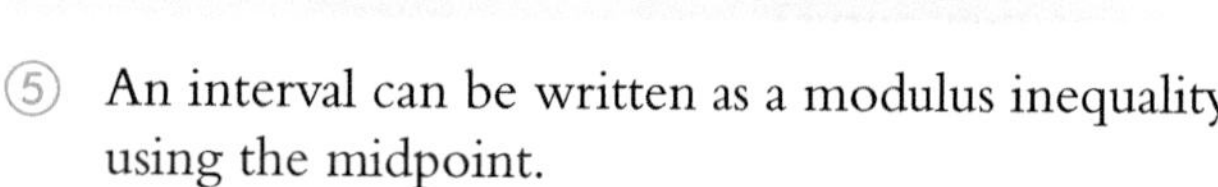

⑤ An interval can be written as a modulus inequality using the midpoint.

Example $1 < x < 11$ has midpoint $x = 6$.
Distance from 6 to the boundary is 5 so $|x - 6| < 5$.

GET IT RIGHT

Solve

a) $|3x - 2| = 4$

b) $|3x - 2| \geqslant 4$.

Hint: Draw a graph to help you.

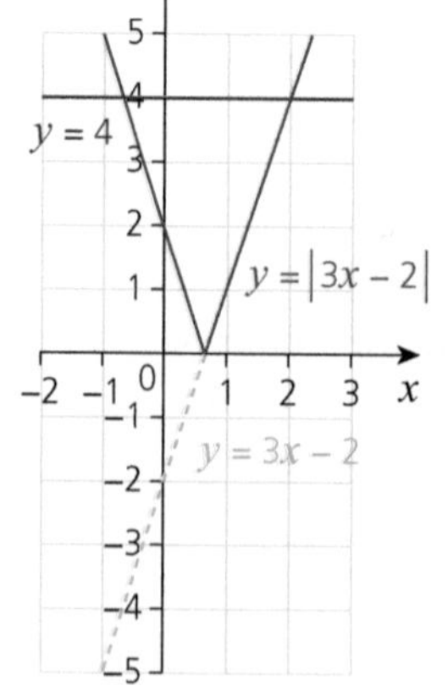

Solution:

a) $3x - 2 = 4$
$\Rightarrow 3x = 6$
$\Rightarrow x = 2$
Or
$3x - 2 = -4$
$\Rightarrow 3x = -2$
$\Rightarrow x = -\frac{2}{3}$

b) $3x - 2$ must be 4 or more or -4 or less.
$3x - 2 \geqslant 4$
$\Rightarrow x \geqslant 2$
Or $3x - 2 \leqslant -4$
$\Rightarrow x \leqslant -\frac{2}{3}$

Hint: To solve $|3x - 2| \geqslant 4$, look for where the red graph is above 4. Remember, two regions need two inequalities!

YOU ARE THE EXAMINER

Which one of these solutions is correct? Where are the errors in the other solution?

Solve $|x+1| < 2x - 3$

PETER'S SOLUTION

$x + 1 < 2x - 3$

$x < 2x - 4$

$4 < x$

$x > 4$

or $x + 1 < -(2x - 3)$

$x + 1 < -2x + 3$

$3x < 2$

$x < \frac{2}{3}$

So $x < \frac{2}{3}$ or $x > 4$

SAM'S SOLUTION

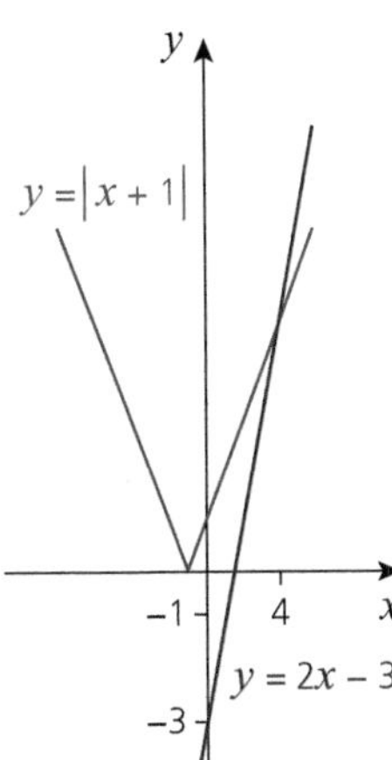

From the graph, I need the point where:

$x + 1 = 2x - 3$

$-x = -4$

$x = 4$

$y = |x+1|$ is above $y = 2x - 3$ when $x > 4$

SKILL BUILDER

1 Solve these equations:

a) $|2x| = 6$ b) $|x+1| = 5$ c) $|x-3| = 8$ d) $|2x-5| = 7$

2 a) Draw the graphs of $y = 2x + 1$, $y = 4$ and $y = |x - 2|$ on the same axes.

b) Use your graphs to help you solve

i) $|x-2| = 4$ ii) $|x-2| < 4$ iii) $|x-2| < 2x + 1$

3 Which of the following could be the equation of the graph below?

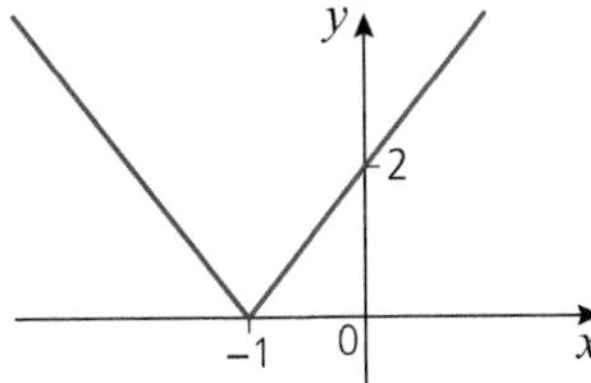

A $y = |2x+1|$ **B** $y = 2|x| + 1$ **C** $y = |x| + 1$

D $y = 2|x+1|$ **E** $y = |x| + 2$

4 Solve the inequality $|3x - 2| \geq 5$.

A $x \geq \frac{7}{3}$ **B** $x \geq \frac{7}{3}$ or $x \geq -1$ **C** $x \leq -1$ or $x \geq \frac{7}{3}$ **D** $-1 \leq x \leq \frac{7}{3}$

5 Solve the equation $3|x+1| = 2x + 5$

A $x = -\frac{8}{5}$ or $x = 2$ **B** $x = 2$ **C** $x = 4$ **D** $x = -\frac{4}{5}$ or $x = 4$ **E** $x = -\frac{2}{5}$ or $x = 2$

6 Write the inequality $-5 < x < 4$ in the form $|x + a| < b$.

A $|x - 0.5| < 4.5$ **B** $|x - 4.5| < -0.5$ **C** $|x + 4.5| < -0.5$ **D** $|x + 0.5| < 4.5$

7 Solve the inequality $|x - 2| > 2x - 1$.

A $x > 1$ **B** $x < 1$ **C** $x < -1$ **D** $-1 < x < 1$

THE LOWDOWN

① A **vector** has magnitude (size) and direction.
Vectors are often shown by arrows.
To write down a vector either use:

- bold (when typed): **a**
- an arrow: $\overrightarrow{AB}$
- underline: $\underline{a}$.

Use the **unit vectors** i and j to break down a **2D vector** into components.
For a **3D vector**, use the unit vectors **i**, **j** and **k**.

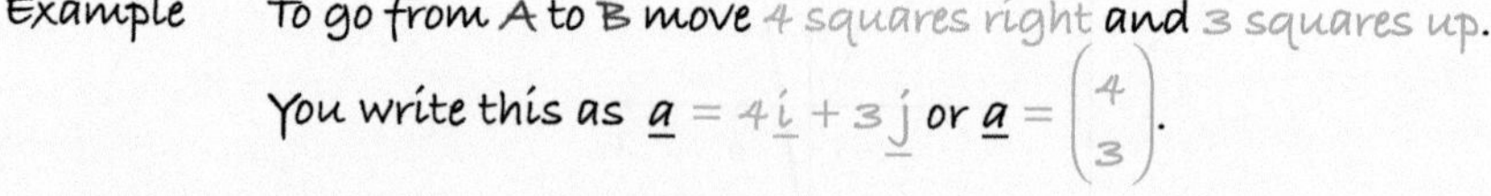

Example To go from A to B move 4 squares right and 3 squares up.
You write this as $\underline{a} = 4\underline{i} + 3\underline{j}$ or $\underline{a} = \begin{pmatrix} 4 \\ 3 \end{pmatrix}$.

② **Equal vectors** have the same magnitude (size) and direction – it doesn't matter where they start and finish.
Parallel vectors are scalar multiples of each other.

Example $\underline{a}$ and $\underline{b}$ are equal vectors.
$\underline{a}$, $\underline{b}$, $\underline{c}$ and d are parallel vectors because

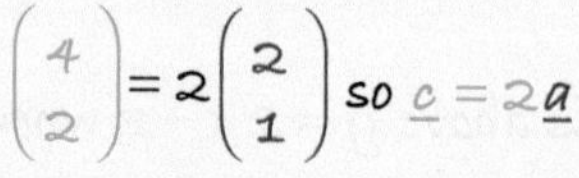

$\begin{pmatrix} 4 \\ 2 \end{pmatrix} = 2\begin{pmatrix} 2 \\ 1 \end{pmatrix}$ so $\underline{c} = 2\underline{a}$

$\begin{pmatrix} 4 \\ 2 \end{pmatrix} = \frac{2}{3}\begin{pmatrix} 6 \\ 3 \end{pmatrix}$ so $\underline{c} = \frac{2}{3}\underline{d}$

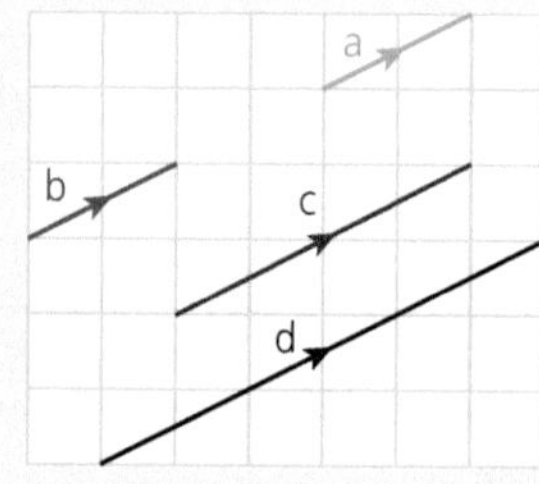

③ **Add/subtract vectors** by adding/subtracting their components.

④ A **position vector** is a vector that **starts** at the origin.

Example The position vector of the point P(2, −3, 1) is
$\overrightarrow{OP} = 2\underline{i} - 3\underline{j} + \underline{k}$

⑤ Use **Pythagoras' theorem** to work out the **magnitude** (size) of a vector.

Example: Rule: $\underline{p} = a\underline{i} + b\underline{j} + c\underline{k} \Rightarrow |\underline{p}| = \sqrt{a^2 + b^2 + c^2}$

In 2D: $\underline{a} = 4\underline{i} + 3\underline{j} \Rightarrow |\underline{a}| = \sqrt{4^2 + 3^2} = \sqrt{25} = 5$

In 3D: $\underline{b} = 2\underline{i} - 3\underline{j} + \underline{k} \Rightarrow |\underline{b}| = \sqrt{2^2 + (-3)^2 + 1^2} = \sqrt{14}$

The **unit vectors** **i**, **j** and **k** have a magnitude (size) of 1 and are at right angles to each other.

Hint: When you multiply a vector by a scalar you multiply each part component.

So $3\begin{pmatrix} 2 \\ -1 \\ 4 \end{pmatrix} = \begin{pmatrix} 3\times 2 \\ 3\times(-1) \\ 3\times 4 \end{pmatrix} = \begin{pmatrix} 6 \\ -3 \\ 12 \end{pmatrix}$

A **scalar** is just a number – it has size only (and no direction).

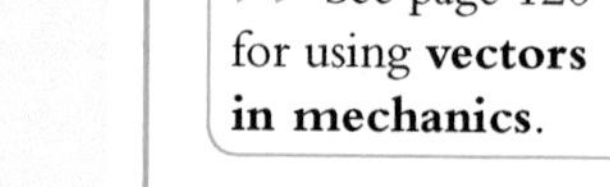

▶▶ See page 120 for using **vectors in mechanics**.

GET IT RIGHT

The points A and B have coordinates (2, 4) and (4, 2).

a) Find the vector $\overrightarrow{AB}$.

b) M is the midpoint of AB. Find the vector $\overrightarrow{OM}$.

Solution:

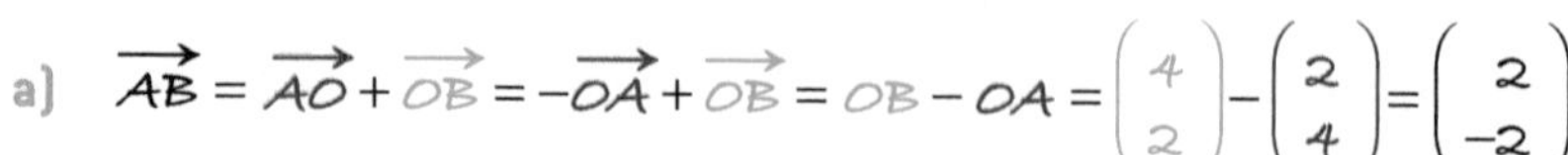

a) $\overrightarrow{AB} = \overrightarrow{AO} + \overrightarrow{OB} = -\overrightarrow{OA} + \overrightarrow{OB} = \overrightarrow{OB} - \overrightarrow{OA} = \begin{pmatrix} 4 \\ 2 \end{pmatrix} - \begin{pmatrix} 2 \\ 4 \end{pmatrix} = \begin{pmatrix} 2 \\ -2 \end{pmatrix}$

b) $\overrightarrow{OM} = \overrightarrow{OA} + \frac{1}{2}\overrightarrow{AB} = \begin{pmatrix} 2 \\ 4 \end{pmatrix} + \frac{1}{2}\begin{pmatrix} 2 \\ -2 \end{pmatrix} = \begin{pmatrix} 2 \\ 4 \end{pmatrix} + \begin{pmatrix} 1 \\ -1 \end{pmatrix} = \begin{pmatrix} 3 \\ 3 \end{pmatrix}$

Imagine an ant walking from A to B: it could go from A to O and then O to B.
If you go in the **opposite direction** to the arrow then the vector is negative.
So $\overrightarrow{AO} = -\overrightarrow{OA}$

Hint: In general, $\overrightarrow{AB} = \mathbf{b} - \mathbf{a}$

YOU ARE THE EXAMINER

Which one of these solutions is correct? Where are the errors in the other solution?

The points P and Q have position vectors $\overrightarrow{OP} = 3\mathbf{i} + \mathbf{k}$ and $\overrightarrow{OQ} = \mathbf{i} - \mathbf{j} + 3\mathbf{k}$.

Find a) $\overrightarrow{PQ}$ and b) $|\overrightarrow{PQ}|$.

MO'S SOLUTION

a) $\overrightarrow{PQ} = \overrightarrow{OQ} - \overrightarrow{OP}$

$$= \begin{pmatrix} 1 \\ -1 \\ 3 \end{pmatrix} - \begin{pmatrix} 3 \\ 0 \\ 1 \end{pmatrix} = \begin{pmatrix} -2 \\ -1 \\ 2 \end{pmatrix}$$

b) $|\overrightarrow{PQ}| = \sqrt{(-2)^2 + (-1)^2 + 2^2}$

$= \sqrt{4 + 1 + 4}$

$= \sqrt{9}$

$= 3$

LILIA'S SOLUTION

a) $\overrightarrow{PQ} = \overrightarrow{OP} - \overrightarrow{OQ}$

$$= \begin{pmatrix} 3 \\ 0 \\ 1 \end{pmatrix} - \begin{pmatrix} 1 \\ -1 \\ 3 \end{pmatrix} = \begin{pmatrix} 2 \\ 1 \\ -2 \end{pmatrix}$$

b) $|\overrightarrow{PQ}| = \sqrt{2^2 + 1^2 + -2^2}$

$= \sqrt{4 + 1 - 4}$

$= \sqrt{1}$

$= 1$

SKILL BUILDER

1 The vectors $\mathbf{p}$ and $\mathbf{q}$ are given by $\mathbf{p} = 2\mathbf{i} - \mathbf{j} + 3\mathbf{k}$ and $\mathbf{q} = 3\mathbf{i} + 2\mathbf{j} - 4\mathbf{k}$.
Find the vector $3\mathbf{p} - 2\mathbf{q}$.

A $\mathbf{j} - \mathbf{k}$ **B** $\mathbf{j} + \mathbf{k}$ **C** $-7\mathbf{j} + 17\mathbf{k}$ **D** $-7\mathbf{j} + \mathbf{k}$

Hint: Work out $3\mathbf{p}$ and $2\mathbf{q}$ first.

Questions 2 and 3 are about three points A(1, 0, 3), B(3, 1, −4) and C(−2, 6, 5).

2 Find the vector $\overrightarrow{BA}$.

A $\begin{pmatrix} 2 \\ 1 \\ -7 \end{pmatrix}$ **B** $\begin{pmatrix} 4 \\ 1 \\ -1 \end{pmatrix}$ **C** $\begin{pmatrix} 3 \\ 0 \\ -12 \end{pmatrix}$ **D** $\begin{pmatrix} -2 \\ -1 \\ 7 \end{pmatrix}$

3 Find the position vector of the midpoint M of BC.

A $\begin{pmatrix} 0.5 \\ 3.5 \\ 0.5 \end{pmatrix}$ **B** $\begin{pmatrix} -2.5 \\ 2.5 \\ 4.5 \end{pmatrix}$ **C** $\begin{pmatrix} 2 \\ 4 \\ -1.5 \end{pmatrix}$ **D** $\begin{pmatrix} -4.5 \\ 8.5 \\ 9.5 \end{pmatrix}$

4 Given $\overrightarrow{OA} = 3\mathbf{i} + 2\mathbf{j}$, $\overrightarrow{OB} = 5\mathbf{i} - 12\mathbf{j}$ and $\overrightarrow{OC} = 4\mathbf{i} - 2\mathbf{j} - 4\mathbf{k}$. Find:

a) $|\overrightarrow{OA}|$ b) $|\overrightarrow{OB}|$ c) $|\overrightarrow{OC}|$ d) $|\overrightarrow{AB}|$

5 The points A, B and C have position vectors $-5\mathbf{i} + \mathbf{j}$, $-3\mathbf{i} - 2\mathbf{j}$ and $4\mathbf{i} - 6\mathbf{j}$ respectively.

a) Find the vector $\overrightarrow{AB}$.

b) Show that AB is parallel to OC

Hint: Show that $\overrightarrow{OC} = k \times \overrightarrow{AB}$.

c) Find the perimeter of the triangle ABC. Give your answer correct to 3 s.f.

Hint: Find the magnitudes of the sides AB, AC and BC and add them together.

d) Find the position vector of M given M is the midpoint of AB.

Hint: Draw a diagram to help you – don't forget to mark the origin O.

Numerical methods

Year 2

THE LOWDOWN

You can use numerical methods to solve equations that you can't solve algebraically.

① **Change of sign method.**

Example You can show that the equation $3^x - 6x = 0$ has a root between 2 and 3 because $3^2 - 6 \times 2 = -3$ and $3^3 - 6 \times 3 = +9$.

As there is a change of sign in the interval [2, 3] the graph must pass through 0, so there is a root.

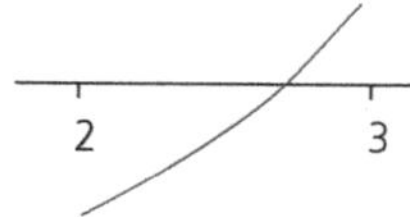

② **Fixed point iteration**. A sequence of answers usually gets close to the root – but it can fail and a different rearrangement may be needed.

Example Follow these steps to solve $x^3 - 5x + 1 = 0$

Step 1 Choose an x to make 'the subject': $x^3 - 5x + 1 = 0$ and rearrange $f(x) = 0$ $5x = x^3 + 1$

into the form $x = g(x)$: $x = \frac{x^3 + 1}{5}$

Step 2 Write down the iterative formula

$x_{n+1} = g(x_n)$: $x_{n+1} = \frac{x_n^3 + 1}{5}$

Step 3 Choose a value for x_1: $x_1 = 2$

Put x_1 into the formula to find x_2: $x_2 = \frac{2^3 + 1}{5} = 1.8$

Put in x_2 to find x_3 and so on....: $x_3 = \frac{1.8^3 + 1}{5} = 1.3664$

Step 4 Wait until the formula keeps giving the same answer to 3 or more d.p. so root is $x = 0.202$ (3 s.f.)

③ **Newton–Raphson method.**

Use the formula $x_{n+1} = x_n - \frac{f(x_n)}{f'(x_n)}$ to find a root of $f(x) = 0$

Watch out!
Change of sign method fails if
- the roots are very close together – so you miss one!
- there is an asymptote – this leads to a false root
- there is a repeated root (no sign change).

Hint: Use the ANS key on your calculator to help you.
Press 2 =
Then
$\frac{\boxed{\text{ANS}}\,\boxed{x^{3}} + 1}{5}$
Keep pressing = until the answer settles down.

▶▶ **Differentiation** on page 62.

$x_{10} = 0.2016...$

GET IT RIGHT

The equation $x^4 - 2x^3 - 2 = 0$ has a root between $x = 2$ and $x = 3$.
Use the Newton–Raphson method, together with $x_1 = 3$, to find the root to 2 d.p.

Solution:

Step 1 Differentiate f(x): $f(x) = x^4 - 2x^3 - 2$
$f'(x) = 4x^3 - 6x^2$

Step 2 Write down the formula: $x_{n+1} = x_n - \frac{x_n^4 - 2x_n^3 - 2}{4x_n^3 - 6x_n^2}$

Step 3 $x_1 = 3$ so $x_2 = 3 - \frac{3^4 - 2 \times 3^3 - 2}{4 \times 3^3 - 6 \times 3^2} = 2.5370...$

$x_3 = 3.3834...$; $x_4 = 2.1993...$; $x_5 = 2.1904...$; $x_6 = x_7 = 2.1903...$

This settles, so root is 2.19 to 2 d.p.

Hint: The **Newton–Raphson** formula is
$x_{n+1} = x_n - \frac{f(x_n)}{f'(x_n)}$

Hint: Make sure you know how to use your calculator efficiently.

YOU ARE THE EXAMINER

Which one of these solutions is correct? Where is the error in the other solution?

The equation $10x^3 - 2x^2 + 1 = 0$ has a root between -1 and 0.

Use a suitable rearrangement to write down an iterative formula.

Use your iterative formula together with $x_1 = 0$ to find the root correct to 1 d.p.

SAM'S SOLUTION

Make the red x the subject:

$10x^3 - 2x^2 + 1 = 0$

$2x^2 = 10x^3 + 1 \Rightarrow x = \sqrt{\frac{10x^3+1}{2}}$

So $x_{n+1} = \sqrt{\frac{10x_n^3+1}{2}}$

$x_1 = 0 \Rightarrow x_2 = \sqrt{\frac{10\times 0^3+1}{2}} = 0.70710..$

$x_3 = 1.50591...,\ x_4 = 4.19228...,$

The numbers are getting bigger and bigger so there isn't a root between -1 and 0.

LILIA'S SOLUTION

Make the red x the subject: $10x^3 - 2x^2 + 1 = 0$

$10x^3 = 2x^2 - 1 \Rightarrow x = \sqrt[3]{\frac{2x^2-1}{10}}$

So $x_{n+1} = \sqrt[3]{\frac{2x_n^2-1}{10}}$

$x_1 = 0 \Rightarrow x_2 = \sqrt[3]{\frac{2\times 0^2-1}{10}} = -0.46415...$

$x_3 = -0.3846...,\ x_4 = -0.4129...,$

$x_5 = -0.4039...$ and so on.

So root is -0.4 correct to 1 d.p.

SKILL BUILDER

1 Use the iterative formula $x_{n+1} = x_n - \frac{x_n^3 + 4x_n - 2}{3x_n^2 + 4}$ together with $x_1 = 0.2$ to find the value of x_4.

Give your answer correct to 5 decimal places.

A $x_4 = 0.48607$ **B** $x_4 = 0.47347$ **C** $x_4 = 0.47354$ **D** $x_4 = 0.48604$

2 Which of these is a correct rearrangement of the equation $3x^3 - x - 6 = 0$?

i $x = \sqrt[3]{\frac{x+6}{3}}$ **ii** $x = \sqrt{\frac{1}{3} + \frac{2}{x}}$ **iii** $x = 3x^3 - 6$

A **i** only **B** **ii** and **iii** **C** **i** and **iii** **D** **i**, **ii**, and **iii**

3 Which of these iterative formulae can be used to find a root of $e^x - x = x^2 + 2$ near $x = 2$?

A $x_{n+1} = e^{x_n} - x_n^2 - 2$ **B** $x_{n+1} = \sqrt{e^{x_n} - x_n - 2}$

C $x_{n+1} = \ln(x_n^2 + x_n + 2)$ **D** All of these

4 Give the Newton–Raphson iterative formula for the equation $2e^{2x} + 3x - 1 = 0$.

A $x_{n+1} = x_n - \frac{2e^{2x_n} + 3x_n - 1}{4e^{2x_n} + 3}$ **B** $x_{n+1} = x_n - \frac{4e^{2x_n} + 3}{2e^{2x_n} + 3x_n - 1}$

C $x_{n+1} = x_n - \frac{2e^{2x_n} + 3x_n - 1}{e^{2x_n} + 3}$ **D** $x_{n+1} = \frac{4e^{2x_n} + 3}{2e^{2x_n} + 3x_n - 1} - x_n$

Hint: The derivative of e^{2x} is $2e^{2x}$ – see page 70 for more about this.

5 **a)** Show that $x^4 - 3x + 1 = 0$ has a root between $x = 1$ and $x = 2$.

b) Use the Newton–Raphson method, together with $x_1 = 2$ to find the root correct to 2 d.p.

6 **a)** Show that $e^x - 5x - 2 = 0$ has a root between $x = 2$ and $x = 3$.

b) Show that $e^x - 5x - 2 = 0$ can be rearranged to give $x = \ln(5x + 2)$.

c) Use an iterative formula based on this equation, together with $x_1 = 1$ to find the root correct to 2 d.p.

Hint: Differentiate the equation first.

The trapezium rule

Year 2

THE LOWDOWN

① To estimate the **area under a curve** divide the area up into trapeziums.

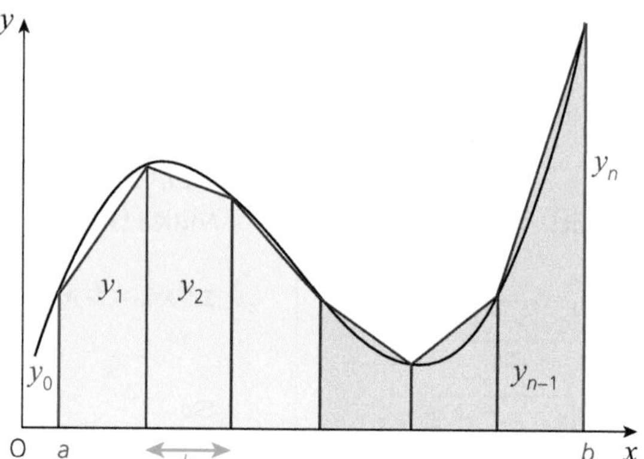

The trapezium rule for the area under the curve between $x = a$ and $x = b$ is:

$$Area \approx \frac{1}{2} \times h \times [y_0 + y_n + 2(y_1 + y_2 + ... + y_{n-1})] \text{ where } h = \frac{b-a}{\text{number of strips}}$$

You can think of this as:

$$\text{Area} \approx \frac{1}{2} \times \text{strip width} \times [\text{ends} + 2 \times \text{middles}]$$

Example You can use the trapezium rule with 4 strips to estimate the area under the curve $y = \sqrt{x-1}$ between $x = 1$ and $x = 5$.

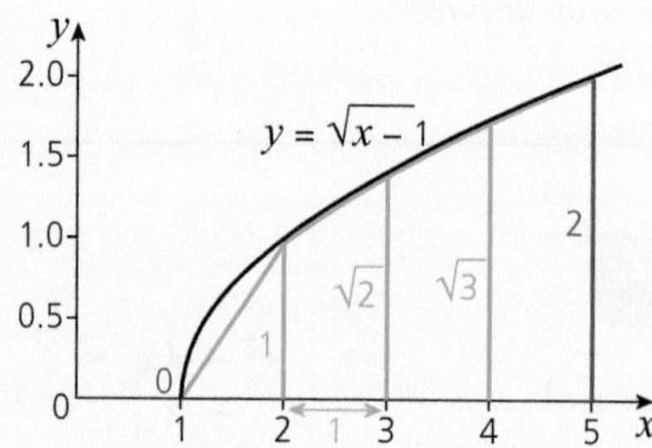

Strip width = 1

End y-values = 0 and 2

Middles = 1, $\sqrt{2}$ and $\sqrt{3}$

$$\text{Area} \approx \frac{1}{2} \times 1 \times [0 + 2 + 2 \times (1 + \sqrt{2} + \sqrt{3})]$$

$$\approx 5.15 \text{ (3 s.f.)}$$

② You can also use rectangles to estimate an **upper and a lower bound** for the area under a curve.

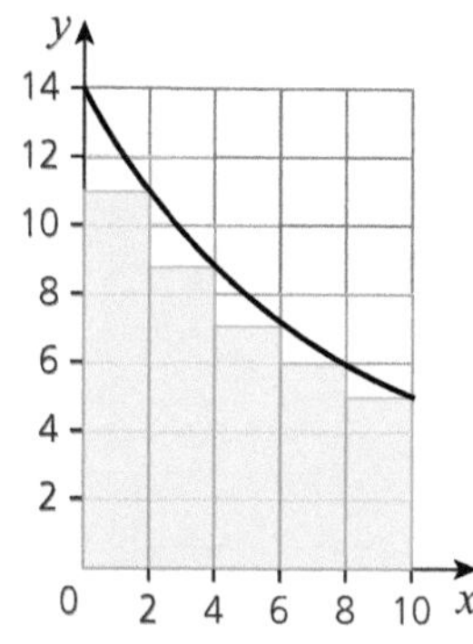

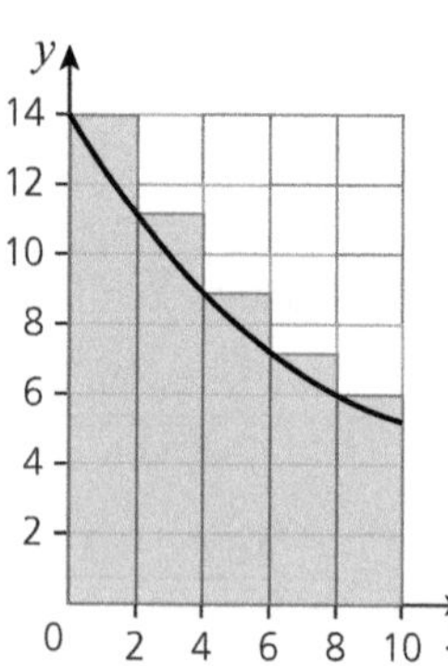

x	y
0	14
2	11.25
4	9
6	7.25
8	6
10	5.25

GET IT RIGHT

Estimate the upper and lower bounds for the area under the curve in **Lowdown 2**.

Solution:

Lower bound = $(2 \times 11.25) + (2 \times 9) + (2 \times 7.25) + (2 \times 6) + (2 \times 5.25) = 77.5$

Upper bound = $(2 \times 14) + (2 \times 11.25) + (2 \times 9) + (2 \times 7.25) + (2 \times 6) = 95$

So $77.5 <$ area under curve < 95

Watch out!
When the sides of the trapeziums lie
- above the curve the result is an overestimate
- below the curve the result is an underestimate.

Hint: To find the height of each trapezium, substitute the x-value into the equation of the curve.

x	y
1	$\sqrt{1-1}$
2	$\sqrt{2-1}$
3	$\sqrt{3-1}$
4	$\sqrt{4-1}$
5	$\sqrt{5-1}$

You can write this as

$$\int_1^5 \sqrt{x-1}\,dx \approx 5.15$$

Hint: The **more strips** you use, the better the estimate.

The y-values in the table give the heights of the rectangles.
The width of each rectangle is 2.

▶▶ **Using integration to find the area under a curve** on page 66.

YOU ARE THE EXAMINER

Which one of these solutions is correct? Where are the errors in the other solution?

Use the trapezium rule with five strips to estimate the area under the curve in Lowdown 2.

Would your answer increase or decrease if you used more strips? Why?

PETER'S SOLUTION

Strip width = 2

Ends = 14 and 5.25

Middles = 11.25, 9, 7.25 and 6

$$A \approx \frac{1}{2} \times 2 \times [14 + 5.25 + 2 \times (11.25 + 9 + 7.25 + 6)]$$

$$\approx 1 \times [19.25 + 2 \times 33.5]$$

$$\approx 86.25$$

This is an overestimate as the trapeziums lie above the curve so if I use more trapeziums my answer will decrease (to get closer to the true area).

SAM'S SOLUTION

Strip width = 2

Ends = 0 and 10

Middles = 2, 4, 6 and 8

$$A \approx \frac{1}{2} \times 2 \times [0 + 10 + 2 \times (2 + 4 + 6 + 8)]$$

$$\approx 1 \times [10 + 2 \times 20]$$

$$\approx 50$$

This is an underestimate as the trapeziums lie below the curve so if I use more trapeziums my answer will increase (to get closer to the true area).

SKILL BUILDER

1 A curve passes through the points given in this table.
Use the trapezium rule to estimate the area between the curve and the x-axis.

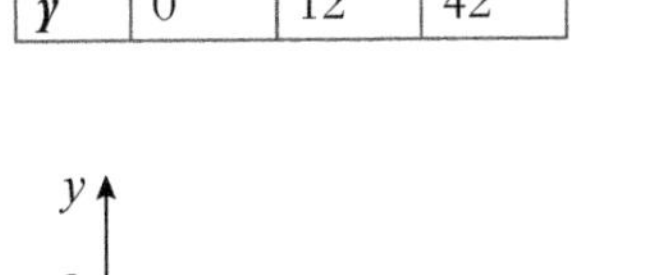

x	1	4	7
y	0	12	42

A 198 **B** 81 **C** 162 **D** 99

2 The graph shows the curve $y = \sqrt{x^2 + 3}$.
By using the method of finding the sum of series of rectangles, find an upper and lower bound for $\int_0^5 \sqrt{x^2 + 3}\,dx$.
Use rectangles of width 1 in your calculations.

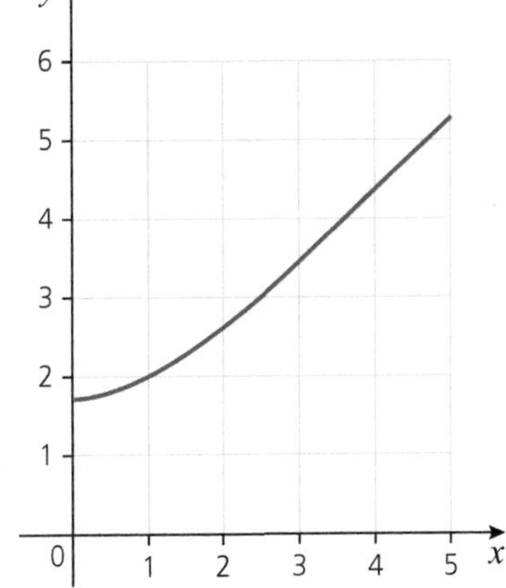

A $14.20 < \int_0^5 \sqrt{x^2 + 3}\,dx < 19.49$

B $17.76 < \int_0^5 \sqrt{x^2 + 3}\,dx < 19.49$

C $14.20 < \int_0^5 \sqrt{x^2 + 3}\,dx < 17.76$

D $11.06 < \int_0^5 \sqrt{x^2 + 3}\,dx < 12.16$

3 **a)** Copy and complete this table for $y = \dfrac{200}{1 + \sqrt{x}}$.

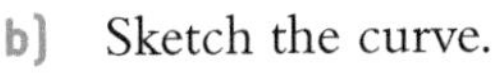

x	1	3	5	7	9	11	13
y							

b) Sketch the curve.

c) Use the trapezium rule with 6 strips to estimate the area under the cuve $y = \dfrac{200}{1 + \sqrt{x}}$ between $x = 1$ and $x = 13$. Give your answer correct to 1 d.p.

d) Is your answer to part b) an overestimate or an underestimate for the area under the curve? Will your answer increase or decrease if you use more strips?
Explain your answer fully.

Trigonometry review

THE LOWDOWN

① Right-angled trigonometry

$\sin\theta = \frac{\text{Opp}}{\text{Hyp}}$ $\cos\theta = \frac{\text{Adj}}{\text{Hyp}}$ $\tan\theta = \frac{\text{Opp}}{\text{Adj}}$

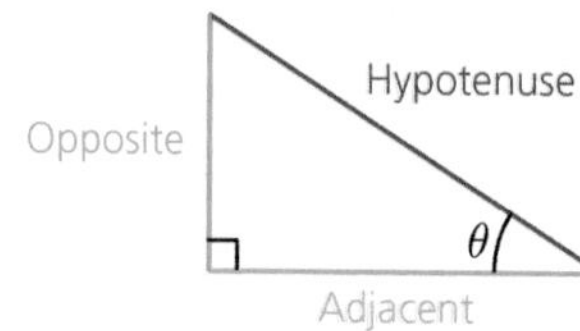

Hint: Remember S^O^H~C^A^H~T^O^A.

② The graph of $y = \sin\theta$

Example To solve $\sin\theta = \dots$

Step 1: Find 1st value from your calculator $\sin\theta = 0.5 \Rightarrow \theta = 30°$

Step 2: 2nd value is $180° - \theta$

$180° - 30° = 150°$

Step 3: $\pm 360°$ to find more possible solutions

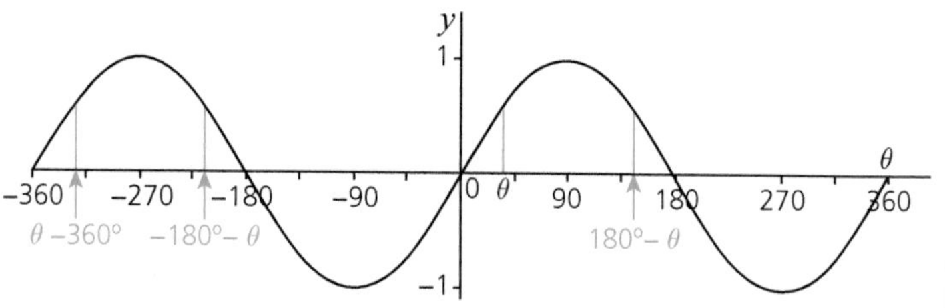

③ The graph of $y = \cos\theta$

Example To solve $\cos\theta = \dots$

Step 1: Find 1st value from your calculator $\cos\theta = 0.5 \Rightarrow \theta = 60°$

Step 2: 2nd value is $360° - \theta$

$360° - 60° = 300°$

Step 3: $\pm 360°$ to find more possible solutions

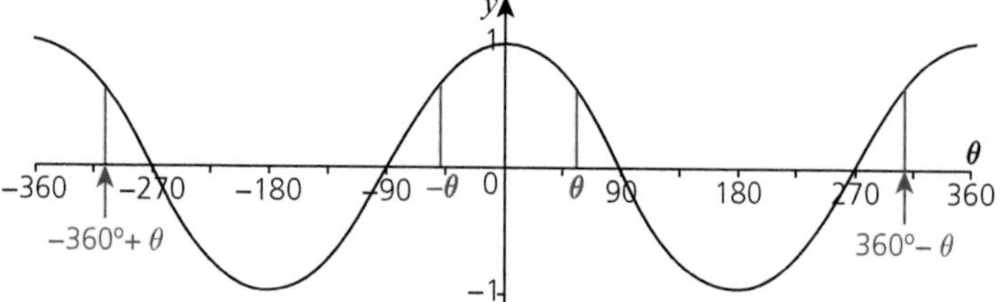

④ The graph of $y = \tan\theta$

Example To solve $\tan\theta = \dots$

Step 1: Find 1st value from your calculator $\tan\theta = -1 \Rightarrow \theta = -45°$

Step 2: $\pm 180°$ to find more possible solutions

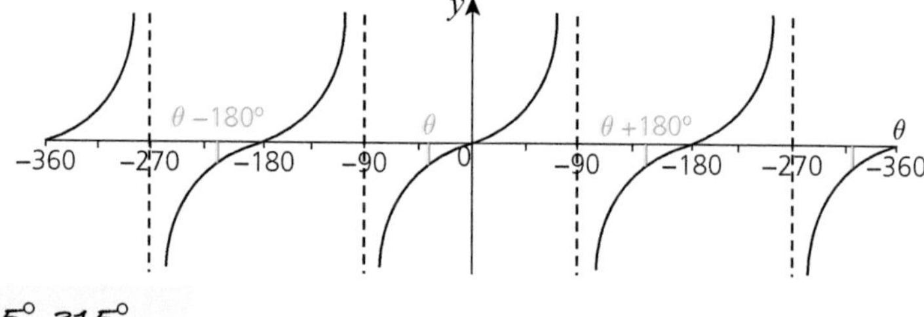

$\dots, -405°, -225°, -45°, 135°, 315°, \dots$

GET IT RIGHT

Solve $4\sin x + 1 = 0$ for $0° \leqslant x \leqslant 360°$.

Solution:

Step 1	Rearrange so you have $\sin x = \dots$	$\sin x = -\frac{1}{4}$
Step 2	Use your calculator to solve:	$x = -14.5°$ (to 1 d.p.)
Step 3	Find the 2nd value:	$x = 180° - (-14.5°) = 194.5°$
Step 4	Find other values:	$x = -14.5° + 360° = 345.5°$
Step 5	List all values **in range** in order:	So $x = 194.5°$ or $345.5°$

▶▶ See page 54 for **working with radians**.

Hint: Give angles correct to 1 d.p.

YOU ARE THE EXAMINER

Which one of these solutions is correct? Where are the errors in the other solution?

Solve $\cos 2x = \frac{\sqrt{2}}{2}$ for $0° \leqslant x \leqslant 360°$.

PETER'S SOLUTION

Let $\cos 2x = \theta$ and solve $\cos\theta = \frac{\sqrt{2}}{2}$

From my calculator $\theta = \cos^{-1}\left(\frac{\sqrt{2}}{2}\right) = 45°$

2nd value is $\theta = 360° - 45° = 315°$

Other values are: $\theta = 45° + 360° = 405°$

and $\theta = 315° + 360° = 675°$

So $\theta = 45°, 315°, 405°, 675°$

So $x = 22.5°, 157.5°, 202.5°$ or $337.5°$

SAM'S SOLUTION

$\cos 2x = \frac{\sqrt{2}}{2} \Rightarrow 2x = 45°$

So $x = 22.5°$

2nd value is $x = 360° - 22.5 = 337.5°$

So $x = 22.5°$ or $337.5°$

SKILL BUILDER

1 You can use these triangles to find exact values of $\sin\theta$, $\cos\theta$ and $\tan\theta$ for the angles 30°, 45° and 60°.

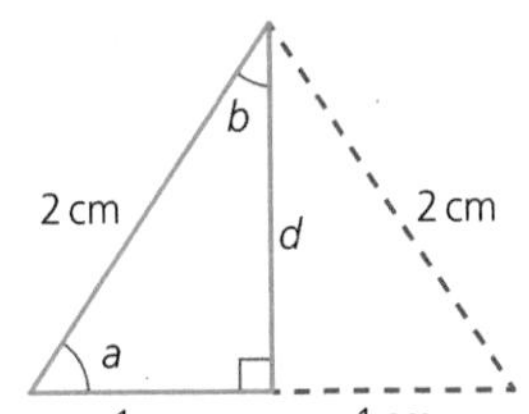

Half of an equilateral triangle of side 2 cm

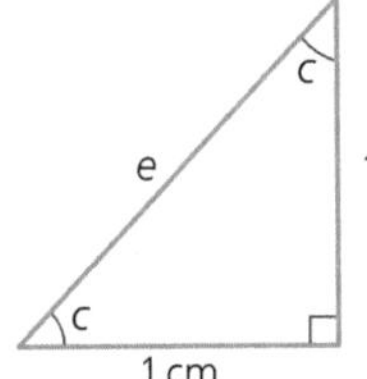

A right-angled isosceles triangle of side 1 cm

Hint: Pythagoras' theorem is $a^2 + b^2 = c^2$ where c is the hypotenuse.

a) Write down the angles a, b and c.

b) Use **Pythagoras' theorem** to find lengths d and e.

c) Use **right-angled trigonometry** and the graphs from the Lowdown to complete the table.

	0°	30°	45°	60°	90°
$\sin\theta$					
$\cos\theta$					
$\tan\theta$					undefined

2 Use the graphs from the Lowdown to help you complete the following.

a) The graph of $y = \sin\theta$ has a period of ___°.

Hint: How often does it repeat?

It has rotational symmetry of order ___ about the ______.

The maximum value of $y = \sin\theta$ is ___ and the minimum value is ___, so ___ $\leqslant \sin\theta \leqslant$ ___.

b) The graph of $y = \cos\theta$ has a period of ___° and it is symmetrical about the ___-axis.

The maximum value of $y = \cos\theta$ is ___ and the minimum value is ___, so ___ $\leqslant \cos\theta \leqslant$ ___.

c) The graph of $y = \tan\theta$ has a period of ___°.

It has rotational symmetry of order ___ about the ______.

$\tan\theta$ is undefined when $\theta = -270°, -90°$, ___°, ___° …, so there are asymptotes at these values.

3 Solve the following equations for $0° \leqslant x \leqslant 360°$.

a) i) $\cos x = \frac{\sqrt{3}}{2}$ ii) $\cos(x + 10°) = \frac{\sqrt{3}}{2}$ iii) $\cos 2x = \frac{\sqrt{3}}{2}$

b) i) $\sin x = \frac{1}{2}$ ii) $\sin x = -\frac{1}{2}$ iii) $\sin^2 x = \frac{1}{4}$

c) i) $5\tan x = 1$ ii) $\tan(x - 30°) = \frac{1}{5}$ iii) $5\tan\left(\frac{x}{2}\right) - 1 = 0$

4 Solve the equation $\sin^2\theta = \sin\theta$ to the nearest degree for $0° \leqslant \theta \leqslant 360°$.

A 0° and 90° **B** 0°, 90°, 180° and 360° **C** 0°, 180° and 360° **D** 90°

Triangles without right angles

GCSE Review/Year 1

THE LOWDOWN

① To label a triangle
- label the vertices with capital letters
- label the opposite sides with the matching small letter.

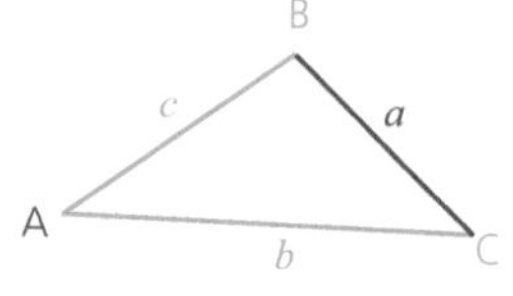

② **Area of a triangle** is $\frac{1}{2}ab\sin C$.

Hint: Use this formula when you know two sides and the angle between them.

Example The area of an equilateral triangle of side 8 cm is

$\frac{1}{2}\times 8\times 8\times \sin 60°$

$= 32\times\frac{\sqrt{3}}{2}$

$= 16\sqrt{3}\ \text{cm}^2$

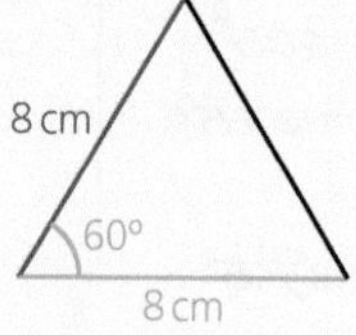

If you know two angles in a triangle then you can use 'angles sum to 180°' to find the third.

③ The **sine rule**
- $\frac{a}{\sin A} = \frac{b}{\sin B} = \frac{c}{\sin C}$ (Use this form to find an unknown side.)
- $\frac{\sin A}{a} = \frac{\sin B}{b} = \frac{\sin C}{c}$ (Flip and use this form to find an unknown angle.)

Hint: Use the sine rule when you know a pair, e.g. a and A.

④ When you use the sine rule OR the area of a triangle formula to find an angle, θ, always check if $180° - \theta$ is also a possible answer. This is called the **ambiguous case**.

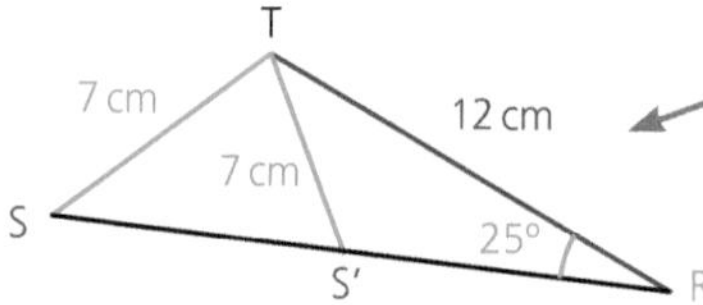

Hint: Always draw a diagram!

Example Triangle RST has sides ST = 7 cm and RT = 12 cm.
Angle R = 25°.
To find angle S use: $\frac{\sin S}{12} = \frac{\sin 25°}{7}$

$\Rightarrow \sin S = \frac{12\sin 25°}{7} = 0.724\ldots$

So $S = 46.4°$ or $S' = 180° - 46.4° = 133.5°$

⑤ The **cosine rule** is
- $a^2 = b^2 + c^2 - 2bc\cos A$ (Use this form to find an unknown side.)
- $\cos A = \frac{b^2 + c^2 - a^2}{2bc}$ (Rearrange to this form to find an unknown angle.)

GET IT RIGHT

Tom walks 4 km from a tower on a bearing of 130°.
Lilia walks 6 km from the tower on a bearing of 070°.
Find the distance between Tom and Lilia.

Solution:

$130° - 70° = 60°$

You know 2 sides and the angle between them so use the cosine rule.

$a^2 = 4^2 + 6^2 - 2\times 4\times 6\cos 60° = 28$

$a = 5.29$

So Tom and Lilia are 5.29 km apart.

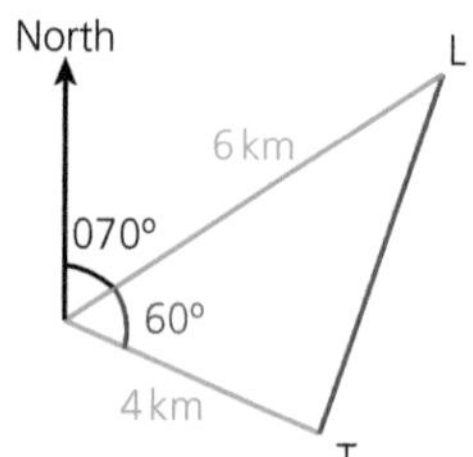

Hint: Use the **cosine rule** when you know:
- **Three sides** and you want **any angle**
- **Two sides** and the **angle between** them and you want the **3rd side**.

Remember: If it isn't a cosine rule question, then it is a sine rule question!

YOU ARE THE EXAMINER

Which one of these solutions is correct?

Where are the errors in the other solution?

In triangle ABC

- AB = 12 cm
- BC = 15 cm
- Angle BCA = 45°.

Work out the size of the angle θ.

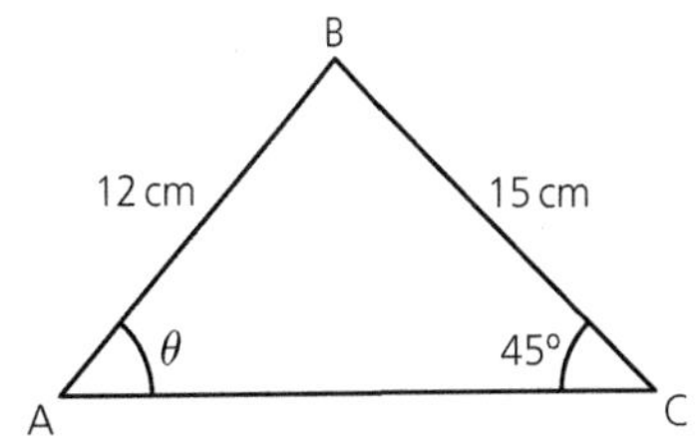

MO'S SOLUTION

$$\frac{\sin\theta}{15} = \frac{\sin 45^\circ}{12}$$

$$\Rightarrow \sin\theta = \frac{15\sin 45^\circ}{12}$$

$$= 0.883...$$

$$\Rightarrow \quad \theta = 62.1^\circ$$

$$\text{Or } \theta = 180^\circ - 62.1^\circ = 117.9^\circ$$

SAM'S SOLUTION

$$\frac{\sin\theta}{12} = \frac{\sin 45^\circ}{15}$$

$$\Rightarrow \sin\theta = \frac{12\sin 45^\circ}{15}$$

$$= 0.565...$$

$$\Rightarrow \quad \theta = 34.4^\circ$$

$$\text{Or } \theta = 180^\circ - 34.4^\circ = 145.6^\circ$$

SKILL BUILDER

1 Work out the area and perimeter of this triangle.

5 cm 125° 8 cm

2 The area of an equilateral triangle is 444 cm^2.
Work out the length of one side.

3 A triangle has sides 5 cm, 7 cm and 9 cm.
Work out the size of its smallest angle.

4 In the triangle ABC, ∠CAB = 37°, ∠ABC = 56° and CB = 4 cm. Find the length of AC.

A 3.24 cm **B** 8.02 cm **C** 5.51 cm **D** 0.18 cm **E** 2.90 cm

5 In the triangle XYZ, XY = 3.8 cm, YZ = 4.5 cm and ∠YZX = 40°. Three of the following statements are false and one is true. Which one is true?

A A possible value for the area of the triangle is exactly 8.55 cm^2.

B The only possible value of ∠XYZ is 90° to the nearest degree.

C You can find the remaining side and angles of the triangle using only the cosine rule.

D The possible values of ∠YXZ are 50° and 130° (to the nearest degree).

6 In the triangle MNP, MN = 5.4 cm, NP = 6 cm and MP = 7 cm. Find angle MNP correct to 3 s.f.

A 48.3° **B** 75.6° **C** 56.1° **D** 1.32°

7 For the triangle given below three of the statements are true and one is false. Which one is false?

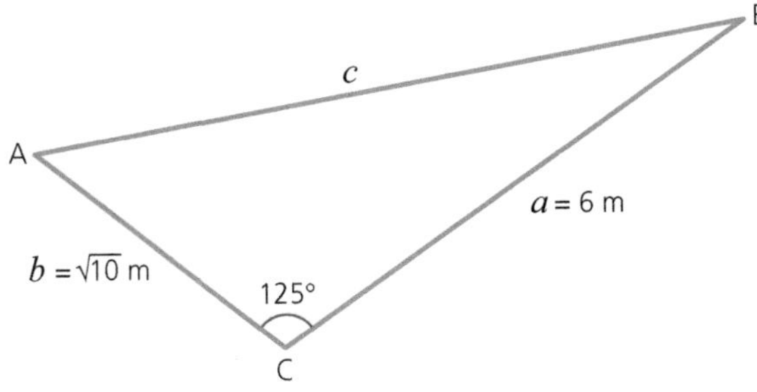

A The area of the triangle is 7.77 m^2 (to 3 s.f.).

B AB is 8.23 m (to 3 s.f.).

C Using only the sine rule you can find the value of c.

D ∠B = 18.34° (to 2 d.p.).

Working with radians

Year 2

THE LOWDOWN

① Radians are used to measure angles. 2π radians = 360°, so 180° = π radians. The symbol for radians is ^{c} or rads so $1.3^c = 1.3$ rads.

- To convert **degrees to radians**, multiply by $\frac{\pi}{180°}$
- To convert radians to degrees, multiply by $\frac{180°}{\pi}$

Example $60°$ is $60° \times \frac{\pi}{180°} = \frac{60\pi}{180} = \frac{\pi}{3}$ radians

$\frac{2\pi}{3}$ radians is $\frac{2\pi}{3} \times \frac{180°}{\pi} = \frac{2 \times 180°}{3} = 120°$

② You can use radians to work out the area and arc length of sectors.

Area of sector $= \frac{1}{2}r^2\theta$ Arc length $= r\theta$

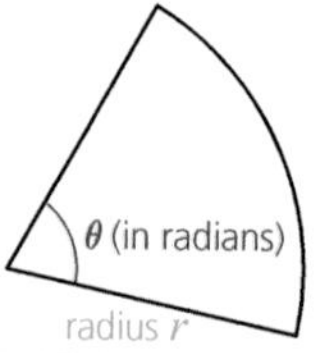

Example When $r = 6$ cm and $\theta = \frac{\pi}{4}$ arc length $= r\theta = 6 \times \frac{\pi}{4} = 4.71$ cm

Perimeter = 4.71 cm + 6 cm + 6 cm = 16.7 cm

Area $= \frac{1}{2}r^2\theta = \frac{1}{2} \times 6^2 \times \frac{\pi}{4} = 14.1$ cm^2 (all to 3 s.f.)

③ You can solve trigonometric equations in radians:

Example Solve $\cos\theta = \frac{\sqrt{3}}{2}$ for $0 \leqslant \theta \leqslant 2\pi$

By calculator: $\theta = \cos^{-1}\left(\frac{\sqrt{3}}{2}\right) = \frac{\pi}{6}$ 2nd value is $2\pi - \frac{\pi}{6} = \frac{11\pi}{6}$

④ Radians can also be used for **small angle approximations**.

When θ **is small and in radians** then

- $\sin\theta \approx \theta$
- $\tan\theta \approx \theta$
- $\cos\theta \approx 1 - \frac{\theta^2}{2}$

Examples $\sin 0.1 \approx 0.1$, $\tan 0.1 \approx 0.1$, $\cos 0.1 \approx 1 - \frac{0.1^2}{2} \approx 0.995$

◀◀ **Trigonometry** on pages 50 and **triangles without right angles** on page 52.
▶▶ **Solving trigonometric equations** on page 56 and **differentiation and integration** on page 70.

Hint: For $\cos\theta$:
- Find the 1st value from your calculator.
- 2nd value is $2\pi - \theta$.
- To find further values you add/subtract 2π.

Use your calculator to check these – you must be in radians!

GET IT RIGHT

Solve a) $\sin\theta = \frac{\sqrt{2}}{2}$ and b) $\sin\theta + \sqrt{3}\cos\theta = 0$ for $0 \leqslant \theta \leqslant 2\pi$.

Solution:

a) Step 1 Find 1st value by calculator: $\theta = \sin^{-1}\left(\frac{\sqrt{2}}{2}\right) = \frac{\pi}{4}$

Step 2 2nd value is $\pi - \frac{\pi}{4} = \frac{3\pi}{4}$

b) Step 1 Divide by $\cos\theta$: $\frac{\sin\theta}{\cos\theta} + \frac{\sqrt{3}\cos\theta}{\cos\theta} = 0$.

Step 2 Use trig identity $\tan\theta = -\sqrt{3}$.

Step 3 Find 1st value by calculator: $\theta = -\frac{\pi}{3}$ not in interval, do not include.

2nd value is $-\frac{\pi}{3} + \pi = \frac{2\pi}{3}$, 3rd value is $\frac{2\pi}{3} + \pi = \frac{5\pi}{3}$.

Watch out! If you see 'π' you should be working in radians.

Watch out! Make sure your calculator is in radians mode.

Hint: To find further values for $\sin\theta$ you add/subtract 2π.

Hint: Add π to find all other values.

YOU ARE THE EXAMINER

Which one of these solutions is correct? Where are the errors in the other solution?

Work out the area of the shaded region.

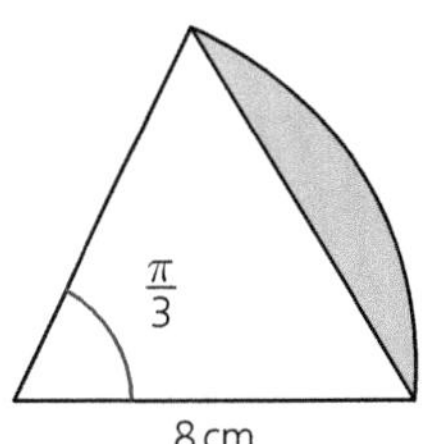

SAM'S SOLUTION

Area shaded region = area of sector − area of Δ

Area of sector $= r\theta = 8 \times \frac{\pi}{3} = 8.377...$

Area of $\Delta = \frac{1}{2}ab\sin C = \frac{1}{2} \times 8^2 \times \sin\left(\frac{\pi}{3}\right)$

$= 0.584...$

Shaded area $= 8.377... - 0.584...$

$= 7.79$ cm² (to 3 s.f.)

LILIA'S SOLUTION

Area shaded region = area of sector − area of Δ

Area of sector $= \frac{1}{2}r^2\theta = \frac{1}{2} \times 8^2 \times \frac{\pi}{3} = \frac{32\pi}{3}$

Area of $\Delta = \frac{1}{2}ab\sin C = \frac{1}{2} \times 8^2 \times \sin\left(\frac{\pi}{3}\right) = 16\sqrt{3}$

Shaded area $= \frac{32\pi}{3} - 16\sqrt{3}$

$= 5.80$ cm² (to 3 s.f.)

SKILL BUILDER

1 Convert these angles from degrees to radians.

a) 30° b) 90° c) 45° d) 210° e) 75° f) 18°

2 Convert these angles from radians to degrees.

a) π b) 4π c) $\frac{\pi}{5}$ d) $\frac{2\pi}{3}$ e) 1^c f) 2^c

3 Without using your calculator, write down the exact values of

a) $\cos\pi$ b) $\sin\left(\frac{\pi}{2}\right)$ c) $\tan\left(\frac{\pi}{6}\right)$ d) $\sin\left(\frac{\pi}{4}\right)$ e) $\cos\left(\frac{\pi}{6}\right)$ f) $\tan\left(\frac{3\pi}{4}\right)$

4 AOB is a sector of a circle with centre O and radius 12 cm.

Given the angle AOB is $\frac{2\pi}{3}$, work out the exact area and perimeter of the sector.

5 Work out the perimeter of the shaded region in 'You are the examiner' at the top of this page.

6 Solve these equations for $0 \leqslant x \leqslant 2\pi$.

a) i) $\cos x = \frac{1}{2}$ ii) $\cos(x + \frac{\pi}{4}) = \frac{1}{2}$ iii) $\cos(2x) = \frac{1}{2}$

b) i) $3\tan x = \sqrt{3}$ ii) $\sqrt{3} + 3\tan x = 0$ iii) $3\tan^2 x = 1$

c) i) $4\sin x - 3 = 0$ ii) $4\sin(2x) - 3 = 0$ iii) $4\sin^2 x - 3 = 0$

Hint: In a) and b) leave your answers in terms of π.

7 Given θ is small and in radians, write down an approximation for the following.

a) $\sin\theta\tan\theta \approx \theta \times \theta = \theta^2$ b) $\frac{\sin\theta}{\tan\theta}$

c) $\frac{\tan^2\theta}{\theta}$ d) $4\sin\theta\tan\theta + 4\cos^2\theta$

Hint: Use the small angle approximations. The first one has been answered for you.

8 Four of the statements below are false and one is true. Which one is true?

A When you convert 540° into radians the answer is 6π.

B When you convert $\frac{7\pi}{15}$ into degrees the answer is 84°.

C $\left(\tan\frac{\pi}{4} - \cos\frac{\pi}{2}\right) = \left(\cos\frac{\pi}{2} - \tan\frac{\pi}{4}\right)$

D $\cos\frac{5\pi}{6}$ is the same as cos 150° and the result is a positive number.

E $\sin(\pi + \theta) = \sin\theta$ is true for all values of θ.

Trigonometric identities

Year 1/2

THE LOWDOWN

1. An equation is true for some values of the variable.
   An **identity** is an equation that is always true.
2. You need to learn these **trigonometric identities**.
   - $\tan\theta \equiv \dfrac{\sin\theta}{\cos\theta}$
   - $\sin^2\theta + \cos^2\theta \equiv 1$
3. The **reciprocal trigonometric functions** are
   - $\operatorname{cosec}\theta \equiv \dfrac{1}{\sin\theta}$
   - $\sec\theta \equiv \dfrac{1}{\cos\theta}$
   - $\cot\theta \equiv \dfrac{1}{\tan\theta} \equiv \dfrac{\cos\theta}{\sin\theta}$
4. Use these identities to
   - simplify an equation to a form that you can solve
   - prove other identities.

Example Start with $\sin^2\theta + \cos^2\theta \equiv 1$ to prove that $\tan^2\theta + 1 \equiv \sec^2\theta$ and $1 + \cot^2\theta \equiv \operatorname{cosec}^2\theta$

Step 1 Divide by $\cos^2\theta$ gives: $\dfrac{\sin^2\theta}{\cos^2\theta} + \dfrac{\cos^2\theta}{\cos^2\theta} \equiv \dfrac{1}{\cos^2\theta}$

So $\tan^2\theta + 1 \equiv \sec^2\theta$

Step 2 Divide by $\sin^2\theta$ gives: $\dfrac{\sin^2\theta}{\sin^2\theta} + \dfrac{\cos^2\theta}{\sin^2\theta} \equiv \dfrac{1}{\sin^2\theta}$

So $1 + \cot^2\theta \equiv \operatorname{cosec}^2\theta$

Hint: Use the 3rd letter of cosec, sec and cot to remember the pairs.

Hint: To prove an identity, work with the more complicated looking side and show it simplifies to give the other side.

Hint: Check out the graphs of these functions.

You need to learn these identities or be able to work them out.

GET IT RIGHT

Solve $2\tan^2\theta + 3\sec\theta = 0$ for $0° \leqslant \theta \leqslant 360°$.

Solution:

$\tan^2\theta + 1 \equiv \sec^2\theta$, so $\tan^2\theta \equiv \sec^2\theta - 1$

$2\tan^2\theta + 3\sec\theta = 2(\sec^2\theta - 1) + 3\sec\theta$

$= 2\sec^2\theta - 2 + 3\sec\theta$

$= 2\sec^2\theta + 3\sec\theta - 2$

So the equation is $2\sec^2\theta + 3\sec\theta - 2 = 0$

Let $x = \sec\theta$, so solve $2x^2 + 3x - 2 = 0$

Factorising gives: $(2x - 1)(x + 2) = 0$

So $x = \dfrac{1}{2} \Rightarrow \sec\theta = \dfrac{1}{2}$

$\Rightarrow \cos\theta = 2$ which is impossible.

Or $x = -2 \Rightarrow \sec\theta = -2$

$\Rightarrow \cos\theta = -\dfrac{1}{2}$

So $\theta = 120°$ or $\theta = 360° - 120° = 240°$

Hint: Use the identity that links $\sec\theta$ and $\tan\theta$.

◀◀ See page 50 **for solving trig equations**.
▶▶ See page 58 for **double angle formulae**.

Watch out: The range $0° \leqslant \theta \leqslant 360°$ tells you to work in degrees.
If the range is $0 \leqslant \theta \leqslant 2\pi$, then the solution is $\theta = \dfrac{2\pi}{3}$ or $\dfrac{4\pi}{3}$.

YOU ARE THE EXAMINER

The steps in Lilia's proof are muddled up. Rewrite the proof in the correct order.

Prove $\text{cosec}^2\theta + \sec^2\theta \equiv \text{cosec}^2\theta\sec^2\theta$.

A: using $\sin^2\theta + \cos^2\theta \equiv 1$

B: $= \dfrac{\cos^2\theta}{\sin^2\theta\cos^2\theta} + \dfrac{\sin^2\theta}{\sin^2\theta\cos^2\theta}$

C: $= \dfrac{1}{\sin^2\theta\cos^2\theta}$

D: $\text{cosec}^2\theta + \sec^2\theta$

E: $= \dfrac{\cos^2\theta + \sin^2\theta}{\sin^2\theta\cos^2\theta}$

F: $= \dfrac{1}{\sin^2\theta} + \dfrac{1}{\cos^2\theta}$

G: as required.

H: $= \text{cosec}^2\theta\sec^2\theta$

SKILL BUILDER

1 Look at this right-angled triangle.

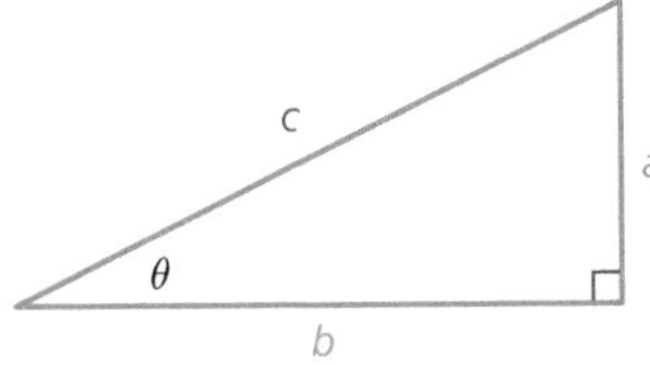

a) Write down expressions in terms of a, b and c for

i) $\sin\theta$ ii) $\cos\theta$ iii) $\tan\theta$ iv) $\text{cosec}\,\theta$ v) $\sec\theta$ vi) $\cot\theta$

b) Use Pythagoras' theorem, $a^2 + b^2 = c^2$, together with your answers to parts i and ii to show that $\sin^2\theta + \cos^2\theta \equiv 1$

2 Given θ is acute and $\cos\theta = \dfrac{5}{13}$, find the exact value of

a) $\sin\theta$ b) $\sec\theta$ c) $\text{cosec}\,\theta$ d) $\cot\theta$

Hint: Draw the right-angled triangle with hypotenuse 13 cm. Find the length of the third side.

3 Find, without using a calculator, the exact value of cosec 240°.

A -2 **B** $-\dfrac{2}{\sqrt{3}}$ **C** $\dfrac{2}{\sqrt{3}}$ **D** $-\dfrac{\sqrt{3}}{2}$

4 Solve these equations for $0° \leqslant x \leqslant 360°$.

a) i) $\sec x = 2$ ii) $\sec(2x) = 2$ iii) $\sec(x - 30°) = 2$

b) i) $\text{cosec}\,x = \dfrac{2}{\sqrt{3}}$ ii) $\text{cosec}\left(\dfrac{1}{2}x\right) = \dfrac{2}{\sqrt{3}}$ iii) $\text{cosec}\left(\dfrac{1}{2}x + 60°\right) = \dfrac{2}{\sqrt{3}}$

c) i) $\cot x = \dfrac{1}{2}$ ii) $3 + 2\cot(2x) = 4$ iii) $4\cot^2 x = 1$

5 Prove these identities.

a) $(1 - \cos\theta)(1 + \cos\theta) \equiv \sin^2\theta$

b) $\cos^2\theta(1 + \tan^2\theta) \equiv 1$

c) $\tan\theta + \cot\theta \equiv \sec\theta\,\text{cosec}\,\theta$

d) $(\sin\theta + \cos\theta)^2 - 1 = 2\sin\theta\cos\theta$

6 a) Prove that $\sin\theta\tan\theta + \cos\theta \equiv \sec\theta$.

Hint: Use $\tan\theta = \dfrac{\sin\theta}{\cos\theta}$ and $\cos\theta = \dfrac{\cos^2\theta}{\cos\theta}$

b) Hence solve $\sin\theta\tan\theta + \cos\theta = \sqrt{2}$ for $0 \leqslant \theta \leqslant 2\pi$.

Compound angles

Year 2

THE LOWDOWN

1. The **compound angle formulae** are
   - $\sin(A+B) = \sin A\cos B + \cos A\sin B$
   - $\sin(A-B) = \sin A\cos B - \cos A\sin B$
   - $\cos(A+B) = \cos A\cos B - \sin A\sin B$
   - $\cos(A-B) = \cos A\cos B + \sin A\sin B$
   - $\tan(A+B) = \dfrac{\tan A + \tan B}{1 - \tan A\tan B}$
   - $\tan(A-B) = \dfrac{\tan A - \tan B}{1 + \tan A\tan B}$

Example: You can use the formulae to expand $\sin(x + 30°)$.

Use $\sin(A+B) = \sin A\cos B + \cos A\sin B$

So $\sin(x+30°) = \sin x\cos 30° + \cos x\sin 30°$

$$= \frac{\sqrt{3}}{2}\sin x + \frac{1}{2}\cos x$$

2. You can use the **compound angle formulae** to give **double-angle formulae**.

Example: Replacing B with A gives

- $\sin(A+A) = \sin A\cos A + \cos A\sin A$

  $\sin(2A) = 2\sin A\cos A$
- $\cos(A+A) = \cos A\cos A - \sin A\sin A$

  $\cos(2A) = \cos^2 A - \sin^2 A$
- $\tan(A+A) = \dfrac{\tan A + \tan A}{1 - \tan A\tan A}$

  $\tan(2A) = \dfrac{2\tan A}{1 - \tan^2 A}$

Hint: These are **identities** – they are true for all values of A and B. Sometimes they are written with the identity symbol $\equiv$. $(A \pm B)$ should not be 90°, 270°, etc. for tan.

Hint: These are in the formula book.

Hint: Remember $\cos 30° = \dfrac{\sqrt{3}}{2}$.

Watch out! You must learn these or derive them in your exam.

◀◀ **Working with radians** on page 54, **trig identities** on page 56 and **surds** on page 6.

GET IT RIGHT

a) Prove that $\cos 2\theta = 2\cos^2\theta - 1$.

b) Hence solve $\cos 2\theta = \cos\theta$ for $0° \leqslant \theta \leqslant 360°$.

Solution:

a) Use the double-angle formula for cos: $\cos(2\theta) = \cos^2\theta - \sin^2\theta$

Replace $\sin^2\theta$ with $(1 - \cos^2\theta)$: $= \cos^2\theta - (1 - \cos^2\theta)$

Remove brackets: $= \cos^2\theta - 1 + \cos^2\theta$

Simplify: $= 2\cos^2\theta - 1$

b) Replace $\cos 2\theta$ with $2\cos^2\theta - 1$ to give $2\cos^2\theta - 1 = \cos\theta$

This is a **quadratic** in $\cos\theta$ so put $= 0$: $2\cos^2\theta - \cos\theta - 1 = 0$

Let $x = \cos\theta$ so: $2x^2 - x - 1 = 0$

Factorise: $(2x+1)(x-1) = 0$

Solve: $x = -\frac{1}{2}$ or $x = 1$

So: $\cos\theta = -\frac{1}{2} \Rightarrow \theta = 120°$ or $360° - 120° = 240°$

or $\cos\theta = 1 \Rightarrow \theta = 0°$ or $360°$

List all the solutions: $\theta = 0°, 120°, 240°$ or $360°$

Hint: Remember $\cos^2\theta + \sin^2\theta = 1$ $\Rightarrow \sin^2\theta = 1 - \cos^2\theta$

Watch out: You can't solve an equation with a mix of $\cos 2\theta$ and $\cos\theta$, so rewrite it in terms of $\cos\theta$ first.

There are 3 useful versions of $\cos 2\theta$

$\cos^2\theta - \sin^2\theta$

$= 2\cos^2\theta - 1$

$= 1 - 2\sin^2\theta$

YOU ARE THE EXAMINER

Why are Sam and Mo's answers different? Can they both be right?

Without using your calculator, find the exact value of $\tan 105°$.

SAM'S SOLUTION

$$\tan(105°) = \tan(60° + 45°)$$

$$\tan(A + B) = \frac{\tan A + \tan B}{1 - \tan A \tan B}$$

$$\tan(60° + 45°) = \frac{\tan 60° + \tan 45°}{1 - \tan 60° \tan 45°}$$

$\tan 60° = \sqrt{3}$ and $\tan 45° = 1$

$$\text{So } \tan(60° + 45°) = \frac{\sqrt{3} + 1}{1 - \sqrt{3} \times 1} = \frac{1 + \sqrt{3}}{1 - \sqrt{3}}$$

MO'S SOLUTION

$$\tan(105°) = \tan(150° - 45°)$$

$$\tan(A - B) = \frac{\tan A - \tan B}{1 + \tan A \tan B}$$

$$\tan(150° - 45°) = \frac{\tan 150° - \tan 45°}{1 + \tan 150° \tan 45°}$$

$\tan 150° = -\frac{\sqrt{3}}{3}$ and $\tan 45° = 1$

$$\tan(150° - 45°) = \frac{-\frac{\sqrt{3}}{3} - 1}{1 - \frac{\sqrt{3}}{3} \times 1}$$

Multiply top and bottom by -3:

$$\tan(150° - 45°) = \frac{\sqrt{3} + 3}{-3 + \sqrt{3}} = \frac{3 + \sqrt{3}}{\sqrt{3} - 3}$$

SKILL BUILDER

1 Isobel says that $\cos(x + 60°) = \cos x + \cos 60°$. Is Isobel correct? Explain your answer fully.

2 Expand and simplify the following expressions.

a) $\cos(x - 45°)$ b) $\sin(x + 60°)$ c) $\tan(x + 45°)$ d) $\tan(x - 45°)$

3 Expand and simplify the following expressions.

a) $\cos\left(x - \frac{\pi}{2}\right)$ b) $\sin(x + \pi)$ c) $\sin\left(x - \frac{\pi}{3}\right)$ d) $\tan\left(x + \frac{\pi}{3}\right)$

Hint: Use radians!

4 a) Expand $\cos(2x)$ by writing $2x$ as $x + x$.

b) Hence show that $\cos(2x) \equiv 2\cos^2 x - 1$ and $\cos(2x) \equiv 1 - 2\sin^2 x$.

Hint: Look at the 'Get it right' box.

5 **Without** using your calculator, find the exact value of

a) i) $\cos 75°$ ii) $\sin 75°$ iii) $\tan 75°$

b) i) $\cos 15°$ ii) $\sin 15°$ iii) $\tan 15°$

Hint: $75° = 45° + 30°$
$15° = 45° - 30°$

6 a) Prove that $\sin(2x) = 2\sin x \cos x$.

b) Hence solve $\sin 2x = \cos x$ for $0° \leqslant x \leqslant 360°$.

Hint: Don't divide by $\cos x$ otherwise you will lose roots – you'll need to factorise!

7 Three of the following statements are false and one is true. Which one is true?

A $\frac{1 - \cos 2\theta}{\sin 2\theta} = \tan\theta$ **B** $\cos(A + B) - \cos(A - B) = 2\sin A \sin B$

C $\sin 3x = 3\sin x \cos x$ **D** $\tan\left(\theta + \frac{\pi}{4}\right) = \tan\theta + 1$

8 Solve the equation $\cos 2\theta = \cos\theta - 1$ for $0 \leqslant \theta \leqslant 2\pi$.

A $60°, 90°, 270°, 300°$ **B** $\frac{\pi}{3}, \frac{\pi}{2}$ **C** $\frac{\pi}{3}, \frac{5\pi}{3}$ **D** $\frac{\pi}{3}, \frac{\pi}{2}, \frac{3\pi}{2}, \frac{5\pi}{3}$

Hint: Look at the 'Get it right' box.

The form $r\sin(\theta + \alpha)$

Year 2

THE LOWDOWN

① You can use the **compound angle formulae** to expand $4\sin(\theta + 60°)$

Example $4\sin(\theta + 60°) = 4\sin\theta\cos 60° + 4\cos\theta\sin 60°$

$$= 4\sin\theta \times \frac{1}{2} + 4\cos\theta \times \frac{\sqrt{3}}{2} = 2\sin\theta + 2\sqrt{3}\cos\theta$$

You can also start with $2\sin\theta + 2\sqrt{3}\cos\theta$ and write it as $4\sin(\theta + 60°)$.

This is called writing $2\sin\theta + 2\sqrt{3}\cos\theta$ **in the form** $r\sin(\theta + \alpha)$.

② There are 4 possible forms for the **r, α formula**

- $a\sin\theta + b\cos\theta = r\sin(\theta + \alpha)$
- $a\sin\theta - b\cos\theta = r\sin(\theta - \alpha)$
- $a\cos\theta + b\sin\theta = r\cos(\theta - \alpha)$
- $a\cos\theta - b\sin\theta = r\cos(\theta + \alpha)$

Example To write $2\sin\theta + 2\sqrt{3}\cos\theta$ in the form $r\sin(\theta + \alpha)$

Step 1 Expand $r\sin(\theta + \alpha) = r\cos\alpha\sin\theta + r\sin\alpha\cos\theta$

Step 2 Compare with $2\sin\theta + 2\sqrt{3}\cos\theta$

So $r\cos\alpha = 2$ ① and $r\sin\alpha = 2\sqrt{3}$ ②

Step 3 Divide: Equation ② ÷ Equation ① $\frac{r\sin\alpha}{r\cos\alpha} = \frac{2\sqrt{3}}{2}$

$\Rightarrow \tan\alpha = \sqrt{3} \Rightarrow \alpha = 60°$

Step 4 Find $r = \sqrt{2^2 + (2\sqrt{3})^2} = \sqrt{4 + 12} = 4$

So $2\sin\theta + 2\sqrt{3}\cos\theta = 4\sin(\theta + 60°)$

Use the **r, α form** when you need to

- sketch the graph $y = a\sin\theta + b\cos\theta$
- solve an equation in the form $a\sin\theta + b\cos\theta = c$
- find maximum and minimum values.

Where $r = \sqrt{a^2 + b^2}$ and $\tan\alpha = \frac{b}{a}$.

Watch out! If the first term has a $\sin\theta$ then use $r\sin(\theta \pm \alpha)$.

If it has a $\cos\theta$ then use $r\cos(\theta \pm \alpha)$.

Hint: Use a right-angled triangle to help you.

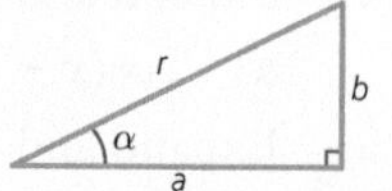

GET IT RIGHT

a) Sketch the graph of $y = 2\sin\theta + 2\sqrt{3}\cos\theta$.

b) Solve $2\sin\theta + 2\sqrt{3}\cos\theta = 2$ for $-180° \leqslant \theta \leqslant 180°$.

Solution:

a) $y = 2\sin\theta + 2\sqrt{3}\cos\theta$ is the same as $y = 4\sin(\theta + 60°)$

Start with graph of $y = \sin\theta$, then carry out (in either order)

- a one-way stretch, scale factor 4, parallel to y-axis
- a translation by vector $\begin{pmatrix} -60° \\ 0 \end{pmatrix}$ (This is 60° to the left.)

From the example in the Lowdown 2.

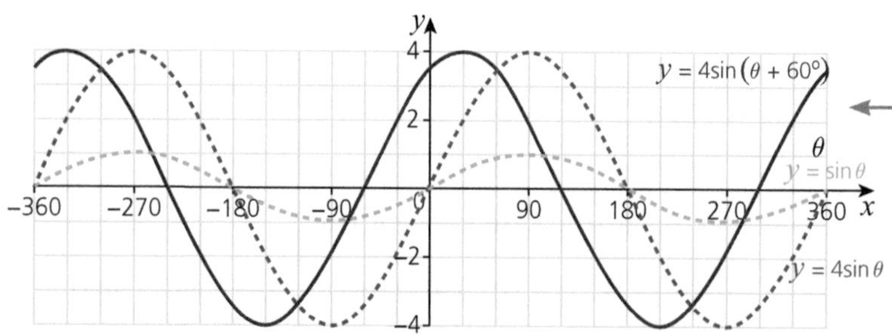

Hint: The graph of $y = \sin\theta$ has a maximum at $(90°, 1)$ and a minimum at $(270°, -1)$.

So the graph of $y = 4\sin(\theta + 60°)$ has a maximum at $(30°, 4)$ and a minimum at $(210°, -4)$.

b) Rewrite $2\sin\theta + 2\sqrt{3}\cos\theta = 2$ as $4\sin(\theta + 60°) = 2$

$$\Rightarrow \sin(\theta + 60°) = \frac{1}{2}$$

Using your calculator: $\theta + 60°$ is $30°$ or $180° - 30° = 150°$

So $\theta = -30°$ or $90°$.

Hint: Subtract 60° from 30° and 150° to find the values of θ.

YOU ARE THE EXAMINER

Which one of these solutions is correct? Where are the errors in the other solution?

Write $\sqrt{2}\cos\theta + \sin\theta$ in the form $r\cos(\theta - \alpha)$ where $0 < \alpha < \frac{\pi}{2}$.

PETER'S SOLUTION

$r\cos(\theta - \alpha) = r\cos\theta\cos\alpha + r\sin\theta\sin\alpha$

Comparing with $\sqrt{2}\cos\theta + \sin\theta$ gives

$r\cos\alpha = \sqrt{2}$ and $r\sin\alpha = 1$

$r = \pm\sqrt{(\sqrt{2})^2 + 1^2} = \pm\sqrt{2+1} = \pm\sqrt{3}$

$\frac{r\sin\alpha}{r\cos\alpha} = \frac{1}{\sqrt{2}} \Rightarrow \tan\alpha = \frac{1}{\sqrt{2}} \Rightarrow \alpha = 45°$

So $\sqrt{2}\cos\theta + \sin\theta = \pm\sqrt{3}\cos(\theta - 45°)$

LILIA'S SOLUTION

$r\cos(\theta - \alpha) = r\cos\theta\cos\alpha + r\sin\theta\sin\alpha$

Comparing with $\sqrt{2}\cos\theta + 1\sin\theta$ gives

$r\cos\alpha = \sqrt{2}$ and $r\sin\alpha = 1$

$r = \sqrt{(\sqrt{2})^2 + 1^2} = \sqrt{2+1} = \sqrt{3}$

$\frac{r\sin\alpha}{r\cos\alpha} = \frac{1}{\sqrt{2}} \Rightarrow \tan\alpha = \frac{1}{\sqrt{2}} \Rightarrow \alpha = 0.615$

So $\sqrt{2}\cos\theta + \sin\theta = \sqrt{3}\cos(\theta - 0.615)$

SKILL BUILDER

1 a) Describe the transformations that map the graph of $y = \sin\theta$ onto

i) $y = 2\sin(\theta + 15°)$ ii) $y = \sqrt{3}\sin(\theta - 50°)$

b) Given $0° \leqslant \theta \leqslant 360°$, write down the coordinates of the minimum and maximum points of the curves in parts a i and ii.

2 a) Expand $r\sin(\theta + \alpha)$.

b) Write $3\sqrt{3}\sin\theta + 3\cos\theta$ in the form $r\sin(\theta + \alpha)$ where $r > 0$ and $0° < \alpha < 90°$.

c) Solve $3\sqrt{3}\sin\theta + 3\cos\theta = 3\sqrt{2}$ for $0° \leqslant \theta \leqslant 360°$.

3 a) Expand $r\cos(\theta + \alpha)$.

b) Write $3\cos\theta - 4\sin\theta$ in the form $r\cos(\theta + \alpha)$ where $r > 0$ and $0° < \alpha < 90°$.

c) Solve $3\cos\theta - 4\sin\theta = 1$ for $0° < \theta < 360°$.

4 What are the values of r and α when $4\cos\theta + 3\sin\theta$ is expressed in the form $r\cos(\theta - \alpha)$, where $r > 0$ and $0 < \alpha < \frac{\pi}{2}$?

A $r = 5, \alpha = 53.1°$ **B** $r = 5, \alpha = 0.644$ **C** $r = 25, \alpha = 0.644$ **D** $r = 5, \alpha = 36.9°$

5 Find the maximum and minimum points of the function $f(\theta) = \sqrt{3}\cos\theta + \sin\theta$ in the range $0 < \theta < 2\pi$.

A $\min\left(\frac{\pi}{6}, 2\right), \max\left(\frac{7\pi}{6}, -2\right)$

B $\min(210°, -2), \max(30°, -2)$

C $\min\left(\frac{7\pi}{6}, -2\right), \max\left(\frac{\pi}{6}, 2\right)$

D $\min\left(\frac{4\pi}{3}, -2\right), \max\left(\frac{\pi}{3}, 2\right)$

Hint: Write $f(\theta)$ in the form $r\cos(\theta - \alpha)$.

6 The graph of $y = \frac{3\sqrt{3}}{2}\cos x - \frac{3}{2}\sin x$ can be obtained from the graph of $y = \cos x$ by a translation and a stretch. Describe the translation and the stretch.

A translation $\begin{pmatrix} -\frac{\pi}{6} \\ 0 \end{pmatrix}$ and one-way stretch, scale factor $\frac{1}{3}$, parallel to the x-axis

B translation $\begin{pmatrix} \frac{\pi}{6} \\ 0 \end{pmatrix}$ and one-way stretch, scale factor 3, parallel to the y-axis

C translation $\begin{pmatrix} \frac{\pi}{6} \\ 0 \end{pmatrix}$ and one-way stretch, scale factor $\frac{1}{3}$, parallel to the x-axis

D translation $\begin{pmatrix} -\frac{\pi}{6} \\ 0 \end{pmatrix}$ and one-way stretch, scale factor 3, parallel to the y-axis

Hint: Write in the form $r\cos(\theta + \alpha)$

Differentiation

Year 1

THE LOWDOWN

① You use **differentiation** to find a formula for the **gradient** of a curve. When you differentiate you find the **gradient function** or the **derivative**. When $y = kx^n$ the **gradient function** is $\frac{dy}{dx} = k \times nx^{n-1}$.

Example $y = 4x^3 + x^2 - 6x + 5 \Rightarrow \frac{dy}{dx} = 12x^2 + 2x - 6$

② You can differentiate twice to find the 2nd derivative.

Example $y = 3x^2 - 5x \Rightarrow \frac{dy}{dx} = 6x - 5 \Rightarrow \frac{d^2y}{dx^2} = 6$

③ You can differentiate functions involving other letters.

Example $A = \frac{4}{3}\pi r^3 \Rightarrow \frac{dA}{dr} = 4\pi r^2$

④ You can substitute an x-coordinate into the gradient function to find the gradient of a curve at a point.

Example To find the gradient of $y = x^2 - 3x$ at $(5, 10)$.

Step 1 Differentiate: $y = x^2 - 3x \Rightarrow \frac{dy}{dx} = 2x - 3$

Step 2 Substitute x into $\frac{dy}{dx}$: When $x = 5$, $\frac{dy}{dx} = 2 \times 5 - 3 = 7$

⑤ You can use the gradient function to find a point with a particular gradient.

Example To find where the gradient of $y = x^2 - 3x$ is 9

Solve $\frac{dy}{dx} = 9$ so $2x - 3 = 9$ giving $x = 6$

At $x = 6$, $y = 6^2 - 3 \times 6 = 18$ so at $(6, 18)$ the gradient of 9.

Hint: You can think of this as 'multiply by the power, then subtract 1 from the power'. You have **differentiated y with respect to x.**

Watch out: If $y = kx$ $\frac{dy}{dx} = k$ and if $y =$ a number then $\frac{dy}{dx} = 0$

Notation: If you start with y and differentiate you get $\frac{dy}{dx}$ and $\frac{d^2y}{dx^2}$.

If you start with $f(x)$ you get $f'(x)$ and $f''(x)$

$f(x) = x^3 - x$ so $f'(x) = 3x^2 - 1$ and $f''(x) = 6x$.

GET IT RIGHT

Find the equation of the

a) tangent and

b) normal to the curve $y = 2x^3 - 4x^2 + 5$ at the point $(2, 5)$.

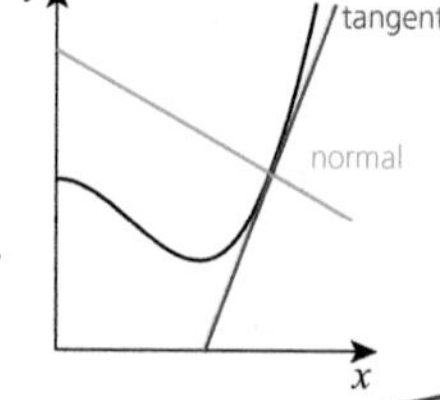

Solution:

a) Step 1 Differentiate: $y = 2x^3 - 4x^2 + 5 \Rightarrow \frac{dy}{dx} = 6x^2 - 8x$

Step 2 Substitute x into $\frac{dy}{dx}$: When $x = 2$, $\frac{dy}{dx} = 6 \times 2^2 - 8 \times 2$

This is the gradient at $(2, 5)$: $\frac{dy}{dx} = 8$

Step 3 Use $y - y_1 = m(x - x_1)$: $y - 5 = 8(x - 2)$ $y = 8x - 11$

b) Step 1 Find gradient of normal: $m_{tangent} = 8 \Rightarrow m_{normal} = \frac{-1}{8}$

Step 2 Use $y - y_1 = m(x - x_1)$: $y - 5 = \frac{-1}{8}(x - 2)$

Step 3 Simplify: $8y - 40 = 2 - x \Rightarrow 8y + x = 42$

Hint: The **gradient of a tangent** is the same as the gradient of the curve at that point.

Hint: The normal is a line at right angles to the tangent.

◀◀ **Equations of parallel and perpendicular lines** on page 34.

▶▶ **Stationary points** on page 64.

YOU ARE THE EXAMINER

Which one of these solutions is correct? Where are the errors in the other solution?

Given $f(x) = \frac{x^5 - 5x^3}{x^2}$, find $f'(x)$ and $f''(2)$.

LILIA'S SOLUTION

$$f'(x) = \frac{5x^4 - 15x^2}{2x}$$

$$f''(x) = \frac{20x^3 - 30x}{2}$$

$$\text{So } f''(2) = \frac{20 \times 2^3 - 30 \times 2}{2} = 50$$

SAM'S SOLUTION

$$\text{Simplify: } f(x) = \frac{x^5}{x^2} - \frac{5x^3}{x^2} = x^3 - 5x$$

$$f'(x) = 3x^2 - 5$$

$$f''(x) = 6x$$

$$\text{So } f''(2) = 6 \times 2 = 12$$

SKILL BUILDER

1 Differentiate the following.

a) $y = 7x^3 - 4x^2 + 5x - 4$

b) $y = 4x^{10} - 3x^9$

c) $A = \pi r^2$

d) $f(x) = 4x^5 - 3x^4 + 2x^3 - x^2$

2 Find the 2nd derivative of each of the equations in question 1.

3 a) Find the gradient of the curve $y = x^3 - 2x^2 + 3$ at the point P(1, 2).

b) Find the equation of the i tangent and ii normal to the curve at P.

4 Differentiate $z = t^4 - 2t^3 - t + 3$.

A $\frac{dz}{dt} = 4t^3 - 6t^2 + 2$ **B** $\frac{dy}{dt} = 4t^3 - 6t^2$ **C** $\frac{dy}{dx} = 4t^3 - 6t^2 - 1$

D $\frac{dy}{dt} = 4t^3 - 5t^2 - 1$ **E** $\frac{dz}{dt} = 4t^3 - 6t^2 - 1$

5 Given $f(x) = 4 - 3x - x^2 + 2x^3$, find $f'(-2)$.

A 31 **B** 25 **C** 29 **D** −23 **E** 21

6 Find the gradient of the curve $y = 3x^2 - 5x - 1$ at the point (2, 1).

A 5 **B** 7 **C** 6 **D** 1 **E** −1

7 Find the coordinates of the point on the curve $y = 4 - 3x + x^2$ at which the tangent to the curve has gradient −1.

A (1, 2) **B** (1, −1) **C** (−2, 14) **D** (−2, −1)

8 Find the gradient of the curve $y = x^5(2x + 1)$ at the point at which $x = -1$.

Hint: Multiply out the bracket first.

A 10 **B** −3 **C** −7 **D** −17 **E** 7

9 Find the value of the second derivative of the curve $y = x^4 - 3x^2 - x + 1$ at the point where $x = -2$.

A −21 **B** 42 **C** −54 **D** 10 **E** −18

10 Find the y-coordinate of the point on the curve $y = x^3 + 6x^2 + 5x - 3$ at which the second derivative is zero.

A −2 **B** −45 **C** 3 **D** −7 **E** 9

11 Find the equation of the normal to the curve $y = x^2 + 7x + 6$ at the point where $x = -2$.

A $y = 3x + 2$ **B** $3y + 10 = x$ **C** $3y + x + 38 = 0$

D $y + 6 = 3x$ **E** $3y + x + 14 = 0$

Stationary points

Year 1 and 2

THE LOWDOWN

① Increasing and decreasing functions

When the gradient function $\frac{dy}{dx}$ is

- positive the function is increasing
- negative the function is decreasing

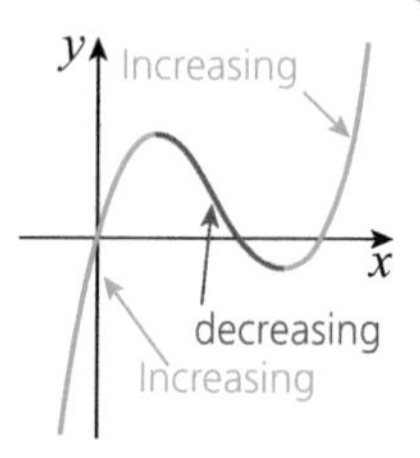

Example $y = x^3$ is an increasing function for all x because $\frac{dy}{dx} = 3x^2$ and $3x^2$ is positive for all values of x.

Hint: Maximum and minimum points are also called turning points or stationary points.

② A **stationary point** is a point on a curve where the gradient is 0, so $\frac{dy}{dx} = 0$.

	Maximum	Minimum	Point of inflection
	0, +, −	−, +, 0	+, +
Before point...	$\frac{dy}{dx} > 0$ (+ve)	$\frac{dy}{dx} < 0$ (−ve)	$\frac{dy}{dx} > 0$ or < 0
At point	$\frac{dy}{dx} = 0$	$\frac{dy}{dx} = 0$	$\frac{dy}{dx} = 0$
...after point	$\frac{dy}{dx} < 0$ (−ve)	$\frac{dy}{dx} > 0$ (+ve)	**Same sign as before**
2nd derivative	$\frac{d^2y}{dx^2} < 0$ (−ve)	$\frac{d^2y}{dx^2} > 0$(+ve)	$\frac{d^2y}{dx^2} = 0$

A **point of inflection** can also look like this:

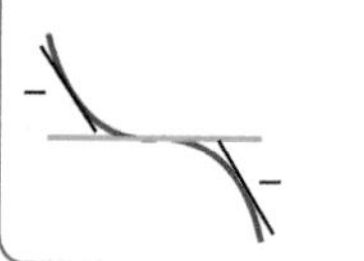

Watch out! If $\frac{d^2y}{dx^2} = 0$, then you **must check** the value of $\frac{dy}{dx}$ on either side of the point.

GET IT RIGHT

Find the coordinates of the stationary points on the curve $y = 2x^3 - 3x^2 - 12x + 3$. Determine the nature of each point.

Hint: 'Determine the nature of a stationary point' means show if it is a maximum, minimum or a point of inflection.

Solution:

Step 1 Differentiate: $\frac{dy}{dx} = 6x^2 - 6x - 12$

Step 2 Solve $\frac{dy}{dx} = 0$: $6x^2 - 6x - 12 = 0 \Rightarrow x = -1$ or $x = 2$

Hint: Use your calculator to solve this.

Step 3 Find the y-coordinates by substituting into the equation of the curve:

At $x = -1$ then $y = 2 \times (-1)^3 - 3 \times (-1)^2 - 12 \times (-1) + 3 = 10$

At $x = 2$ then $y = 2 \times 2^3 - 3 \times 2^2 - 12 \times 2 + 3 = -17$

Step 4 Differentiate again: $\frac{d^2y}{dx^2} = 12x - 6$

Step 5 Substitute in x values:

At $x = -1$: $\frac{d^2y}{dx^2} = 12 \times (-1) - 6 = -18 < 0$ so maximum at $(-1, 10)$

At $x = 2$: $\frac{d^2y}{dx^2} = 12 \times 2 - 6 = 18 >$ so minimum at $(2, -17)$

YOU ARE THE EXAMINER

Which one of these solutions is correct? Where are the errors in the other solution?

Find and classify the stationary point(s) of the curve $y = x^4$.

SAM'S SOLUTION

At a stationary point: $\frac{dy}{dx} = 4x^3 = 0$

$4x^3 = 0 \Rightarrow x = 0$

Find y-coordinate:

$y = x^4$ so when $x = 0$, $y = 0$

At $x = -1$: $\frac{dy}{dx} = 4 \times (-1)^3 = -4 < 0$

At $x = 1$: $\frac{dy}{dx} = 4 \times 1^3 = 4 > 0$

Gradient is −ve, 0, +ve

So $(0, 0)$ is a minimum.

MO'S SOLUTION

At a stationary point: $\frac{dy}{dx} = 4x^3 = 0$

$4x^3 = 0 \Rightarrow x = 0$

Find y-coordinate:

$y = x^4$ so when $x = 0$, $y = 0$

$\frac{d^2y}{dx^2} = 12x^2$: At $x = 0$: $\frac{d^2y}{dx^2} = 0$

Hence $(0, 0)$ is a point of inflection.

SKILL BUILDER

1 Find the coordinates of the stationary point(s) on the curve $y = x^3 - 3x^2 - 9x + 11$.

A $(1, 0)$ and $(-3, -16)$ **B** $(-1, 0)$ and $(3, 0)$ **C** $(-1, 16)$ and $(3, -16)$

D $(1, -12)$ and $(-3, 36)$ **E** $(1, 0)$ only

2 Find the coordinates of the stationary point(s) on the curve $y = 3x^4 + 2x^3 + 1$.

A $\left(-\frac{1}{2}, \frac{15}{16}\right)$ only **B** $(0, 1)$ and $\left(-\frac{1}{2}, \frac{9}{16}\right)$ **C** $\left(-\frac{1}{2}, \frac{9}{16}\right)$ only

D $(0, 1)$ and $\left(-\frac{1}{2}, \frac{15}{16}\right)$ **E** $(0, 1)$ and $\left(\frac{1}{2}, \frac{23}{16}\right)$

3 Find the range of values of x for which the function $y = x^3 + 2x^2 + x$ is a decreasing function of x.

Hint: Find the stationary points and sketch the graph.

A $x < -1$ and $x > -\frac{1}{3}$ **B** $x \leqslant -1$ and $x \geqslant -\frac{1}{3}$ **C** $x < -2$

D $-1 \leqslant x \leqslant -\frac{1}{3}$ **E** $-1 < x < -\frac{1}{3}$

4 Find and classify the stationary point(s) of the curve $y = x^3 - 6x^2 + 12x$.

5 A rectangular sheet of sides 24 cm and 15 cm has four equal squares of sides x cm cut from the corners.

The sides are then turned up to makes an open rectangular box.

Find an expression in terms of x for the volume of the rectangular box.

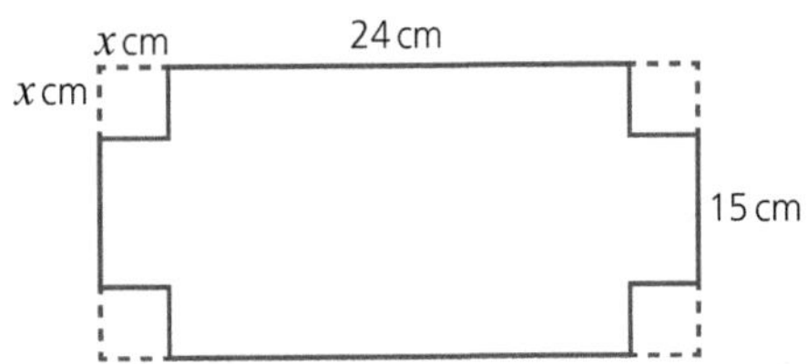

Hint: Find the dimensions of the base of the box in terms of x.

A $4x^2 - 78x + 360$ **B** $360x$

C $4x^3 - 78x^2 + 360x$ **D** $2x^3 - 39x^2 + 180x$

6 Find the value of x so that the volume of the box in question 5 is a maximum.

A 3 **B** 10 **C** 9.75 **D** 2.55

Integration

Year 1

THE LOWDOWN

① When you **differentiate** $y = x^2 + 6$, $y = x^2 + 2$ or $y = x^2 - 1$ then $\frac{dy}{dx} = 2x$.

Integration is the reverse of differentiation, $\int 2x\,dx$ means integrate $2x$.

So when you integrate $2x$ the result is $x^2 + c$.

The rule for integrating a power of x is

$\int kx^n dx = \frac{k}{n+1}x^{n+1} + c$ where c is the constant of integration.

Example $\int(3x^4 + x - 4)dx = \frac{3}{5}x^5 + \frac{1}{2}x^2 - 4x + c$

Hint: When you differentiate a constant it disappears, so add in a constant c when you integrate.

② A **definite integral** has **limits** which you substitute in to the integrated function.

Example To evaluate $\int_{-2}^{3}(x^2 - 2x)dx$

Integrate: $= \left[\frac{1}{3}x^3 - x^2\right]_{-2}^{3}$

Substitute in the limits: $= \left(\frac{1}{3}\times 3^3 - 3^2\right) - \left(\frac{1}{3}\times(-2)^3 - (-2)^2\right)$

Work out the answer: $= (9 - 9) - \left(-\frac{8}{3} - 4\right) = 6\frac{2}{3}$

Hint: Add 1 to the power and divide by the **new** power.

Watch out! You don't need 'c' for definite integration as it cancels out.

③ **Area under the curve** $y = f(x)$ between $x = a$ and $x = b$ is $\int_a^b f(x)\,dx$

Example $A = \int_1^2 (x^3 + 5)dx$

$= \left[\frac{1}{4}x^4 + 5x\right]_1^2$

$= \left(\frac{1}{4}\times 2^4 + 5\times 2\right) - \left(\frac{1}{4}\times 1^4 + 5\times 1\right)$

$= (4 + 10) - \left(\frac{1}{4} + 5\right)$

$= 8.75$ square nits

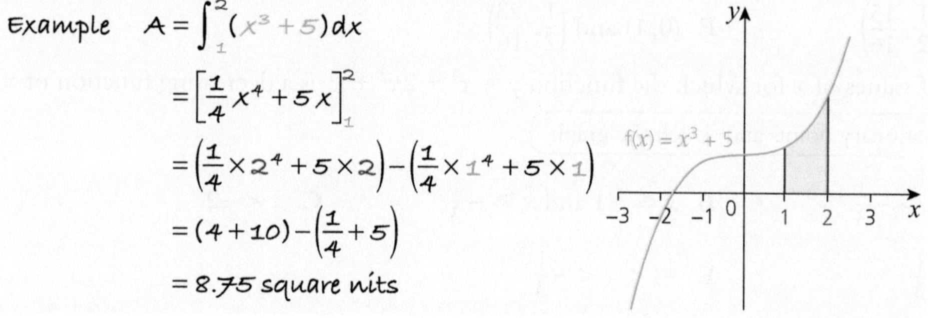

Watch out! An area under the x-axis is negative.

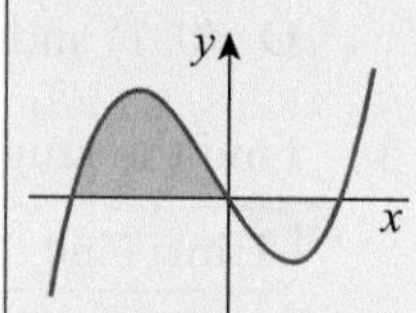

If the curve goes under the x-axis you need to work out the area in 2 parts.

GET IT RIGHT

The gradient function for a curve is $\frac{dy}{dx} = 6x^2 - 3x + 5$.

Find the equation of the curve given it passes through the point (2, 17).

Solution:

Step 1 Integrate: $y = \int(6x^2 - 3x^1 + 5)dx = \frac{6}{3}x^3 - \frac{3}{2}x^2 + 5x + c$

So $y = 2x^3 - \frac{3}{2}x^2 + 5x + c$ (Don't forget $+ c$)

Step 2 The curve passes through (2, 17) so substitute $x = 2$ and $y = 17$ into the equation of the curve to find c.

$17 = 2\times 2^3 - \frac{3}{2}\times 2^2 + 5\times 2 + c \Rightarrow 17 = 16 - 6 + 10 + c \Rightarrow c = -3$

So the equation of the curve is $y = 2x^3 - \frac{3}{2}x^2 + 5x - 3$

Hint: Add 1 to the power and then divide by the new power.

Remember:

kx ⟵ integrate / differentiate ⟶ k

★ YOU ARE THE EXAMINER

Which one of these solutions is correct? Where are the errors in the other solution?

The diagram shows the curve $y = x^2 - 5x$.

Find the total area of the shaded regions.

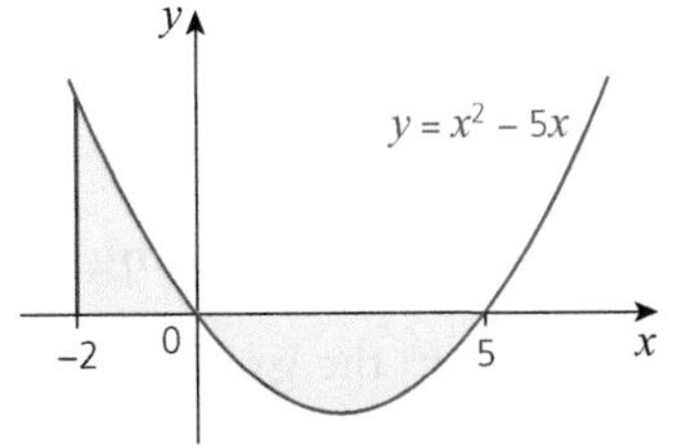

MO'S SOLUTION

$$\int_{-2}^{5}(x^2 - 5x)\,dx$$

$$= \left[\frac{1}{3}x^3 - \frac{5}{2}x^2\right]_{-2}^{5}$$

$$= \left(\frac{1}{3}\times 5^3 - \frac{5}{2}\times 5^2\right) - \left(\frac{1}{3}\times(-2)^3 - \frac{5}{2}\times(-2)^2\right)$$

$$= \left(-20\frac{5}{6}\right) - \left(-12\frac{2}{3}\right) = -8\frac{1}{6}$$

Area can't be negative so area is $8\frac{1}{6}$ sq. units

NASREEN'S SOLUTION

Area of region 1 $= \int_{-2}^{0}(x^2 - 5x)\,dx$

$$= \left[\frac{1}{3}x^3 - \frac{5}{2}x^2\right]_{-2}^{0} = 12\frac{2}{3}$$

Area of region 2 $= \int_{0}^{5}(x^2 - 5x)\,dx$

$$= \left[\frac{1}{3}x^3 - \frac{5}{2}x^2\right]_{0}^{5} = -20\frac{5}{6}$$

Total area $= 12\frac{2}{3} + 20\frac{5}{6} = 33.5$ sq. units

✓ SKILL BUILDER

1 Find

a) $\int(x^3 - 2)\,dx$ b) $\int(t^4 - 3t^3 + 5)\,dt$ c) $\int(5y^4 - 2y^2 - 1)\,dy$

2 Evaluate these.

a) $\int_1^3(2x - 1)\,dx$ b) $\int_{-1}^{2}(x^2 - x)\,dx$ c) $\int_{-3}^{0}(4x^3 + 2x + 3)\,dx$

3 Find $\int 2x^3 dx$.

A $\frac{2x^3}{3} + c$ **B** $\frac{x^4}{2} + c$ **C** $6x^2 + c$ **D** $\frac{2x^4}{3} + c$

4 Find $\int(3x^2 + 1)\,dx$.

A $x^3 + c$ **B** $6x + c$ **C** $x^3 + 1 + c$ **D** $x^3 + x + c$

5 A curve has gradient given by $\frac{dy}{dx} = 4x$ and passes through the point $(-2, 3)$. What is the equation of the curve?

A $y = 4x + 11$ **B** $y = 2x^2 + 11$ **C** $y = 2x^2 - 20$ **D** $y = 2x^2 - 5$

6 A curve has gradient given by $\frac{dy}{dx} = 6x^2 - 6x$ and passes through the point $(2, -1)$. What is the equation of the curve?

A $y = 2x^3 - 3x^2 - 5$ **B** $y = 12x - 6$ **C** $y = 6x^3 - 6x^2 - 23$ **D** $y = 2x^3 - 3x^2 + 7$

7 A curve has gradient given by $\frac{dy}{dx} = x^2 - 2x + 1$ and passes through the point $(-1, -2)$. Where does the curve intercept the y-axis?

A $y = -1\frac{2}{3}$ **B** $y = -2$ **C** $y = -1$ **D** $y = \frac{1}{3}$

8 Find the total area of the shaded regions for each of the following graphs.

a)

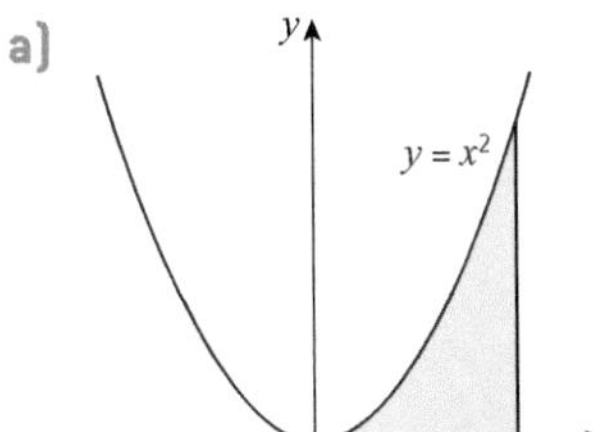

b)

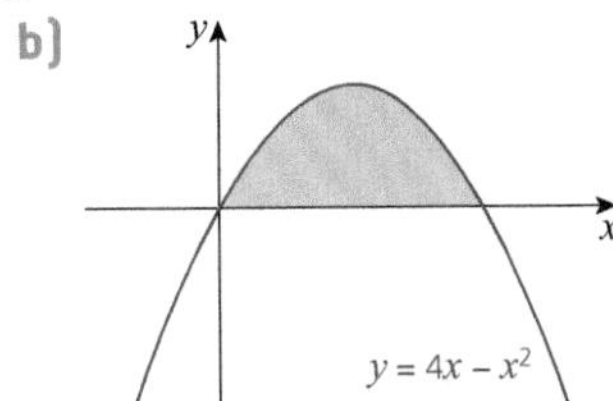

c)

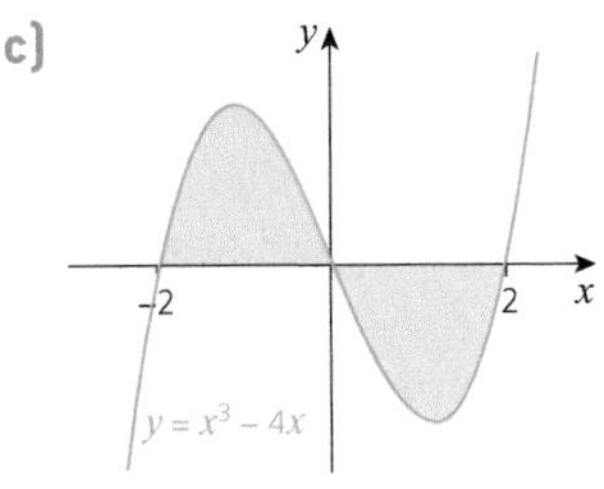

Extending the rules

Year 1

THE LOWDOWN

① The rule for **differentiating** a power of x can be used for any value of n.

When $y = kx^n$ the **gradient function** is $\frac{dy}{dx} = k \times nx^{n-1}$.

Think: Multiply by the power and then subtract 1 from the power.

Example To differentiate $y = 6\sqrt[3]{x} - \frac{2}{x}$

Step 1 Rewrite as powers of x: $y = 6x^{\frac{1}{3}} - 2x^{-1}$

Step 2 Differentiate: $\frac{dy}{dx} = 6 \times \frac{1}{3}x^{-\frac{2}{3}} - 2 \times (-1)x^{-2}$

Step 3 Tidy up: $\frac{dy}{dx} = \frac{2}{\sqrt[3]{x^2}} + \frac{2}{x^2}$

② The rule for **integrating** a power of x can be used for any value of n except $n = -1$.

$\int kx^n dx = \frac{k}{n+1}x^{n+1} + c$ where c is the constant of integration.

Think: Add 1 to the power and then divide by the new power.

Example To find $\int\left(\frac{3}{x^2} - \sqrt[3]{x^2}\right)dx$

Step 1 Rewrite as powers of x: $\int\left(3x^{-2} - x^{\frac{2}{3}}\right)dx$

Step 2 Integrate: $= 3 \times \frac{x^{-1}}{-1} - \frac{x^{\frac{5}{3}}}{5/3} + c$

Step 3 Tidy up: $= \frac{-3}{x} - \frac{3}{5}x^{\frac{5}{3}} + c$

Watch out! n could be negative or a fraction.

Hint: Remember $\sqrt[n]{x} = x^{\frac{1}{n}}$ and $\frac{1}{x^n} = x^{-n}$

Hint: You can leave this as $\frac{dy}{dx} = 2x^{-\frac{2}{3}} + 2x^{-2}$

Watch out! The rule doesn't work for $n = -1$ because you can't divide by 0.

$\int \frac{1}{x}dx = \int x^{-1}dx \neq \frac{x^0}{0}$ ✗

Hint: Flip the fraction in the denominator.

GET IT RIGHT

a) Find the equation of the tangent to the curve $y = 3\sqrt{x}$ at the point (4, 6).

b) Find the area enclosed by the tangent, the curve and y-axis.

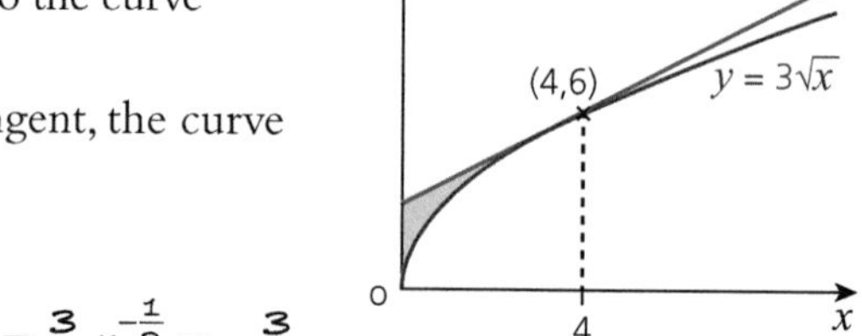

Solution:

a) Differentiate: $y = 3x^{\frac{1}{2}} \Rightarrow \frac{dy}{dx} = \frac{3}{2}x^{-\frac{1}{2}} = \frac{3}{2\sqrt{x}}$

When $x = 4$, $\frac{dy}{dx} = \frac{3}{2\sqrt{4}} = \frac{3}{4}$

Equation of tangent at (4, 6) is $y - 6 = \frac{3}{4}(x - 4) \Rightarrow y = \frac{3}{4}x + 3$

b) Area = Area under line − area under curve

$= \int_0^4 (\frac{3}{4}x + 3)dx - \int_0^4 3x^{\frac{1}{2}}dx$

$= [\frac{3}{8}x^2 + 3x]_0^4 - [2x^{\frac{3}{2}}]_0^4$

$= \{(\frac{3}{8} \times 4^2 + 3 \times 4) - (0)\} - \{(2 \times 4^{\frac{3}{2}}) - (0)\}$

$= 18 - 16 = 2$ square units

◀◀ **Laws of indices** on page 6, **differentiation** on page 62 and **integration** on page 66.

▶▶ **Further calculus** on pages 70-83.

▶▶ See **variable acceleration** for using calculus in mechanics on page 114.

★ YOU ARE THE EXAMINER

Which one of these solutions is correct? Where are the errors in the other solution?

Given $f'(x) = \sqrt[3]{x} + \dfrac{1}{3x^2}$, find f(x) and f''(x).

PETER'S SOLUTION

$$f'(x) = x^{\frac{1}{3}} + 3x^{-2}$$
$$\Rightarrow f(x) = \int (x^{\frac{1}{3}} + 3x^{-2})\,dx$$
$$= \frac{3}{4}x^{\frac{4}{3}} + \frac{3}{-3}x^{-1}$$
$$= \frac{3}{4}x^{\frac{4}{3}} - \frac{1}{x}$$
$$f''(x) = \frac{1}{3}x^{-\frac{2}{3}} - 6x^{-3}$$
$$= \frac{1}{3\sqrt{x^3}} - \frac{6}{x^3}$$

MO'S SOLUTION

$$f'(x) = x^{\frac{1}{3}} + \frac{1}{3}x^{-2}$$
$$\Rightarrow f(x) = \int (x^{\frac{1}{3}} + \frac{1}{3}x^{-2})\,dx$$
$$= \frac{3}{4}x^{\frac{4}{3}} + \frac{1}{-3}x^{-1} + c$$
$$= \frac{3}{4}x^{\frac{4}{3}} - \frac{1}{3x} + c$$
$$f''(x) = \frac{1}{3}x^{-\frac{2}{3}} + \frac{1}{3}\times(-2)x^{-3}$$
$$= \frac{1}{3\sqrt[3]{x^2}} - \frac{2}{3x^3}$$

✓ SKILL BUILDER

1 Differentiate the function $y = \dfrac{1}{3x^5}$.

A $\dfrac{dy}{dx} = -\dfrac{15}{x^6}$ **B** $\dfrac{dy}{dx} = -\dfrac{5}{3x^4}$

C $\dfrac{dy}{dx} = -\dfrac{15}{x^4}$ **D** $\dfrac{dy}{dx} = -\dfrac{5}{3x^6}$

E $\dfrac{dy}{dx} = \dfrac{1}{15x^4}$

2 Find the value of the second derivative of the curve $y = \dfrac{1}{3x}$ at the point where $x = 3$.

A $\dfrac{2}{81}$ **B** $-\dfrac{2}{81}$ **C** $\dfrac{2}{9}$ **D** $-\dfrac{1}{27}$ **E** $-\dfrac{2}{9}$

3 Find the gradient of the function $y = x - 2 + \dfrac{3}{x} - \dfrac{2}{x^2}$ at the point where $x = -2$.

A $-\dfrac{1}{4}$ **B** $-\dfrac{5}{4}$ **C** $-\dfrac{3}{4}$ **D** $\dfrac{7}{2}$ **E** -4

4 Find the coordinates of the point(s) at which the graph $y = x - \dfrac{1}{x}$ has gradient 5.

A $\left(\frac{1}{2}, -\frac{3}{2}\right)$ **B** $\left(2, \frac{3}{2}\right)$ and $\left(-2, -\frac{3}{2}\right)$

C $\left(2, \frac{3}{2}\right)$ **D** $\left(-\frac{1}{2}, \frac{3}{2}\right)$

E $\left(\frac{1}{2}, -\frac{3}{2}\right)$ and $\left(-\frac{1}{2}, \frac{3}{2}\right)$

5 Find the equation of the normal to the curve $y = \dfrac{1}{x}$ at the point where $x = 2$.

A $y + x = 4$ **B** $2y = 8x - 15$

C $2y + 8x = 17$ **D** $4y = x$

E $2y = 8x - 17$

6 Find the x-coordinate of the stationary point of the curve $y = x - \sqrt{x} + 2$.

Use the second derivative to identify its nature.

A $x = \dfrac{1}{4}$; min **B** $x = \dfrac{1}{4}$; max

C $x = \sqrt{2}$; min **D** $x = \dfrac{1}{\sqrt{2}}$; min

E $x = \dfrac{1}{\sqrt{8}}$; max

7 Find $\displaystyle\int_2^4 \left(x^3 + \frac{1}{x^3}\right)dx$.

A $56\dfrac{9}{64}$ **B** $60\dfrac{3}{32}$ **C** $60\dfrac{15}{1024}$ **D** $59\dfrac{29}{32}$

8 Find $\displaystyle\int \sqrt[3]{x}\,dx$.

A $-\dfrac{1}{2x^2} + c$ **B** $x^{\frac{1}{3}} + c$

C $\dfrac{1}{3\sqrt[3]{x^2}} + c$ **D** $\dfrac{3\left(\sqrt[3]{x}\right)^4}{4} + c$

9 The gradient of a curve is given by $\dfrac{dy}{dx} = \dfrac{3}{x^4}$.

The curve passes through $(1, 2)$.

Find the equation of the curve.

A $y = 3 - \dfrac{1}{x^3}$ **B** $y = -\dfrac{12}{x^5}$

C $y = 1\dfrac{1}{8} - \dfrac{1}{x^3}$ **D** $y = 2\dfrac{3}{5} - \dfrac{3}{5x^5}$

Calculus with other functions

Year 2

THE LOWDOWN

1. You need to know the following standard results for **differentiation**.

- $y = \sin kx \Rightarrow \frac{dy}{dx} = k\cos kx$
- $y = \cos kx \Rightarrow \frac{dy}{dx} = -k\sin kx$
- $y = \tan kx \Rightarrow \frac{dy}{dx} = k\sec^2 kx$

Example $y = \sin 3x - \cos 2x \Rightarrow \frac{dy}{dx} = 3\cos 3x + 2\sin 2x$

- $y = e^{kx} \Rightarrow \frac{dy}{dx} = ke^{kx}$
- $y = \ln x \Rightarrow \frac{dy}{dx} = \frac{1}{x}$

Example $y = \ln(7x) \Rightarrow y = \ln 7 + \ln x \Rightarrow \frac{dy}{dx} = 0 + \frac{1}{x} = \frac{1}{x}$

2. You need to know the following standard results for **integration**.

- $\int \sin kx\, dx = -\frac{1}{k}\cos kx + c$
- $\int \cos kx\, dx = \frac{1}{k}\sin kx + c$
- $\int \sec^2 kx\, dx = \frac{1}{k}\tan kx + c$

Example

$$\int_0^{\frac{\pi}{2}} \sin 3\theta\, d\theta = \left[-\frac{1}{3}\cos 3\theta\right]_0^{\frac{\pi}{2}} = \frac{1}{3}\left(-\cos 3\left(\frac{\pi}{2}\right) + \cos 3(0)\right) = \frac{1}{3}(0+1) = \frac{1}{3}$$

- $\int \frac{1}{x}dx = \ln|x| + c$ and $\int \frac{k}{x}dx = k\ln|x| + c$
- $\int e^{kx}\, dx \Rightarrow \frac{dy}{dx} = \frac{1}{k}e^{kx}$

Example $\int (\frac{8}{x} + \ln 2)dx = \int 8 \times \frac{1}{x} + \ln 2\, dx = 8\ln|x| + x\ln 2 + c$

Watch out! You must use radians for trig functions.

▶▶ **Trigonometry** on page 50 and **logs and exponentials** on page 22.
▶▶ **Differentiation** on pages 72–75 and **integration** on pages 76–80.

Hint: To help with the signs, differentiate down the list, integrate up the list.

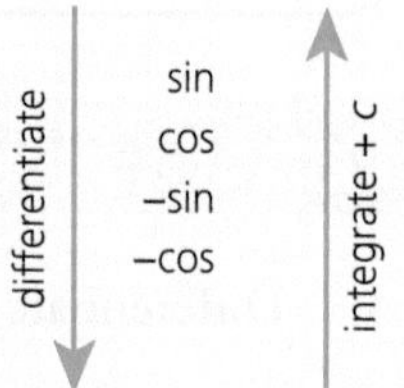

Hint: Integration gives you the fraction of $\frac{1}{k}$.

Watch out! You can't have the log of a negative number so you need the modulus of x.

Hint: ln 2 is just a number!

GET IT RIGHT

Find the x-coordinate of the stationary point of the curve $y = e^{2x} - e^x$.

Solution:

$\frac{dy}{dx} = 2e^{2x} - e^x$

At a stationary point $\frac{dy}{dx} = 0 \Rightarrow 2e^{2x} - e^x = 0$

Factorise: $e^x(2e^x - 1) = 0 \Rightarrow e^x = \frac{1}{2} \Rightarrow x = \ln\frac{1}{2} = -\ln 2$

Hint: Take ln of both sides to undo e.

Hint: $\ln\frac{1}{2} = \ln 2^{-1} = -\ln 2$

YOU ARE THE EXAMINER

Which one of these solutions is correct? Where is the error in the other solution?

Find the area of the shaded region.

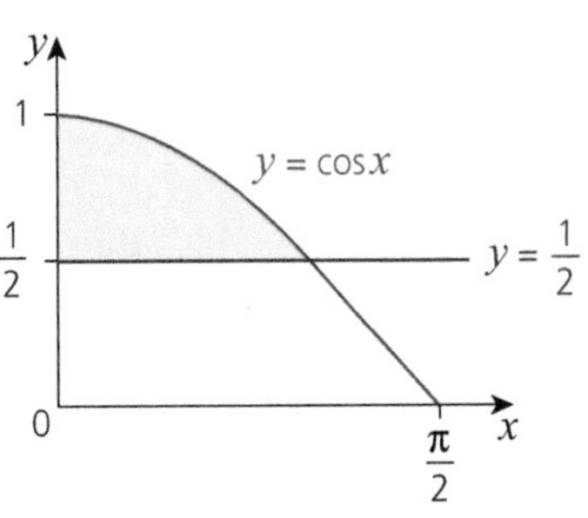

MO'S SOLUTION

$\cos x = 0.5 \Rightarrow x = \frac{\pi}{3}$

Shaded region $= \int_0^{\frac{\pi}{3}} \cos x\,dx - \int_0^{\frac{\pi}{3}} \frac{1}{2}\,dx$

$= [\sin x]_0^{\frac{\pi}{3}} - \left[\frac{1}{2}x\right]_0^{\frac{\pi}{3}}$

$= \left\{\left(\frac{\sqrt{3}}{2}\right) - (0)\right\} - \left\{\left(\frac{1}{2} \times \frac{\pi}{3}\right) - (0)\right\}$

$= \frac{\sqrt{3}}{2} - \frac{\pi}{6}$

LILIA'S SOLUTION

$\cos x = 0.5 \Rightarrow x = 60°$

Shaded region $= \int_0^{60} \cos x\,dx - \int_0^{60} \frac{1}{2}\,dx$

$= [\sin x]_0^{60} - \left[\frac{1}{2}x\right]_0^{60}$

$= \left\{\left(\frac{\sqrt{3}}{2}\right) - (0)\right\} - \left\{\left(\frac{1}{2} \times 60\right) - (0)\right\}$

$= \frac{\sqrt{3}}{2} - 30$

SKILL BUILDER

1 Find f'(x).

a) $f(x) = \sin 2x + \cos 3x$
b) $f(x) = \tan 2x$
c) $f(x) = 4\sin\left(\frac{x}{2}\right) - 3\cos\left(\frac{x}{6}\right)$

2 Differentiate:

a) $y = e^{5x} - e^{-x} + e^{x}$
b) $y = \frac{1}{e^{3x}} + e$
c) $y = \ln(4x) + \ln 2$

3 Find:

a) $\int \left(\sin 2x + \cos\left(\frac{x}{4}\right)\right) dx$
b) $\int \sec^2(5x)\,dx$
c) $\int_{\frac{\pi}{3}}^{\frac{\pi}{6}} (3\cos x - 2\sin x)\,dx$

4 Find:

a) $\int (e^{2x} + 3e^{x})\,dx$
b) $\int \frac{5}{x}\,dx$
c) $\int_1^2 \frac{6e^{3x}}{e^{x}}\,dx$

5 Given $f(x) = \sin x - \cos x$, find $f''\left(\frac{\pi}{3}\right)$.

A $\frac{\sqrt{3}-1}{2}$ **B** $\frac{1-\sqrt{3}}{2}$ **C** $\frac{\sqrt{3}+1}{2}$ **D** $-\sin x + \cos x$

6 Find the equation of the normal to the curve $f(x) = \sin x + 2\cos x$ at $x = \frac{\pi}{2}$.

A $y + 2x = 1 + \pi$ **B** $2y + 2x = 4 + \pi$ **C** $4y + 2x = 4 + \pi$ **D** $4y - 2x = 4 - \pi$

7 Which of the following is the area of the shaded region?

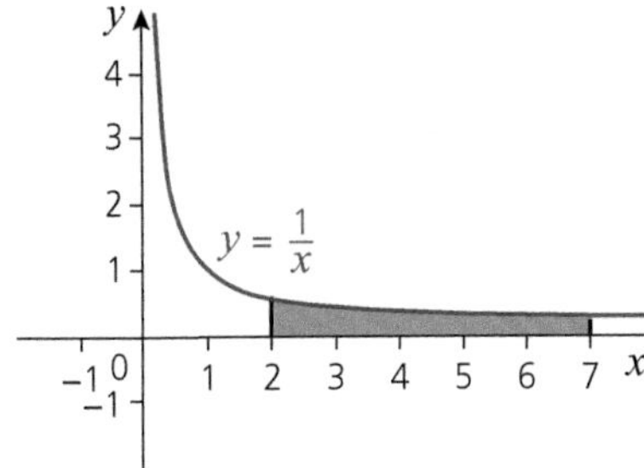

A $\ln 5$ **B** $\ln 3.5$ **C** $\frac{1}{4} - \frac{1}{49}$

D $\frac{\ln 7}{\ln 2}$ **E** 0.544

8 The shaded region in this graph is formed by the curves $y = e^{3x}$ and $y = e^{2x}$ and the line $x = 2$.

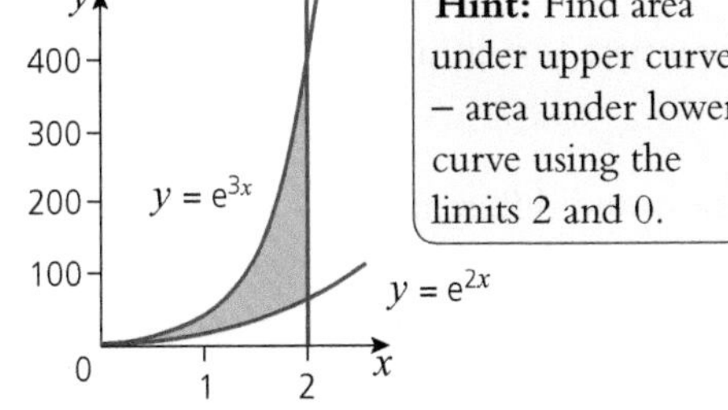

Hint: Find area under upper curve − area under lower curve using the limits 2 and 0.

Find the area of the region.

A $\frac{1}{3}e^6 - \frac{1}{2}e^4 + \frac{1}{6}$ **B** $\frac{1}{3}e^6 - \frac{1}{2}e^4$

C $e^2 - 1$ **D** $3e^6 - 2e^4 - 1$

The chain rule

Year 2

THE LOWDOWN

① The chain rule says $\frac{dy}{dx} = \frac{dy}{du} \times \frac{du}{dx}$.

Example Differentiate $y = (2x-5)^6$

Step 1 Let u = the 'inside' function so $u = 2x - 5$

and $y = u^6$

Step 2 Differentiate: $u = 2x - 5 \Rightarrow \frac{du}{dx} = 2$

$y = u^6 \Rightarrow \frac{dy}{du} = 6u^5$

Step 3 Multiply: $\frac{dy}{dx} = \frac{dy}{du} \times \frac{du}{dx}$

$= 6u^5 \times 2 = 12u^5$

Step 4 Rewrite in terms of x: $\frac{dy}{dx} = 12(2x-5)^5$

Example Differentiate $y = e^{\sin x}$

Let $u = \sin x \Rightarrow \frac{du}{dx} = \cos x$ and $y = e^u \Rightarrow \frac{dy}{du} = e^u$

So $\frac{dy}{dx} = \cos x \times e^u = \cos x\, e^{\sin x}$

② The chain rule leads to a handy shortcut:

- $y = \ln|f(x)| \Rightarrow \frac{dy}{dx} = \frac{f'(x)}{f(x)}$

③ You can use the chain rule to solve problems with **connected rates of change**.

Example The surface area of a circular oil spill increases at $40\,m^2 s^{-1}$.

So $A = \pi r^2 \Rightarrow \frac{dA}{dr} = 2\pi r$ and $\frac{dA}{dt} = 40$

Chain rule gives $\frac{dA}{dt} = \frac{dA}{dr} \times \frac{dr}{dt} = 2\pi r \times \frac{dr}{dt} = 40$

So $\frac{dr}{dt} = \frac{40}{2\pi r} = \frac{20}{\pi r}\, m s^{-1}$

When $r = 5\,m$, the radius is increasing at a rate of $\frac{20}{\pi \times 5} = 1.27\, m s^{-1}$.

Watch out! You must give your answer in terms of x, not u.

Hint: You can turn the gradient function upside down to find $\frac{dx}{dy}$ in terms of x.

Watch out! If you are not 100% confident, it is safer to use the chain rule.

Hint: You use the **chain rule** to differentiate a function of a function.

Hint: Differentiate the formula linking the two variables, and put that in the middle of the '**chain**'.

◀◀ **Functions** on page 18.
▶▶ **Parametric equations** on page 82.

GET IT RIGHT

Differentiate $y = \ln(\sin x)$.

Solution:

Define u and differentiate: $u = \sin x \Rightarrow \frac{du}{dx} = \cos x$

and $y = \ln u \Rightarrow \frac{dy}{du} = \frac{1}{u}$

Use the chain rule: $\frac{dy}{dx} = \frac{dy}{du} \times \frac{du}{dx} = \frac{1}{u} \times \cos x = \frac{\cos x}{\sin x} = \cot x$

YOU ARE THE EXAMINER

Which one of these solutions is correct? Where are the errors in the other solution?

Differentiate $y = \dfrac{1}{(1-3x^2)^5}$.

PETER'S SOLUTION

$u = 1 - 3x^2 \Rightarrow \dfrac{du}{dx} = -6x$

$y = \dfrac{1}{u^5} = u^{-5} \Rightarrow \dfrac{dy}{du} = -5u^{-4}$

Use the chain rule: $\dfrac{dy}{dx} = \dfrac{dy}{du} \times \dfrac{du}{dx}$

$\dfrac{dy}{dx} = -5u^{-4} \times (-6x)$

$= \dfrac{6x}{5(1-3x^2)^4}$

NASREEN'S SOLUTION

$u = 1 - 3x^2 \Rightarrow \dfrac{du}{dx} = -6x$

$y = \dfrac{1}{u^5} = u^{-5} \Rightarrow \dfrac{dy}{du} = -5u^{-6}$

Use the chain rule: $\dfrac{dy}{dx} = \dfrac{dy}{du} \times \dfrac{du}{dx}$

$\dfrac{dy}{dx} = -5u^{-6} \times (-6x)$

$= \dfrac{30x}{(1-3x^2)^6}$

SKILL BUILDER

1 Use the chain rule to differentiate the following.

a) $y = (3x-2)^4$ b) $y = \dfrac{1}{(3x-2)^7}$ c) $y = \sqrt{3x-2}$

d) $y = \sin(3x-2)$ e) $y = \ln(3x-2)$ f) $y = e^{3x-2}$

Hint: Let $u = 3x - 2$.

2 Use the chain rule to differentiate the following.

a) $y = (4x^2 - 5x + 1)^{10}$ b) $y = e^{3x^2}$ c) $y = \sqrt{x^4 - 3x^2}$

d) $y = (e^{4x} - 1)^5$ e) $y = \cos^2 x$ f) $y = \ln(\cos x)$

Hint: For **e)** and **f)**: let $u = \cos x$.

3 A metal cube of side x cm expands as it is heated at a constant rate of $0.02\,\text{cm}^3\text{s}^{-1}$.

a) Find V and $\dfrac{dV}{dx}$. b) Find $\dfrac{dx}{dt}$ at the time when $x = 4$ cm.

Hint: $\dfrac{dV}{dt} = \dfrac{dV}{dx} \times \dfrac{dx}{dt}$

4 Find the derivative of $y = (2x^2 - 3)^3$.

A $3(2x^2-3)^2$ **B** $12x(2x^2-3)^3$ **C** $12x(2x^2-3)^2$ **D** $6x(2x^2-3)^2$

5 You are given that $y = \dfrac{1}{(1-2x)^3}$. Find $\dfrac{dy}{dx}$.

A $-\dfrac{3}{(1-2x)^4}$ **B** $\dfrac{6}{(1-2x)^4}$ **C** $\dfrac{6}{(1-2x)^2}$ **D** $-\dfrac{6}{(1-2x)^4}$ **E** $\dfrac{6}{(1-2x)^3}$

6 You are given that $f(x) = 3\ln(5x-1)$. Find the exact value of $f'(2)$.

A $\dfrac{1}{3}$ **B** $15\ln 9$ **C** $\dfrac{5}{3}$ **D** 15

7 Which of the following is the gradient function of $y = \sqrt{\sin 2x}$?

A $\dfrac{1}{2}(\sin 2x)^{-1/2}$ **B** $\dfrac{1}{2}(\sin 2x)^{-1/2}\cos 2x$ **C** $\dfrac{\cos 2x}{\sqrt{\sin 2x}}$ **D** $-\dfrac{\cos 2x}{\sqrt{\sin 2x}}$ **E** $\sqrt{\sin 2x}\cos 2x$

Product and quotient rules

Year 2

THE LOWDOWN

① The **product rule** says:

When $y = uv$ then $\frac{dy}{dx} = u\frac{dv}{dx} + v\frac{du}{dx}$

Example To differentiate $y = x^3e^{2x}$ then

$u = x^3$ $\frac{du}{dx} = 3x^2$

$v = e^{2x}$ $\frac{dv}{dx} = 2e^{2x}$

So $\frac{dy}{dx} = x^3 \times 2e^{2x} + e^{2x} \times 3x^2$

$= 2x^3e^{2x} + 3x^2e^{2x}$

② The **quotient rule** says:

When $y = \frac{u}{v}$ then $\frac{dy}{dx} = \frac{v\frac{du}{dx} - u\frac{dv}{dx}}{v^2}$

Example To differentiate $y = \frac{x^2}{3x-2}$ then

$u = x^2$ $\frac{du}{dx} = 2x$

$v = 3x - 2$ $\frac{dv}{dx} = 3$

So $\frac{dy}{dx} = \frac{(3x-2) \times 2x - x^2 \times 3}{(3x-2)^2}$

$= \frac{2x(3x-2) - 3x^2}{(3x-2)^2} = \frac{3x^2 - 4x}{(3x-2)^2}$

◀◀ **Differentiation** on page 62 and 72.
▶▶ **Integration by parts** on page 78 and **using calculus in mechanics** on page 114.

Hint: Multiply the expressions linked by the green arrows and **add them together.**

Watch out! Multiply the expressions linked by the **arrows** – start with the **black arrow** and then the green arrow. The top line must be in the right order!

GET IT RIGHT

Show that when $y = \tan x$ then $\frac{dy}{dx} = \sec^2 x$.

Solution:

$y = \tan x = \frac{\sin x}{\cos x}$

$u = \sin x$ $\frac{du}{dx} = \cos x$

$v = \cos x$ $\frac{dv}{dx} = -\sin x$

So $\frac{dy}{dx} = \frac{\cos x \times \cos x - \sin x \times (-\sin x)}{(\cos x)^2}$

$= \frac{\cos^2 x + \sin^2 x}{\cos^2 x}$

$= \frac{1}{\cos^2 x}$

$= \sec^2 x$ as required.

Watch out! Make sure the top line is in the right order – start with differentiating the top! The quotient rule is given to you in your formula book – make sure you use it!

Hint: Remember $\sin^2 x + \cos^2 x \equiv 1$.

YOU ARE THE EXAMINER

Which one of these solutions is correct? Where are the errors in the other solution?

Differentiate $y = \sqrt{x}\cos 2x$.

NASREEN'S SOLUTION

$u = \sqrt{x}$ $\quad \frac{du}{dx} = \frac{1}{2}x^{\frac{1}{2}}$

$v = \cos 2x$ $\quad \frac{dv}{dx} = \frac{1}{2}\sin 2x$

$\frac{dy}{dx} = \sqrt{x} \times \frac{1}{2}\sin 2x + \cos 2x \times \frac{1}{2}x^{\frac{1}{2}}$

$= \frac{1}{2}\sqrt{x}\sin 2x + \frac{1}{2}\sqrt{x}\cos 2x$

PETER'S SOLUTION

$u = x^{\frac{1}{2}}$ $\quad \frac{du}{dx} = \frac{1}{2}x^{-\frac{1}{2}}$

$v = \cos 2x$ $\quad \frac{dv}{dx} = -2\sin 2x$

$\frac{dy}{dx} = x^{\frac{1}{2}} \times (-2\sin 2x) + \cos 2x \times \frac{1}{2}x^{-\frac{1}{2}}$

$= -2\sqrt{x}\sin 2x + \frac{1}{2\sqrt{x}}\cos 2x$

SKILL BUILDER

1 Complete the working to differentiate:

a) $y = \sqrt{x}(x+4)^3$

b) $y = \frac{3x+2}{2x^2-1}$.

a) $u =$ ______ $\frac{du}{dx} =$ ______

$v =$ ______ $\frac{dv}{dx} =$ ______

So $\frac{dy}{dx} =$ ____ × ____ + ____ × ____

Tidy up to give $\frac{dy}{dx} =$ ______ + ______

b) $u =$ ______ $\frac{du}{dx} =$ ______

$v =$ ______ $\frac{dv}{dx} =$ ______

So $\frac{dy}{dx} = \frac{\_\_\_\_ \times \_\_\_\_ - \_\_\_\_ \times \_\_\_\_}{(\_\_\_\_)^2}$

Tidy up to give $\frac{dy}{dx} = \frac{\_\_\_\_ - \_\_\_\_}{\_\_\_\_}$

2 Use the product rule to differentiate the following.

a) $y = xe^x$ b) $y = x^2\sin x$ c) $y = e^{2x}\ln x$

3 Use the quotient rule to differentiate the following.

a) $y = \frac{2x-3}{4x+1}$ b) $y = \frac{\sin x}{(x^2+1)}$ c) $y = \frac{\cos x}{e^x + x}$

4 a) Given that $f(x) = \cos x \sin x$ find $f'(\pi)$.

b) Given that $f(x) = \frac{\cos x}{\sin x}$ find $f'(\frac{\pi}{6})$.

5 The equation of a curve is $y = \frac{x}{2x+1}$. Three of the following statements are false and one is true. Which one is true?

A The gradient function is $\frac{dy}{dx} = \frac{1}{2}$.

B The gradient function is positive for all values of x.

C The gradient at $x = 2$ is $\frac{1}{25}$.

D The gradient at $x = -2$ is $-\frac{1}{9}$.

6 Given that $y = \frac{x}{2 + 3\ln x}$, find the value of $\frac{dy}{dx}$ when $x = 1$.

A $-\frac{1}{4}$ **B** $\frac{1}{2}$ **C** $\frac{1}{4}$ **D** $\frac{1}{3}$

7 Find the gradient of the curve $y = \frac{\cos 2x}{x}$ at the point where $x = \pi$.

A $-\frac{2}{\pi}$ **B** $-\frac{1}{\pi^2}$ **C** $\frac{1}{\pi^2}$ **D** 1

Integration by substitution

Year 2

THE LOWDOWN

① You can use a **substitution** to simplify an integral.

Example To find $\int x(x^2+3)^4\,dx$

Step 1 Write down the substitution: Let $u = x^2 + 3$

Step 2 Differentiate: $\frac{du}{dx} = 2x$

Step 3 Find dx: $\frac{1}{2x}du = dx$

Step 4 Substitute into the integral: $\int x(u)^4 \frac{1}{2x}du$

Step 5 Simplify (x cancels!) $= \int \frac{1}{2}u^4\,du$

Step 6 Integrate $= \frac{1}{10}u^5 + c$

Step 7 Write in terms of x $= \frac{1}{10}(x^2+3)^5 + c$

② When a fraction has a 'top' that is the **derivative** of the '**bottom**' use this.

$$\int \frac{f'(x)}{f(x)}dx = \ln|f(x)| + c$$

Sometimes you have to 'fix' the top to get it in the right form.

Example In the integral $\int \frac{3x-4}{3x^2-8x}dx$ the derivative of the bottom is $6x-8$ so the top of $3x-4$ is 'half' what you want it to be.

$$\int \frac{3x-4}{3x^2-8x}dx = \frac{1}{2}\int \frac{6x-8}{3x^2-8x}dx$$

Now the top is the derivative of the bottom so

$$\frac{1}{2}\int \frac{6x-8}{3x^2-8x}dx = \frac{1}{2}\ln|3x^2-8x| + c$$

Hint: This is like the chain rule in reverse – so choose u to be the inside function.

Watch out! Sometimes you are still left with an x at this point – just make x the subject of your expression for u and substitute that in as well.

Hint: When the 'top' is the **derivative** of the '**bottom**' then the answer is ln | *bottom* | + *c*.

Hint: Fix the integral by doubling the top and then halving.

GET IT RIGHT

Use the substitution $u = 2x^2$ to find $\int_0^1 8xe^{2x^2}\,dx$.

Solution:

Differentiate u: $u = 2x^2 \Rightarrow \frac{du}{dx} = 4x$

Make dx the subject: $\frac{1}{4x}du = dx$

Substitute into integral and simplify: $\int 8xe^{2x^2}dx = \int 8xe^u \frac{1}{4x}du$

$= \int 2e^u du$

Integrate and rewrite in terms of x $= 2e^u + c = 2e^{2x^2} + c$

Now use the limits: $\int_0^1 8xe^{2x^2}dx = \left[2e^{2x^2}\right]_0^1 = 2e^2 - 2e^0 = 2e^2 - 2$

◀◀ **Differentiating other functions** on page 70 and **partial fractions** on page 30.
▶▶ **Further calculus** on page 80.

Watch out: The limits are from $x = 0$ to $x = 1$ so you need to rewrite the integral in terms of x before you can use them.

YOU ARE THE EXAMINER

Which one of these solutions is correct? Where are the errors in the other solution?

Find $\int\left(\frac{2}{4x+1}+\frac{1}{x^2}\right)dx$.

NASREEN'S SOLUTION

$$\int\left(\frac{2}{4x+1}+\frac{1}{x^2}\right)dx = \int\left(\frac{1}{2}\times\frac{4}{4x+1}+x^{-2}\right)dx$$
$$= \frac{1}{2}\ln|4x+1| - x^{-1} + c$$
$$= \frac{1}{2}\ln|4x+1| - \frac{1}{x} + c$$

PETER'S SOLUTION

$$\int\left(\frac{2}{4x+1}+\frac{1}{x^2}\right)dx = 2\ln|4x+1| + \ln x^2 + c$$

SKILL BUILDER

1 Use the rule $\int\frac{f'(x)}{f(x)}dx = \ln|f(x)| + c$ to integrate the following:

a) i) $\int\frac{2x-6}{x^2-6x-1}dx$ ii) $\int\frac{4x-12}{x^2-6x-1}dx$ iii) $\int\frac{x-3}{x^2-6x-1}dx$

b) i) $\int\frac{2e^{2x}}{e^{2x}+5}dx$ ii) $\int\frac{e^{2x}}{e^{2x}+5}dx$ iii) $\int\frac{5e^{2x}}{e^{2x}+5}dx$

c) i) $\int\frac{\cos x}{\sin x}dx$ ii) $\int\frac{\cos x}{3\sin x}dx$ iii) $\int\frac{\sin x}{\cos x}dx$

2 Use integration by substitution to find:

a) $\int(2x+1)^3\,dx$ let $u = 2x+1$ b) $\int\sqrt{4x+3}\,dx$ let $u = 4x+3$

3 Find $\int\frac{6x^2}{1+4x^3}dx$.

Two of these answers are correct. Which two are correct?

A $\frac{1}{2}\ln|1+4x^3| + c$ **B** $2\ln|1+4x^3| + c$ **C** $\frac{2x^2}{x+x^4} + c$

D $2x^3 + \frac{3}{2}\ln|x| + c$ **E** $\frac{1}{2}\ln|2(1+4x^3)| + c$

4 Find $\int\frac{x}{\sqrt{1+x^2}}dx$. You may wish to use the substitution $u = 1+x^2$.

A $2\sqrt{1+x^2} + c$ **B** $\sqrt{1+x^2} + c$

C $u^{\frac{1}{2}} + c$ **D** $\frac{1}{2}\ln(1+x^2) + c$

5 Find the exact value of $\int_0^1\frac{2e^x}{1+e^x}dx$. You may wish to use the substitution $u = 1+e^x$.

A $2 - 2\ln 2$ **B** $\frac{2e}{1+e} - 2$ **C** $\frac{1}{2}\ln\left(\frac{1+e}{2}\right)$ **D** $2\ln\left(\frac{1+e}{2}\right)$

6 The diagram shows a sketch of the curve $y = 10x(2x-1)^3$.
Find the area of the shaded region using the substitution $u = 2x-1$.

A $\frac{1}{4}$ **B** 0.055

C $\frac{1}{8}$ **D** $-\frac{1}{8}$

y
$y = 10x(2x-1)^3$
0
x

Hint: The curve cuts the x-axis when $x = 0$ and when $(2x-1) = 0$ so $x = \frac{1}{2}$.

Integration by parts

Year 2

THE LOWDOWN

① You can integrate some products such as

- $\int 3x\cos x\,dx$
- $\int xe^x\,dx$
- $\int 2x\sin x\,dx$

by using the rule for **integration by parts:**

$$\int u\frac{dv}{dx}dx = uv - \int v\frac{du}{dx}dx$$

Example To find $\int 3x\cos x\,dx$

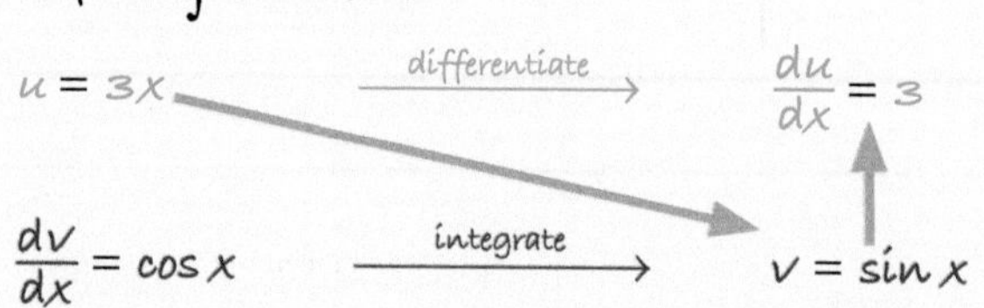

Substitute in the formula and then integrate:

$$\int 3x\cos x\,dx = 3x\sin x - \int \sin x \times 3\,dx$$
$$= 3x\sin x - (-3\cos x) + c$$
$$= 3x\sin x + 3\cos x + c$$

② You can also use **integration by parts** for **definite integrals**.
The rule is

$$\int_a^b u\frac{dv}{dx}dx = [uv]_a^b - \int_a^b v\frac{du}{dx}dx$$

Hint: Use integration by parts when you have x with e^x, $\cos x$ or $\sin x$.

Hint: Use LATE to decide which part to call u and which to call $\frac{dv}{dx}$: (u first)
Logs ($\ln x$)
Algebra (e.g. x)
Trig
Exponential

Hint: Multiply the expressions linked by the green arrows and subtract.

GET IT RIGHT

Find the **exact** value of $\int_0^1 4xe^{2x}\,dx$.

Solution:

Choose u to be $4x$ as x is simpler when you differentiate it. (LATE)

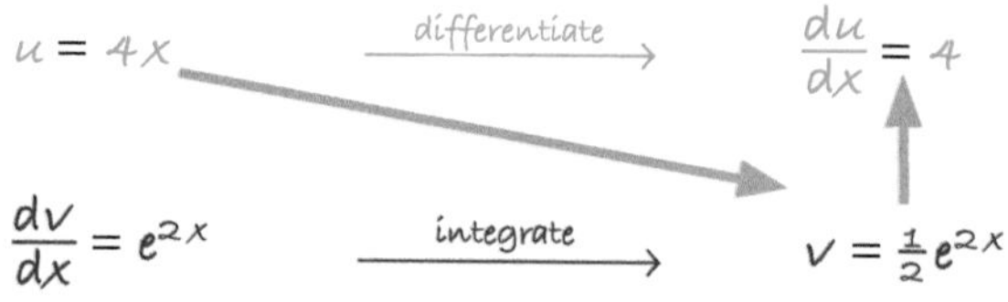

Substitute in the formula: $\int_0^1 4xe^{2x}\,dx = \left[4x \times \frac{1}{2}e^{2x}\right]_0^1 - \int_0^1 \frac{1}{2}e^{2x} \times 4\,dx$

Tidy up $= \left[2xe^{2x}\right]_0^1 - \int_0^1 2e^{2x}\,dx$

Integrate $= \left[2xe^{2x}\right]_0^1 - \left[e^{2x}\right]_0^1$

Substitute in the limits $= (2 \times 1 \times e^{2\times 1} - 2 \times 0 \times e^{2\times 0}) - (e^{2\times 1} - e^{2\times 0})$
$= (2e^2 - 0) - (e^2 - 1)$

Leave your answer in **exact** form: $= e^2 + 1$

Watch out! Choose u to be the function that becomes **simpler** when you differentiate it.

Watch out! Choose $\frac{dv}{dx}$ to be the function of e^x, $\cos x$ or $\sin x$.

Hint: Remember that when you integrate $2e^{2x}$ you get $2 \times \frac{1}{2}e^{2x}$.

Hint: Remember that $e^0 = 1$ as anything to the power 0 is 1.

YOU ARE THE EXAMINER

Which one of these solutions is correct? Where are the errors in the other solution?

To integrate $\int x \ln x \, dx$ you should choose u to be ___ and $\frac{dv}{dx}$ to be ____.

LILIA'S SOLUTION

Let $u = \ln x$ and $\frac{dv}{dx} = x$

MO'S SOLUTION

Let $u = x$ and $\frac{dv}{dx} = \ln x$

SKILL BUILDER

1 Complete the working to integrate each of the following by parts.

a) $\int xe^x \, dx$

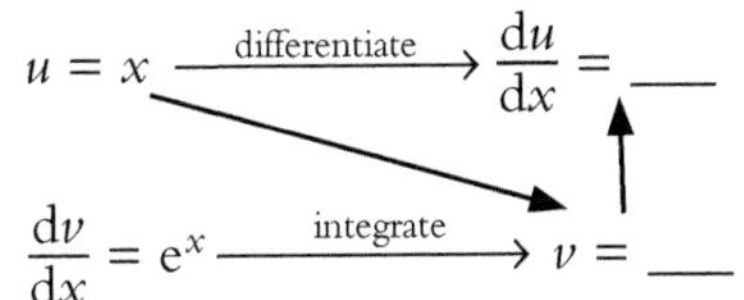

$\int xe^x \, dx = [\_\_\_\_] - \int \_\_\_\_ \, dx$

$= \_\_\_\_\_\_\_\_$

b) $\int x \ln x \, dx$

$u = \ln x$ —differentiate→ $\frac{du}{dx} =$ ___

$\frac{dv}{dx} = x$ —integrate→ $v =$ ___

$\int x \ln x \, dx = [\_\_\_\_] - \int \_\_\_\_ \, dx$

$= \_\_\_\_\_\_\_\_$

2 Use integration by parts to find:

a) $\int x \cos x \, dx$ b) $\int 9xe^{3x} \, dx$ c) $\int 2x \sin 2x \, dx$

3 Find $\int (x + 1) \sin x \, dx$.

A $-\left(\frac{1}{2}x^2 + x\right)\cos x + c$

B $(x + 1)\cos x + \sin x + c$

C $-(x + 1)\cos x + \sin x + c$

D $(x + 1)\cos x - \sin x + c$

4 Find $\int_0^1 xe^{2x} \, dx$.

A $\frac{e^2}{4}$

B $2(2 - e^2)$

C $\frac{1}{2}xe^{2x} - \frac{1}{4}e^2 + \frac{1}{4}$

D $\frac{1}{4}(e^2 + 1)$

5 The diagram shows part of the curve $y = x \sin 2x$.
Find the exact area of the shaded region which lies between the curve, the line $x = \frac{\pi}{3}$ and the x-axis.

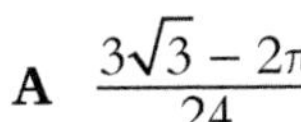

A $\frac{3\sqrt{3} - 2\pi}{24}$

B $\frac{\pi + 6\sqrt{3}}{2}$

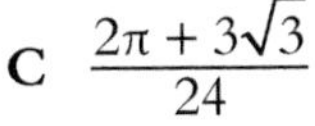

C $\frac{2\pi + 3\sqrt{3}}{24}$

D $\frac{\pi^2}{72}$

6 Find $\int x\ln 2x dx$.

A $\frac{1}{2}x^2\ln 2x - \frac{1}{4}x^2 + c$

B $\frac{1}{2}x^2\ln 2x - \frac{1}{8}x^2 + c$

C $\ln 2$

D $1 - \ln x + c$

Further calculus

Year 2

THE LOWDOWN

① You can differentiate y with respect to x by using **implicit differentiation.**

Example To differentiate y^3 with respect to x

Step 1 Let $z = y^3 \Rightarrow \frac{dz}{dy} = 3y^2$

Step 2 Use the chain rule $\frac{dz}{dx} = \frac{dz}{dy} \times \frac{dy}{dx} = 3y^2 \frac{dy}{dx}$

When you differentiate

- y^3 with respect to y you get $3y^2$
- y^3 with respect to x you get $3y^2 \frac{dy}{dx}$.

Example Given $y^3 + 4e^y = 3x^2$ find $\frac{dy}{dx}$

Differentiating gives $3y^2 \frac{dy}{dx} + 4e^y \frac{dy}{dx} = 6x$

Make $\frac{dy}{dx}$ the subject: $\frac{dy}{dx}(3y^2 + 4e^y) = 6x \Rightarrow \frac{dy}{dx} = \frac{6x}{3y^2 + 4e^y}$

② A **differential equation** involves a derivative such as $\frac{dy}{dx}$.

Example To find the general solution of $\frac{dy}{dx} = 2x + 1$

Integrate both sides $\int dy = \int (2x + 1)dx \Rightarrow y = x^2 + x + c$

You may need to **separate the variables** first so **like terms** are on the same side.

Example To solve $\frac{dy}{dx} = \frac{x}{2y}$ given that when $x = \sqrt{2}$, $y = 3$

Rearrange and integrate: $\int 2y\,dy = \int x\,dx \Rightarrow y^2 = \frac{1}{2}x^2 + c$

Substitute x and y values to find c: $3^2 = \frac{1}{2}\sqrt{2}^2 + c$

So $9 = 1 + c \Rightarrow c = 8$

So the solution is $y^2 = \frac{1}{2}x^2 + 8$

◀◀ **Chain rule** on page 72 and **differentiating other functions** on page 70.
▶▶ **Using calculus in mechanics** on page 114.

Hint: When you differentiate a y term differentiate as normal and then multiply by $\frac{dy}{dx}$.

The graph of $y^3 + 4e^y = 3x^2$ has a **turning point** when $\frac{dy}{dx} = 0 \Rightarrow 6x = 0$, i.e. at $x = 0$.

This is called the **general solution** as c could be any value. You only need a constant on **one side**.

This is called a **particular solution** as you have found c.

GET IT RIGHT

Find the **general solution** of the differential equation $\frac{dy}{dx} = 2xy$.

Solution:

Rearrange: $\frac{dy}{dx} = 2xy \Rightarrow \int \frac{1}{y}dy = \int 2x\,dx$

Integrate: $\Rightarrow \ln|y| = x^2 + c$

Take exponentials of both sides: $y = e^{x^2 + c}$

Use laws of indices to rewrite it as: $y = e^c e^{x^2} = Ae^{x^2}$

Hint: Remember $\ln y$ and e^y are inverses, so one undoes the other.

Hint: e^c is just a constant so you can just call it A.

YOU ARE THE EXAMINER

Which one of these solutions is correct? Where are the errors in the other solution?

Use the **product rule** to differentiate $5x^3y^2$ with respect to x.

SAM'S SOLUTION

$u = 5x^3$, $\frac{du}{dx} = 15x^2$

$v = y^2$, $\frac{dv}{dx} = 2y$

So $\frac{dy}{dx} = 5x^3 \times 2y + y^2 \times 15x^2$

$= 10x^3y + 15x^2y^2$

MO'S SOLUTION

$u = 5x^3$, $\frac{du}{dx} = 15x^2$

$v = y^2$, $\frac{dv}{dx} = 2y\frac{dy}{dx}$

So the derivative of $5x^3y^2$ is

$5x^3 \times 2y\frac{dy}{dx} + y^2 \times 15x^2 = 10x^3y\frac{dy}{dx} + 15x^2y^2$

SKILL BUILDER

1 Differentiate the following with respect to x.

a) i) x^5 ii) y^5 iii) 5 iv) $x^5 + y^5 + 5$

b) i) e^{3x} ii) e^{2y} iii) $3y$ iv) $e^{3x} + e^{2y} + 3y$

c) i) $\cos 2x$ ii) $\sin 3y$ iii) $\tan y$ iv) $\cos 2x + \sin 3y + \tan y$

d) i) $2x^6$ ii) y^3 iii) $3x^2y^3$

e) i) e^{2x} ii) $\sin y$ iii) $e^{2x}\sin y$

Hint: For d iii and e iii: Use the product rule.

2 Find $\frac{dy}{dx}$ when $y^2 = e^x + \sin y$.

A $\frac{e^x}{2y - \cos y}$ **B** $\frac{e^x}{2y + \cos y}$ **C** $\frac{e^x + \cos y}{2y}$ **D** $\frac{2y - e^x}{\cos y}$

3 The curve shown is defined implicitly by the equation $y^2 + 2 = x^3 + y$.

Find the gradient of the curve at the point (2, 3).

A $\frac{12}{5}$ **B** 9

C 2 **D** −6

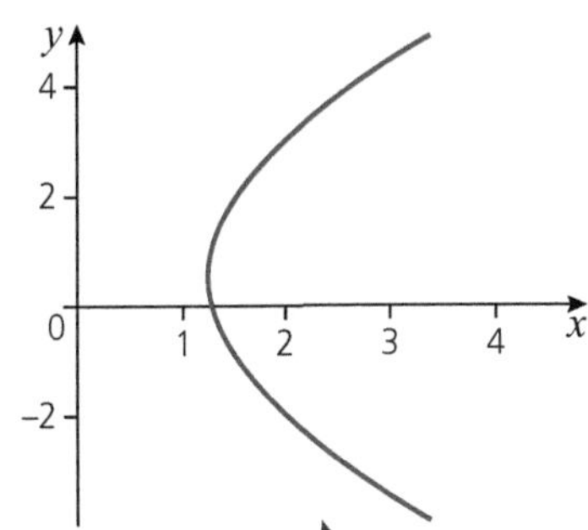

Hint:
Step 1 Differentiate
Step 2 Substitute $x = 2$ and $y = 3$
Step 3 Rearrange to make $\frac{dy}{dx}$ the subject.

4 Which of the following is the general solution of the differential equation $\frac{dy}{dx} = \sqrt{x} + 3$?

A $y = \frac{3}{2}x^{\frac{3}{2}} + 3x + c$ **B** $y = x^{\frac{3}{2}} + 3x + c$

C $y = \frac{2}{3}x^{\frac{3}{2}} + 3x + c$ **D** $y = \frac{2}{3}x^{\frac{3}{2}} + c$

5 a) Find the general solution to the differential equation $\frac{dy}{dx} = \frac{6x^2}{y^3}$.

b) Find the particular solution given that at $x = 2$, $y = 3$.

6 Solve the differential equation $\frac{dy}{dx} = 3x^2y$ given when $x = 0$, $y = 10$.

Hint: Look at the 'Get it Right' box.

7 a) Find the general solution to the differential equation $\frac{dy}{dx} = \frac{4\cos 2x}{e^y}$.

b) Find the particular solution given when $x = \frac{\pi}{2}$, $y = 0$.

Parametric equations

Year 2

THE LOWDOWN

① The equation of a curve can be written in

- **cartesian form,** where one equation links x and y
- **parametric form,** where x and y are both given in terms of a 3rd variable such as t or θ.

Example — A curve is defined by the parametric equations $x = t + 3$ and $y = t^2 - 4t$
When $t = 1$, $x = 1 + 3 = 4$
and $y = 1^2 - 4 \times 1 = -3$
so the curve passes through the point $(4, -3)$.

Hint: You can find the coordinates of a point on the curve by substituting the value of t into the parametric equation.

② You can **convert** between parametric form and cartesian form.

Example — Make t the subject of one equation: $x = t + 3 \Rightarrow t = x - 3$
Substitute into the other equation: $y = t^2 - 4t$

$\Rightarrow y = (x-3)^2 - 4(x-3)$

$= x^2 - 6x + 9 - 4x + 12$

$= x^2 - 10x + 21$

Hint: So this curve is a **quadratic.**

③ You can use the **chain rule** to find the gradient of a curve written parametrically.

The gradient function is $\frac{dy}{dx} = \frac{dy}{dt} \times \frac{dt}{dx} \Rightarrow \frac{dy}{dx} = \frac{\frac{dy}{dt}}{\frac{dx}{dt}}$

◀◀ **Chain rule** on page 72, **circles** on page 36 and **trigonometric identities** on page 56.

Example — To find the stationary points of the curve: $x = t^2 + t$, $y = t^3 - 12t$

Step 1 Differentiate: $\frac{dy}{dt} = 3t^2 - 12$ and $\frac{dx}{dt} = 2t + 1$

Step 2 Use the chain rule: $\frac{dy}{dx} = \frac{3t^2 - 12}{2t + 1}$

Step 3 Solve $\frac{dy}{dx} = 0$: $\frac{3t^2 - 12}{2t + 1} = 0 \Rightarrow 3t^2 - 12 = 0$

So $t^2 = 4 \Rightarrow t = \pm 2$

Hint: The only way a fraction can equal 0 is if the top line equals 0.

Step 4 Find the coordinates:

$t = 2$: $x = 2^2 + 2 = 6$, $y = 2^3 - 12 \times 2 = -16 \Rightarrow (6, -16)$

$t = -2$: $x = (-2^2) + (-2) = 2$, $y = (-2)^3 - 12 \times (-2) = 16 \Rightarrow (2, 16)$

Hint: Think of the trig identity that links the trig functions given in the equations.

GET IT RIGHT

Find the cartesian equation of the curve with parametric equations $x = 5\cos\theta$, $y = 5\sin\theta$ and sketch the curve.

Solution:

$x = 5\cos\theta \Rightarrow \cos\theta = \frac{x}{5}$ and $y = 5\sin\theta \Rightarrow \sin\theta = \frac{y}{5}$

Use the identity $\cos^2\theta + \sin^2\theta + \equiv 1$

$\left(\frac{x}{5}\right)^2 + \left(\frac{y}{5}\right)^2 = 1 \Rightarrow \frac{x^2}{25} + \frac{y^2}{25} = 1 \Rightarrow x^2 + y^2 = 25$

This is a circle centre (0, 0) and radius 5.

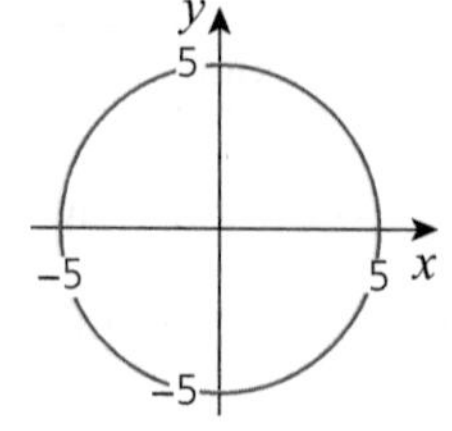

Hint: Think of the trig identity that links the trig functions given in the equations.

A **circle** with **centre** (a, b) radius r has parametric equations
$x = a + r\cos\theta$
$y = b + r\sin\theta$

YOU ARE THE EXAMINER

Which one of these solutions is correct? Where are the errors in the other solution?

A curve has the parametric equations $x = 3\ln t$, $y = e^{2t} + t$.

Find an expression for $\frac{dy}{dx}$ in terms of t.

SAM'S SOLUTION

$\frac{dy}{dt} = 2e^{2t} + 1$ and $\frac{dx}{dt} = \frac{3}{t}$

$\frac{dy}{dx} = \frac{dy}{dt} \times \frac{dt}{dx} = (2e^{2t} + 1) \times \frac{t}{3}$

$= \frac{1}{3}t(2e^{2t} + 1)$

PETER'S SOLUTION

$\frac{dy}{dt} = 2e^{2t} + 1$ and $\frac{dx}{dt} = \frac{1}{3t}$

So $\frac{dy}{dx} = \frac{\left(\frac{dy}{dt}\right)}{\left(\frac{dx}{dt}\right)} = \frac{2e^{2t} + 1}{\frac{1}{3t}} = \frac{2e^{2t} + 1}{3t}$

SKILL BUILDER

1 A curve has the parametric equations $x = 4t - 1$, $y = t^2 + 3t$.

a) Find the coordinates of the point where $t = 3$.

b) Find an expression for $\frac{dy}{dx}$ in terms of t.

c) Find the coordinates of the stationary point.

d) Find the cartesian equation of the curve.

Hint: $\frac{dy}{dx} = 0$ at a stationary point.

2 A curve has the parametric equations $x = 4\cos t$, $y = 4\sin t$.

a) Find the coordinates of the point where $t = \frac{\pi}{6}$

b) Find an expression for $\frac{dy}{dx}$ in terms of t.

c) Find the gradient at the point where $t = \frac{\pi}{6}$.

Hint: Use the gradient function $\frac{dy}{dx}$.

d) Find the equation of the i) tangent and ii) normal to the curve at the point where $t = \frac{\pi}{6}$.

Hint: Use your answers for parts a) and c) and $y - y_1 = m(x - x_1)$.

e) Find the cartesian equation of the curve.

3 Find the cartesian equation of the curve given by the parametric equations $x = 2t^2$, $y = 4(t - 1)$.

A $y = \pm\sqrt{8x - 16}$ **B** $y = -4 \pm \sqrt{8x}$ **C** $y = -4 \pm \sqrt{2x}$ **D** $y = -1 \pm \sqrt{8x}$

4 A curve is defined parametrically by $x = 3 + 5\cos\theta$, $y = 5\sin\theta - 2$.

Three of the following statements are false and one is true.

Which statement is true?

A The curve is a circle centre (3, 2) and radius 5.

B The curve cuts the y-axis at the point where $\theta = 0.41$ radians.

C There is only one point on the curve with x-coordinate 3.

D At the point (8, −2), $\theta = 0$.

5 A curve has parametric equations $x = 4t$, $y = 1 - \frac{1}{t}$. Find the value of $\frac{dy}{dx}$ when $t = 3$

A $\frac{1}{36}$ **B** 36 **C** $\frac{4}{9}$ **D** $-\frac{1}{36}$

Sampling and displaying data 1

Year 1

THE LOWDOWN

① **Surveys**

Polly surveys students at her college about how long they study each week.

She could ask every student (census) or a take sample, say 10%, of students.

Here are some ways she could take a sample.

- **Simple random sample:** Cut up a college register (**sampling frame**).
- **Systematic sample:** Use the register of the whole college, choose one of the first 10 names and ask every following 10th student on the list.
- **Cluster sample:** Ask everyone in 10% of tutor groups.
- **Stratified sample:** Ask separate samples from Y12 and Y13.
- **Opportunity sample:** Go into the library and ask everyone who is there.
- **Self-selecting sample:** Put up a poster and ask students to message her how long they study each week.

② **Stem-and-leaf diagrams**

- Split each number into two parts, for example tens and units. Those with the same stem go together.
- Provide a key so you can read the data.

Example: Working through the list, Polly puts the values 12, 24, 9, 12, 21, 10, 28, 8, 25, 15 into the diagram then redraws.

Key 1|2 means 12 hours

0	9	8			
1	2	2	0	5	
2	4	1	8	5	

0	8	9		
1	0	2	2	5
2	1	4	5	8

③ **Averages: Mode and median**

The mode is the value that occurs most often (there may be more than one).

The median is the value below which half the values lie.

For a list with n values, write the values in order and use the $\frac{n+1}{2}$th value.

Example: The median is the 5.5th value, between 12 and 15 so 13.5 hours.

Hint: The sample should be representative, not biased.

Hint: It will be hard to find the people on your list.

Hint: These are likely to be biased samples.

Hint: It is usual to redraw with the 'leaves' in order of size.

▶▶ See page 86 for **displaying data 2** and page 88 for **quartiles, mean and standard deviation**.

Hint: This is the middle value, or the number halfway between the middle two if you have an even number of values, when they are written in size order.

GET IT RIGHT

The stem-and-leaf diagram shows Polly's data for year 12 and year 13 students. Compare the modes and medians for the two data sets.

				Year 12			Year 13				
			5	0	**0**	8	9				
6	6	3	1	0	**1**	0	1	2	3	5	8
	3	1	1	1	**2**	3	3	7	9		

Key: 1|**2**|3 means 21 hours for Year 12 and 23 hours for Year 13.

Hint: The data on the left of the centre column should be read from right to left.

Hint: The mean may be the best average to consider here.

Solution:

Step 1 Look for repeated values.

Mode: Y12, 21 hours, Y13, 23 hours so the mode is higher for Y13.

Step 2 Count the values in each list.

For Y12 $\frac{11+1}{2}$ = 6th value 16 hours.

For Y13 $\frac{12+1}{2}$ = 6.5th value (between 13 and 15) so 14 hours.

So the median is higher for Y12.

YOU ARE THE EXAMINER

Which one of these solutions is correct? Where are the errors in the other solution?

The stem-and-leaf diagram shows the lengths, in cm, of petals from a plant.

Find the mode and the median.

1	7 9
2	2 2 7 7 7
3	3 5 5 6 7 8 8 9
4	0 4

Key: **1**|7 represents 1.7 cm

LILIA'S SOLUTION

The mode is 2.7 cm and the median is 3.5 cm.

MO'S SOLUTION

The mode is 7 and the median is 5.

SKILL BUILDER

1 A student is investigating how students in his college use their leisure time. He starts with a list of all students at the college and looks at male and female students separately. He chooses equal numbers of male students and female students at random. Which sampling method is the student using?

A Opportunity **B** Quota **C** Simple random **D** Stratified **E** Systematic

2 A market researcher wants to know what users of a particular post office think of the facilities available. Which of the following is likely to be the least biased sampling method?

A Choose everyone who is in the post office at noon on Monday.

B Put an advert in the local paper and ask volunteers to get in touch if they use the post office and want to give their opinions about the facilities.

C Visit the post office at twelve times during one week, two on each day Monday to Saturday, and choose every fifth person who comes into the post office during half an hour.

D Ask post office staff to give a letter to 20 men and 20 women who use the post office. The letter asks the person who receives it to complete a questionnaire and give it back to the post office staff.

E Put up a sign that can be seen by everyone entering the post office asking them to get in touch with the market researcher to give their opinion about the facilities. Choose a simple random sample from those who get in touch.

3 Write down the advantages and disadvantages of each type of sampling technique listed in the lowdown.

4 a) Draw an ordered stem-and-leaf diagram for this list.

52, 41, 42, 58, 35, 67, 65, 52, 57, 54, 45, 58, 38, 68, 35, 64

b) Find the mode and the median.

5 The stem-and-leaf diagram shows life expectancy (in years) the countries of two continents. Compare the modes and medians for Africa and Asia.

Africa		Asia
9 8 4 3	**5**	
9 6 6 1 1 0	**6**	5 6 9 9 9
6 6 4 2 2	**7**	0 0 1 1 3 4 6 6 6 6 8
	8	4 4 4

Key: 0|**6**|5 means 60 years for Africa and 65 years for Asia.

Displaying data 2

GCSE/Year 1

THE LOWDOWN

① **Histograms**

A **histogram** is used to display grouped continuous data.

Each group (class) is represented by a rectangle whose area represents frequency. The height of each rectangle is the frequency density.

$$\text{Frequency density} = \frac{\text{frequency}}{\text{class width}}$$

Remember, when a histogram is drawn for you:

class width × height = **frequency**.

② The **modal class** is the class with the highest frequency density.

③ **Skew**

Histograms may be symmetric. 'Skew' describes the asymmetry of other histograms.

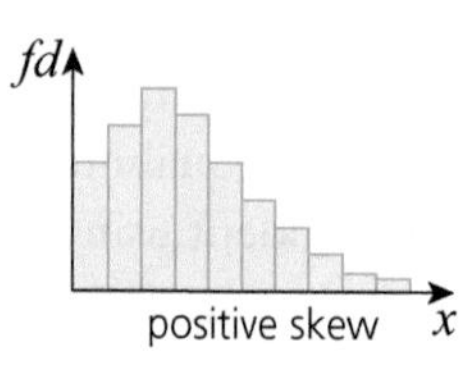

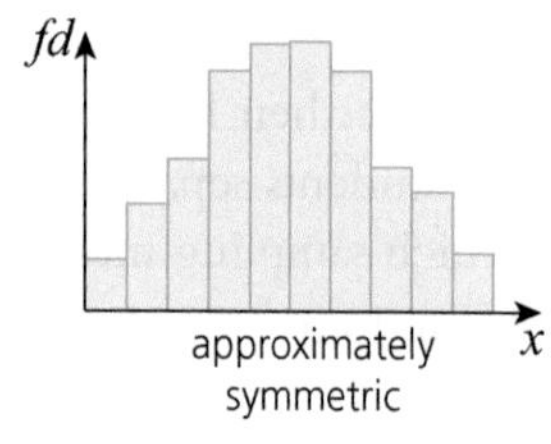

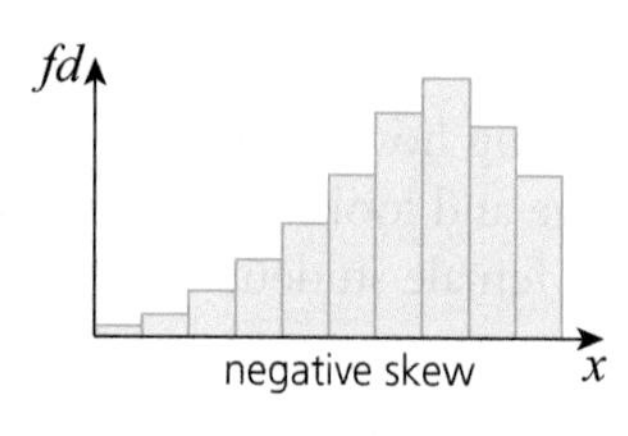

Watch out! This may not be the class with the highest frequency.

▶▶ **Boxplots** on page 90.

GET IT RIGHT

Polly has done a survey of 186 students from her college to find out how long they study each week. Draw a histogram to show her results and comment on the shape of the distribution.
State the modal class.

Time (h)	$0 \leqslant t < 5$	$5 \leqslant t < 10$	$10 \leqslant t < 20$	$20 \leqslant t < 25$	$25 \leqslant t < 30$	$30 \leqslant t < 40$
Frequency	6	12	43	30	38	57

Hint: The first group do less than 5 hours, the next 5 or more but less than 10 hours, etc.

Solution:

Step 1 Find the width of each group.

Step 2 Calculate frequency density.

$$\text{Frequency density} = \frac{\text{frequency}}{\text{class width}}$$

Step 3 Draw axes and then draw each rectangle – there should not be gaps between the rectangles.

Time (h)	Class width	Frequency density
$0 \leqslant t < 5$	5	1.2
$5 \leqslant t < 10$	5	2.4
$10 \leqslant t < 20$	10	4.3
$20 \leqslant t < 25$	5	6
$25 \leqslant t < 30$	5	7.6
$30 \leqslant t < 40$	10	5.7

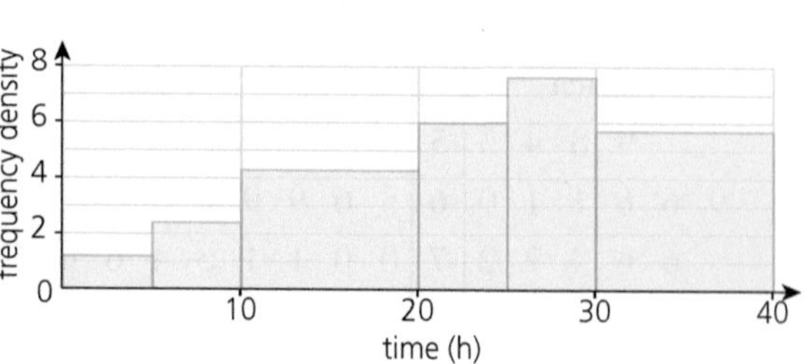

The histogram shows that the data has a negative skew.
The modal class is $25 \leqslant t < 30$.

YOU ARE THE EXAMINER

Which one of these solutions is correct? Where are the errors in the other solution?

The histogram shows the results of a survey of students and the number of hours they spend studying.

How many students took part in the survey?

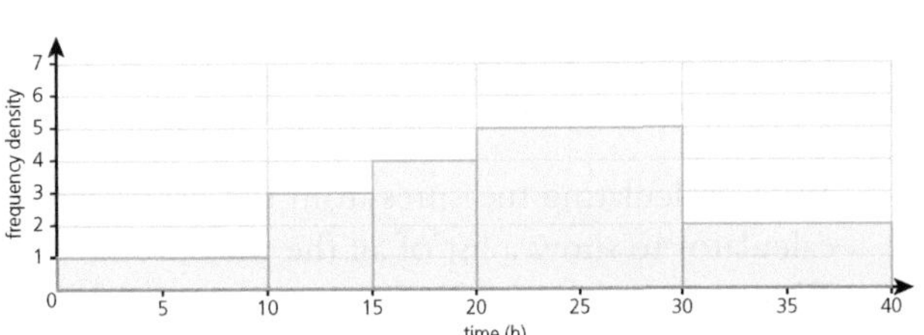

LILIA'S SOLUTION

Reading values from the horizontal axis:

$1 + 3 + 4 + 5 + 2 = 15$

15 students in the survey

PETER'S SOLUTION

Frequency is the area of each rectangle

$1 \times 10 + 3 \times 5 + 4 \times 5 + 5 \times 10 + 2 \times 10 = 115$

115 students in the survey

SKILL BUILDER

1 Aarav collects data on the ages of the children at a park.

Age (years)	0–4	5–8	9–10	11–13	14–18
Frequency	9	12	10	14	2

Hint: The 0–4 class includes all the children up to their 5th birthday so has a class width of 5.

Draw a histogram of the data and comment on the shape of the distribution.

2 The table shows the speed of 140 cars passing through a village.

The values are rounded to the nearest whole number.

Speed (mph)	**Lower boundary**	**Upper boundary**	**Class width**	**Frequency**	**Frequency density**
20–29				70	
30–34				40	
35–44				27	
45–60				3	

Hint: the first lower boundary is the lowest speed (to 1 d.p.) that rounds up to 20, so it is 19.5; the first upper boundary is the same as the second lower boundary.

a) Complete the table and draw a histogram.

b) Comment on the shape of the histogram.

3 The table and histogram (both incomplete) show the ages of passengers on a plane.

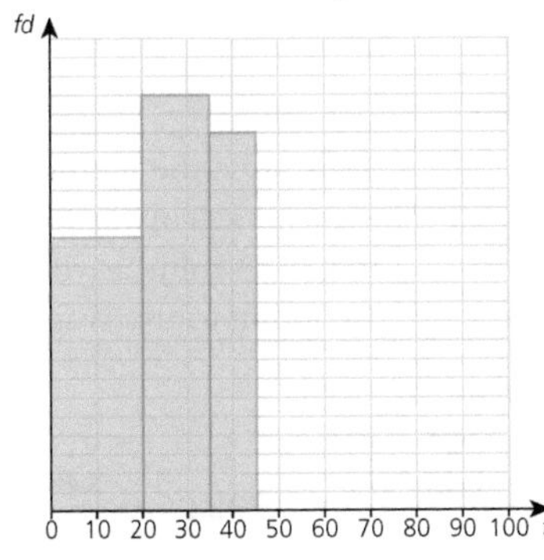

Age (years)	**Frequency**	**Frequency density**
$0 \leqslant x < 20$		
$20 \leqslant x < 35$		
$35 \leqslant x < 45$	20	2
$45 \leqslant x < 65$	20	
$65 \leqslant x < 90$		0.16

a) Use the $35 \leqslant x < 45$ class in the table to determine the scale on the histogram.

b) Read the frequency density from the histogram for the first two classes.

c) Complete the table and the histogram.

Averages and measures of spread

GCSE/Year 1

THE LOWDOWN

① When calculating measures from the data, use the Statistics menu on your calculator to show a list of all the measures. If you only know n, the total $\sum x$ and sum of squares $\sum x^2$, use the formulae.

② The **mean** of a set of values is given by the formula $\bar{x} = \frac{\sum x}{n}$.

Hint: The mean is the total divided by how many numbers there are.

③ **Measures of spread**

The **range** is the difference between the largest and the smallest values.

The **median** divides the data into two halves.

The **median** of each half is a quartile. The **interquartile range** is the difference between the upper and lower quartiles.

The **variance** s^2 uses S_{xx}, the sum of the squares of the differences between each value and the mean.

$$S_{xx} = \sum (x_i - \bar{x})^2 = \sum x_i^2 - \frac{\left(\sum x_i\right)^2}{n} = \sum x_i^2 - n\bar{x}^2$$

An easier to use alternative for this variance is $\frac{\sum x_i^2}{n} - \bar{x}^2$

For MEI use $s^2 = \frac{1}{n-1} S_{xx}$ otherwise use variance $= \frac{S_{xx}}{n}$

The **standard deviation** (sd) is the square root of the variance.

Hint: Your calculator will give values for two versions of standard deviation. The smaller of the two uses n and the larger uses $n - 1$. For large values of n they are very similar.

④ **Outliers** are values that are unusually large or small.

A value not between $\bar{x} - 2 \times sd$ and $\bar{x} + 2 \times sd$ is an outlier.

A value not between $LQ - 1.5 \times IQR$ and $UQ + 1.5 \times IQR$ is an outlier.

⑤ **Deciding which measures to use**

The **mean, mode** and **median** are often similar in value, so any can be used.

If the data has a strong skew, the **median** and interquartile range may be better.

Outliers have the biggest effect on the **mean** and the range.

When there are outliers, other measures may represent the data better.

◀◀ See page 84 for **mode** and **median**.

▶▶ See page 92 for **calculations from a frequency table.**

GET IT RIGHT

Sean collects test marks from 30 students. He calculates $\sum x = 1657$ and $\sum x^2 = 99309$.

a) Calculate the mean and the standard deviation.

b) Find the range of values for which the mark be considered an outlier.

Solution:

a) Step 1 Calculate the mean: $\frac{\sum x}{n} = \frac{1657}{30} = 55.23$ (2 d.p.)

Hint: Use the memory on your calculator for the exact value as you are going to use it again.

Step 2 Use the formula to calculate the variance:

$$\frac{\sum x^2}{n} - \bar{x}^2 = \frac{99309}{30} - 55.2333^2$$

$$= 259.6$$

Step 3 Take the square root: standard deviation $= \sqrt{259.6} = 16.1$

b) $\bar{x} + 2 \times sd = 55.23 + 2 \times 16.1 = 87.5$

and $\bar{x} - 2 \times sd = 55.23 - 2 \times 16.1 = 23.0$

A mark is an outlier if $x < 23.0$ or $x > 87.5$

For MEI calculate S_{xx} first.

$$S_{xx} = \sum x_i^2 - \frac{\left(\sum x_i\right)^2}{n} = 99309 - \frac{1657^2}{30}$$

$$s = \sqrt{\frac{S_{xx}}{n-1}} = \sqrt{\frac{7787.367}{29}} = 16.39$$

$\bar{x} + 2 \times sd = 55.23 + 2 \times 16.39 = 88.01$

$\bar{x} - 2 \times sd = 55.23 - 2 \times 16.39 = 22.45$

A mark is an outlier if $x < 22.45$ or $x > 88.01$

YOU ARE THE EXAMINER

Which one of these solutions is correct? Where are the errors in the other solutions?

A shop records the sales made to its customers (five men and eight women) over one hour. The mean amount spent by the five men is £45.20 and the mean amount for the eight women is £59.75. Find the mean for all the customers.

SAM'S SOLUTION

Mean is £45.20 + £59.75 = £104.95

PETER'S SOLUTION

$$\frac{£45.20 + £59.75}{2} = £52.475$$

So the mean is £52

LILIA'S SOLUTION

Total for men is 5 × 45.20 = 226

Total for women is 8 × 59.75 = 478

Total of sales is 226 + 478 = 704

Number of customers is 5 + 8 = 13

Mean is $\frac{704}{13}$ = 54.154 so £54.15

SKILL BUILDER

1 A park keeper weighs 10 grey squirrels that visit his feeding station. His results, in grams, are 570, 578, 561, 554, 562, 529, 568, 568, 549, 590.

a) Find the mean, mode and median.

b) Find the range, interquartile range and the standard deviation.

c) Use the mean and standard deviation to determine which values would be outliers.

d) Use the quartiles to determine which values would be outliers.

2 Sharon gets £8 per week pocket money. She asks her mother for more money. Her mother says she already gets more than the average for her friends. Sharon asks her friends and they get £5, £5, £5, £6, £6.50, £7, £7, £8, £10, £15, £20. Sharon says she gets less than average. Who is right?

3 Gary and Henry find a website that states that the average family in the UK has 1.7 children. Gary argues this must be wrong because there has to be a whole number of children. Henry argues that 170 children will move into the village when 100 new family homes are built there. Comment on their arguments.

4 Colin and Ahmed record their times in seconds in several 100 m races. For Colin the times are 14.2, 14.0, 11.9, 12.9, 16.3, 14.3, 11.1, 13.7, 15.1. For Ahmed the summary data is $n = 10, \sum x = 145.3, \sum x^2 = 2124.21$. Compare the two sets of times.

5 Ten donations to a charity one day have a mean of £27.50.

a) Find the total amount given.

b) One person gave £120. Find the mean for the other nine donations.

Hint: Find the total for the other nine donations.

c) An 11th person wants to give a gift that will bring up the mean donation to £30. How much would they have to give?

Hint: Work out what the total would have to be for the 11 donations together.

Cumulative frequency graphs and boxplots

GCSE/ Year 1

THE LOWDOWN

① **A cumulative frequency graph** shows the number less than each value.

Example Draw a cumulative frequency curve for Polly's data from page 86.

Time (hours)	Frequency
$0 \leqslant t < 5$	6
$5 \leqslant t < 10$	12
$10 \leqslant t < 20$	43
$25 \leqslant t < 30$	30
$30 \leqslant t < 35$	38
$35 \leqslant t < 40$	57

Less than	Cumulative frequency
5	6
10	18
20	61
25	91
30	129
40	186

Hint: You can do without a new table if you add the cumulative frequency column to the first table.

Hint: You must remember to plot the point at the **end** of the interval.

② You can **estimate the median** by drawing a horizontal line to the curve from $\frac{186}{2} = 93$ and read the value for time (25.3).

Repeat at $\frac{186}{4} = 46.5$ for lower quartile (16.6).

and $\frac{3}{4} \times 186 = 139.5$ for upper quartile (31.8).

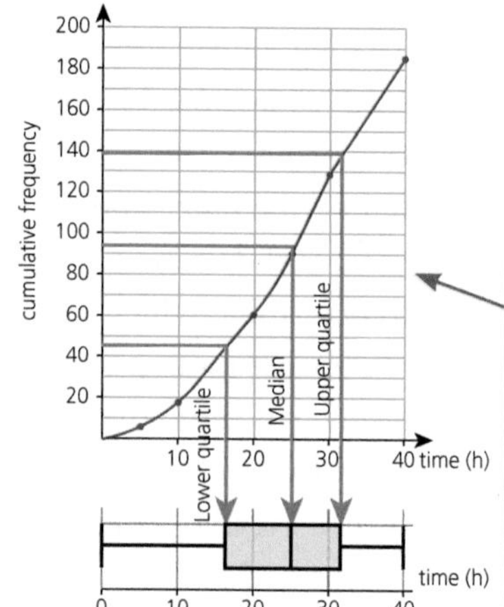

③ **Boxplot**

The box represents the middle 50% of the population from the lower quartile to the upper quartile.

The median is a vertical line through the box.

The horizontal lines ('whiskers') extend to the minimum and maximum values.

④ **Recognising skew** from a boxplot

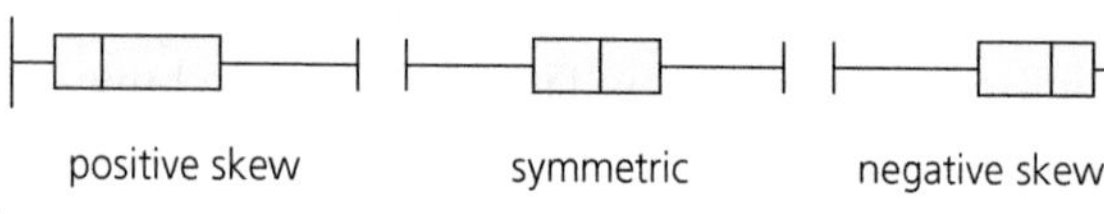

◀◀ **Skew** on page 86.

▶▶ See page 92 for **linear interpolation**.

When the data is symmetric, the cumulative frequency curve has an S-shape.

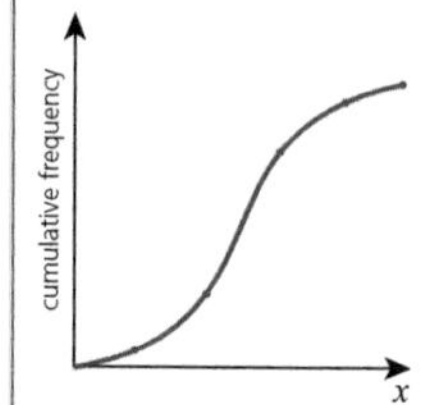

GET IT RIGHT

These are the test marks for 15 students:

14, 22, 35, 42, 44, 45, 49, 53, 54, 54, 57, 58, 60, 61, 71.

a) Find the median and the interquartile range.

b) Determine whether any values are outliers.

c) Draw a boxplot showing the outliers as crosses.

Solution:

a) The median is the 8th value, 53.

LQ is the 4th value, 42; UQ is the 12th value 58; so IQR = 58 − 42 = 16

b) For outliers: values are below LQ − 1.5 × IQR = 42 − 1.5 × 16 = 18 so only the minimum 14 is an outlier at the bottom end, OR above UQ + 1.5 × IQR = 58 + 1.5 × 16 = 82 so no more outliers.

c)

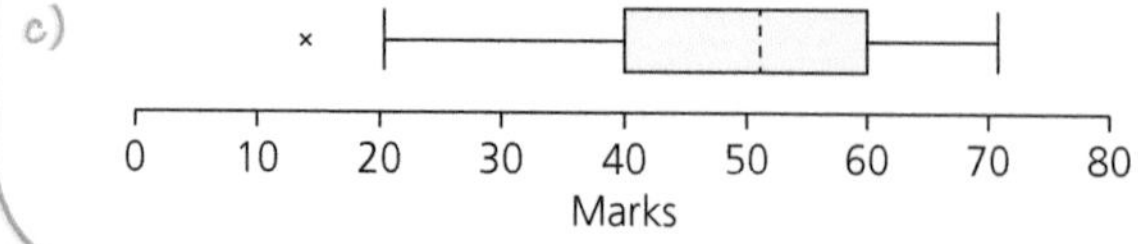

YOU ARE THE EXAMINER

Whose statements are correct?

The box plots below show the numbers of infant deaths per 1000 children born for a sample of countries in Africa (above) and Asia (below).

Comment on the information that can be found in the diagrams.

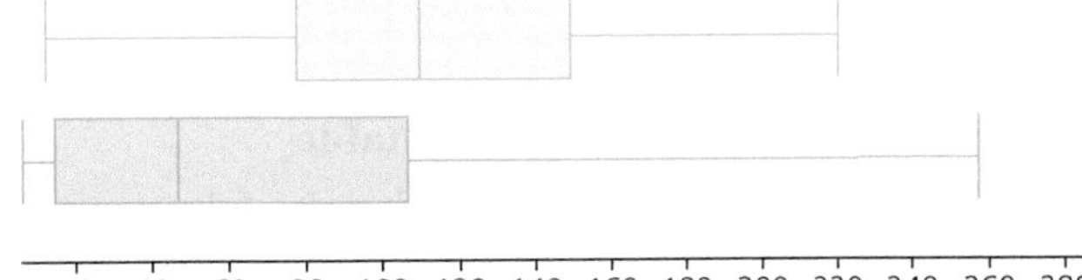

NASREEN'S SOLUTION

Of the countries in the sample, the country with the higher rate of infant deaths is in Asia.

SAM'S SOLUTION

On average there are more deaths per 1000 children in African countries than Asian countries.

MO'S SOLUTION

The country in the world with the lowest rate of infant deaths is in Asia.

LILIA'S SOLUTION

There are more infant deaths in Africa than Asia.

SKILL BUILDER

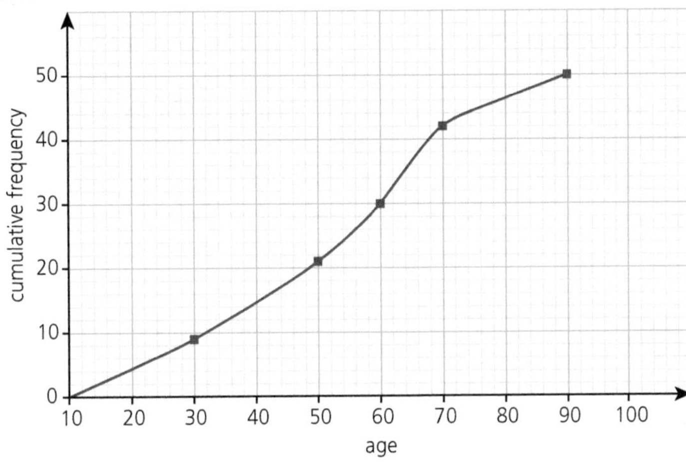

1 Which of the frequency tables below shows the data corresponding to the cumulative frequency curve?

A	
Age	**Frequency**
10–30	9
31–50	21
51–60	30
61–70	42
71–90	50

B	
Age	**Frequency**
10–29	9
30–49	12
50–59	9
60–69	12
70–89	8

C	
Age	**Frequency**
20–40	9
40–55	12
55–65	9
65–80	12
80–100	8

D	
Age	**Frequency**
11–30	9
31–50	21
51–60	30
61–70	42
71–90	50

2 Find the median and quartiles for the data set in the graph above. Draw a box plot for the data.

3 The table shows the lengths of horse chestnut tree leaves, in cm.

a) Draw the cumulative frequency graph and estimate the median and the lower and upper quartile.

Length (cm)	$4 \leqslant x < 8$	$8 \leqslant x < 15$	$15 \leqslant x < 20$	$20 \leqslant x < 25$	$25 \leqslant x < 30$	$30 \leqslant x < 35$
Frequency	3	10	28	48	55	60

b) Draw a boxplot. **Hint:** There are no leaves less than 4 cm long.

Grouped frequency calculations

Year 1

THE LOWDOWN

① **Calculations from a frequency table**

Change your calculator settings so that the values and their frequencies are used.

Example The table shows the lengths of words in a child's story book.

No. of letters	1	2	3	4	5	6
Frequency	4	6	14	18	12	1

Typing the values and their frequencies, from the calculator the mean is 3.6 standard deviation is $\sigma_x = 1.34$ ($s_x = 1.35$ for MEI).

② **Calculations from a grouped frequency table**

Example The table shows the heights of 100 children in a club. You can estimate the mean and standard deviation by using the midpoint of the class to represent the whole class.

Height (cm)	$100 \leqslant h < 110$	$110 \leqslant h < 120$	$120 \leqslant h < 135$	$135 \leqslant h < 140$
Frequency	13	33	40	14
Midpoint	105	115	127.5	142.5

Example Estimate for mean 122.55
standard deviation $\sigma_x = 11.15$ ($s_x = 11.21$ for MEI).

③ **Linear interpolation** can be used to estimate the median without using the midpoint for the class.

Median = lower class boundary plus fraction of the class width (see 'Get it right', below).

◀◀ See page 88 for **mean and standard deviation** and page 84 for **median**.

Hint: In some books you will see the formulae include the frequencies, e.g.: $\sum x, \sum x_i, \sum xf$ are all used to indicate the total.

Hint: Draw a diagram to show what fraction of the class width needs adding to the lower class boundary.

GET IT RIGHT

The table shows the cumulative frequencies for the heights of 100 children in the example above.

Height (cm)	$100 \leqslant h < 110$	$110 \leqslant h < 120$	$120 \leqslant h < 135$	$135 \leqslant h < 150$
Frequency	13	33	45	9
Cumulative	13	46	91	100

Use linear interpolation to estimate the median height.

Solution:

Step 1 The median is the $\frac{100-1}{2} = 50.5$th value, which is in the $120 \leqslant h < 135$ class.

Step 2 The median is the $50.5 - 46 = 4.5$th person of 45 in the $120 \leqslant h < 135$ class, which is $\frac{1}{10}$ of the way through the children in that class.

Step 3 The median is more than the lower boundary of $120 \leqslant h < 135$.

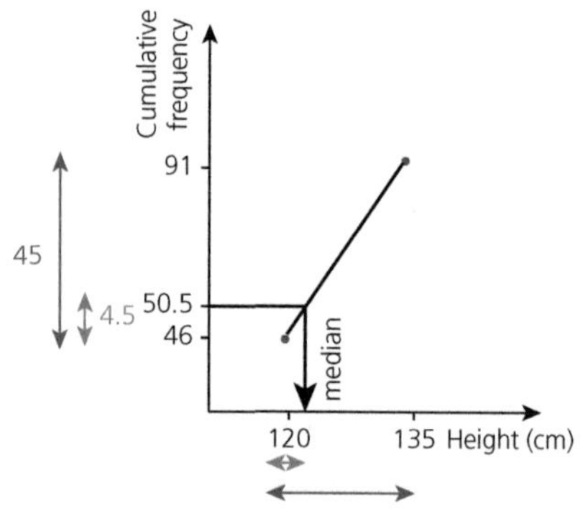

The extra is $\frac{1}{10}$ of the class width $(135 - 120 = 15)$ – as if the 45 children were equally spaced between 120 cm and 135 cm.

The median is $120 + \frac{4.5}{45} \times \text{class width} = 120 + \frac{1}{10} \times 15 = 121.5$ cm.

YOU ARE THE EXAMINER

Which of these solutions is correct? Where are the errors in the other three solutions?

The table shows the heights of 20 children in a playground.

Calculate the mean height.

Height (m)	$1.00 \leqslant h < 1.20$	$1.20 \leqslant h < 1.40$	$1.40 \leqslant h < 1.50$
Frequency	4	11	5

LILIA'S SOLUTION

$$\frac{4 \times 1.20 + 11 \times 1.40 + 5 \times 1.50}{20} = \frac{27.7}{20}$$
$$= 1.385$$

NASREEN'S SOLUTION

$$\frac{4 \times 1.10 + 11 \times 1.30 + 5 \times 1.45}{20} = \frac{25.95}{20}$$
$$= 1.2975$$

PETER'S SOLUTION

$$\frac{4 \times 1.00 + 11 \times 1.20 + 5 \times 1.40}{20} = \frac{24.2}{20}$$
$$= 1.21$$

MO'S SOLUTION

$$\frac{4 + 11 + 5}{3} = 6.67$$

SKILL BUILDER

1 The table shows the lengths of words in a paragraph of an A level Maths textbook.

Letters	2	3	4	5	6	7	8	9	10
Frequency	10	11	9	3	2	5	2	4	2

a) Calculate the mean, mode, median, standard deviation and range.

b) Compare with the values for the children's story book in Example 1 in the Lowdown.

2 A number of students are asked how many school dinners they ate in the past week with the following results.

Number of dinners eaten	0	1	2	3	4	5
Frequency	7	10	10	12	21	18

Four of the statements below are false and one is true. Which one is true?

A The median is 3.5.

B The mode is 10.

C The number of students in the sample was 6.

D The number of students in the sample was 15.

E The median is 2.5.

3 The waist measurements of a sample of boys are shown below.

Waist (cm)	50–59	60–69	70–79	80–89	90–109
Frequency	2	45	80	19	7

Four of the statements below are false and one is true. Which one is true?

A The mode is 80 cm.

B The mean is 30.6 cm.

C A good estimate of the mean is 74.5 cm.

D The median is 74.5 cm.

E The median is somewhere in the group 70–79 cm.

4 The weights of a sample of two species of capuchin monkey are given in the table.

Use the midpoints and your calculator to estimate the mean and standard deviation for both species.

Weight (kg)	$2.6 \leqslant w < 2.8$	$2.8 \leqslant w < 3.0$	$3.0 \leqslant w < 3.2$	$3.2 \leqslant w < 3.5$	$3.5 \leqslant w < 3.8$
White-faced	0	1	6	34	14
Kaapori	7	42	42	8	0

5 Use linear interpolation to find the median for the data sets in question 4.

6 Compare the weights of the two species of capuchin monkey in question 4.

Probability

THE LOWDOWN

① Here are two ways to present the **same data** about adults on a small beach.

	With hat (H)	Without hat (H')
Men (M)	14	21
Women (W)	18	27

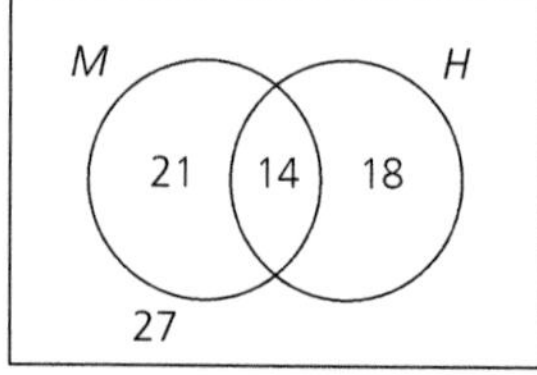

Experiment: Select a person at random from the beach.

An **event** is a subset of the set of possible outcomes.

Events: male, female, wears hat, does not wear hat, etc.

② **Combining events**

Complement M' not a man

Intersection $M \cap H$ man wearing a hat

Union $M \cup H$ either a man or person wearing a hat (or both)

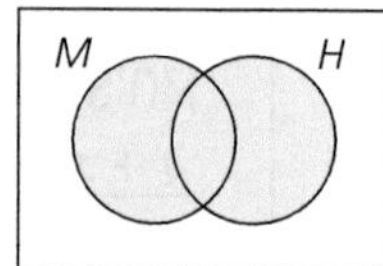

$M' \cap H'$ neither a man nor wearing a hat

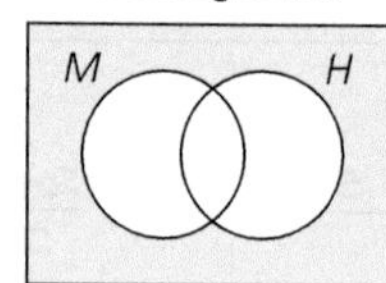

③ **Calculate the probability** of event A from

$$P(A) = \frac{\text{no of ways } A \text{ can happen}}{\text{no of equally likely outcomes}}$$

Example a) $P(\text{woman}) = P(W) = \frac{18+27}{80} = \frac{9}{16}$

b) $P(\text{man with hat}) = P(M \cap H) = \frac{14}{80} = \frac{7}{40}$

The event W could also be written as M' (not man).

④ **Mutually exclusive** events cannot happen at the same time.

Example: $P(\text{man and woman}) = P(M \cap H) = 0$

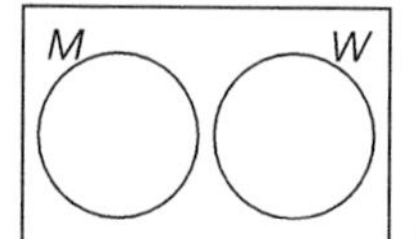

⑤ If two events are **not** mutually exclusive, adding their probabilities counts some possibilities twice – those in the intersection must be subtracted from the total.

Example $P(\text{person is either a man or is wearing a hat})$

$P(M \cup H) = P(M) + P(H) - P(M \cap H)$

$= \frac{35}{80} + \frac{32}{80} - \frac{14}{80} = \frac{53}{80}$

Hint: You can also work this out by adding the numbers 21 + 14 + 18 from the Venn diagram and dividing by 80.

⑥ Two events are **independent** if they **do not** affect each other.

$P(A \cap B) = P(A) \times P(B)$

▶▶ **Conditional probability** on page 96.

GET IT RIGHT

Show that the events M and H are independent.

Solution:

Step 1 Find the probability of the intersection:

$$P(M \cap H) = P(\text{man with hat}) = \frac{\text{no of men with hats}}{80} = \frac{14}{80} = \frac{7}{40}$$

Hint: Use the numbers in the table to work this out.

Step 2 Find the two probabilities: $P(M) = \frac{35}{80}$ and $P(H) = \frac{32}{80}$

Step 3 Multiply these two probabilities:

$$P(M) \times P(H) = \frac{35}{80} \times \frac{32}{80} = \frac{1120}{6400} = \frac{7}{40}$$

Step 4 Compare answers and make a conclusion:

$$P(M) \times P(H) = \frac{7}{40} = P(M \cap H)$$

So the events are independent as the probabilities are the same.

YOU ARE THE EXAMINER

Which one of these solutions is correct? Where are the errors in the other solution?

There are 80 children's pictures in a gallery.

32 feature hearts, 15 feature spots.

11 pictures have hearts and spots.

Draw a Venn diagram and find the probability that a picture chosen has neither hearts nor spots.

MO'S SOLUTION

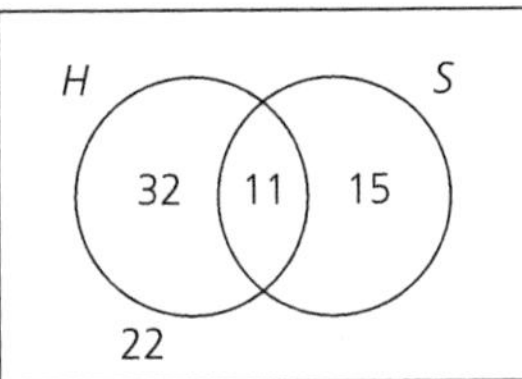

$P(H' \cup S') = \frac{22}{80}$

PETER'S SOLUTION

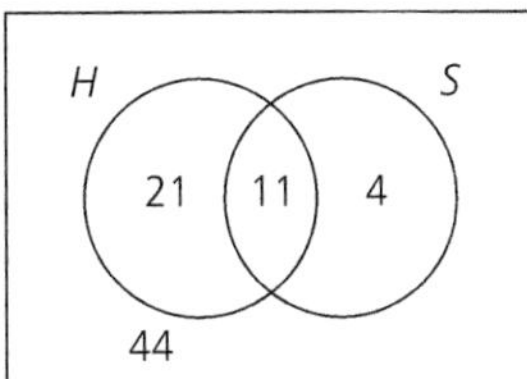

$P(H' \cap S') = \frac{44}{80}$

SKILL BUILDER

1 Raffle tickets numbered 277 to 575 inclusive are put into a large container and one ticket is taken out at random. What is the probability of a number divisible by 5 being chosen?

A $\frac{1}{5}$ **B** $\frac{30}{149}$ **C** $\frac{60}{299}$ **D** $\frac{59}{298}$ **E** $\frac{59}{299}$

2 Some gardeners have phoned in to ask for their garden to appear on TV. The number of these gardens having certain features is shown in the Venn diagram. One of these gardens will be chosen at random to be in the TV show. What is the probability that a garden with both a vegetable patch and a pond will be chosen?

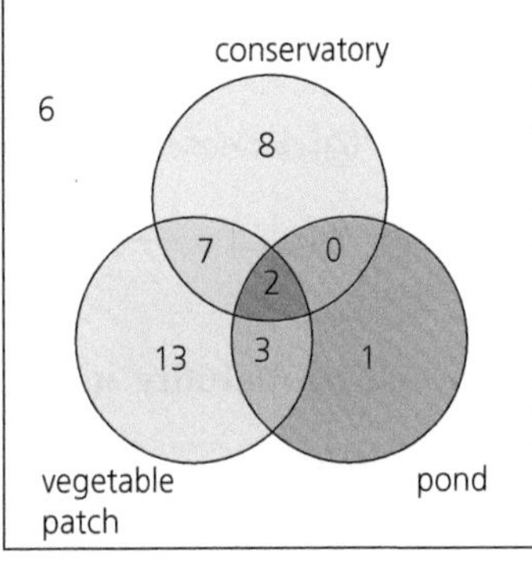

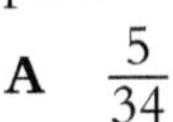

A $\frac{5}{34}$ **B** $\frac{13}{20}$ **C** $\frac{1}{8}$

D $\frac{3}{40}$ **E** $\frac{7}{20}$

3 The two-way table shows information of a survey in a playground. Show the same information as a Venn diagram showing the sets A and B.

	Has a bag (B)	**Has no bag**
Adults (A)	21	38
Children	18	23

4 The Venn diagram shows how many books on a shelf are hardback (H) and how many are fiction (F).

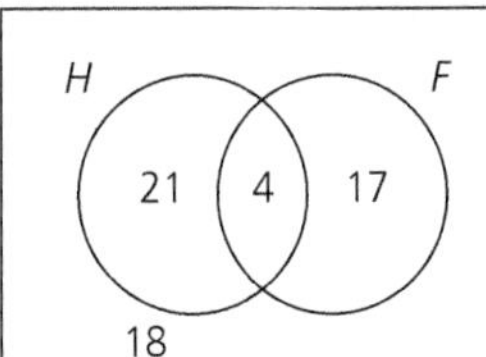

a) Show the information as a two-way table.

b) What sort of books are described by $H \cap F$?
Calculate the probability a book chosen at random is of that type.

5 Use the information in the two-way table to calculate:

	B	***B'***
A	35	29
A'	47	9

a) $P(A)$ and $P(B)$.

b) $P(A \cap B)$

c) Show the events are not independent.

Conditional probability

Year 1

◀◀ **Probability** on page 94.
▶▶ **Binomial distribution** on page 100.

THE LOWDOWN

① **Conditional probability** is the probability that an event happens **given that** another event has already happened.

② **Conditional probability from tree diagrams**

Fei asks Mo: 'What is the probability that you walk to college?'

He replies, 'It depends on whether it's raining – 10% if it's raining and 80% if not.'

You write the probability of walking *given that* it is not raining as $P(W|R') = 80\% = 0.8$

Fei finds out there are 156 out of 365 rain days in a year in the UK.

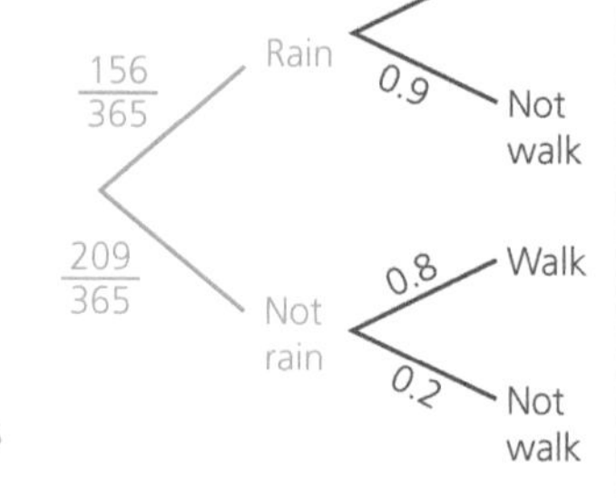

Example The probability that Mo walks is

$$\left(\frac{156}{365}\times 0.1\right)+\left(\frac{209}{365}\times 0.8\right) = 0.501$$

Hint: The probabilities on the second part of the tree diagram are the conditional probabilities.

You multiply along the branches and add the answers for the two branches.

③ **Conditional probability using a two-way table**

The two-way table shows information about a class.

	Long hair (*L*)	Short hair (*L'*)	Total
Boys (*B*)	3	14	17
Girls (*G*)	11	1	12
Total	14	15	29

Example: P(girl chosen has long hair)

$$= P(L|G) = \frac{\text{no. girls with long hair}}{\text{no. of girls}} = \frac{11}{12}$$

Watch out! The probability of a person with long hair being a girl is not the same.

④ **Conditional probability using the formula**

Notice that $P(B|A) = \dfrac{\text{no. of ways } A \text{ and } B \text{ both happen}}{\text{no. of ways } A \text{ can happen}}$

Dividing by the total number of outcomes gives $P(B|A) = \dfrac{P(A\cap B)}{P(A)}$

Hint: For independent events the probability of *A* is the same whether *B* has happened or not. $P(A|B) = P(A|B')$ $= P(A)$

GET IT RIGHT

The probability Mia visits her Gran on a weekday is 0.05 and at the weekend it is 0.45. Find the probability that it is a weekday, given that Mia visits her Gran.

Solution:

Step 1 Draw a tree diagram:

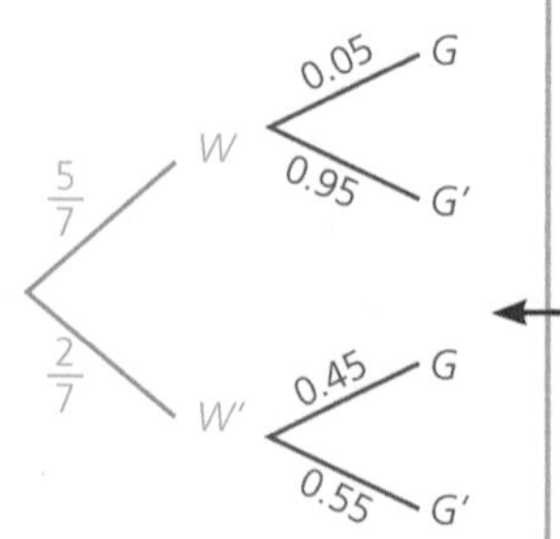

Step 2 Find the probability of both conditions:

$$P(W\cap G) = \frac{5}{7}\times 0.05 = \frac{1}{28}$$

Step 3 Find the probability of Mia visiting Gran:

$$P(G) = \frac{5}{7}\times 0.05 + \frac{2}{7}\times 0.45 = \frac{23}{140}$$

Step 4 Divide: $P(W|G) = \dfrac{P(W\cap G)}{P(G)} = \dfrac{1/28}{23/140} = \dfrac{5}{23}$

Hint: Write down what the meaning of the letters you are using – here *W* for a weekday and *G* for visiting Gran. The question asks for $P(W|G)$.

YOU ARE THE EXAMINER

Which one of these solutions is correct?
Where are the errors in the other solution?

The Venn diagram shows the probabilities for events A and B.
Calculate the probability $P(A|B)$.

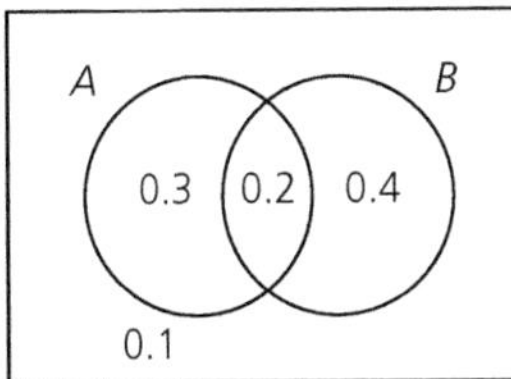

LILIA'S SOLUTION

$$P(A|B) = \frac{P(A \cap B)}{P(A)} = \frac{0.2}{0.3} = \frac{2}{3}$$

PETER'S SOLUTION

$$P(A|B) = \frac{P(A \cap B)}{P(B)} = \frac{0.2}{0.2 + 0.4} = \frac{1}{3}$$

SKILL BUILDER

1 A jar contains eight red discs and five green discs. Two discs are picked at random. The first one is not replaced. What is the probability that the second disc is red, given that the first disc is green?

A $\frac{1}{2}$ **B** $\frac{2}{3}$ **C** $\frac{7}{12}$ **D** $\frac{8}{13}$

E It is impossible to work out without more information.

2 A researcher is conducting a survey by asking questions in the street. She keeps a record of the number of people who were willing to answer her questions and of those who were not. The results are shown in the table below. Assuming that these results are typical, what is the probability that someone is willing to answer given that the person is female?

	Male	**Female**
Willing to answer	51	81
Not willing to answer	70	61

A $\frac{81}{263}$ **B** $\frac{132}{263}$ **C** $\frac{142}{263}$ **D** $\frac{81}{132}$ **E** $\frac{81}{142}$

3 The two-way table shows information about a sample of passengers at an airport.

	With hold luggage	**Without hold luggage**
Male	58	120
Female	55	67

Calculate the probability that a passenger chosen at random:

a) has hold luggage

b) is a male with hold luggage

c) has hold luggage, given that they are male

d) is male, given that they have hold luggage.

4 The Venn diagram shows the probabilities of letters being first class (F) and delivered on the next working day (N).

Find the probability that:

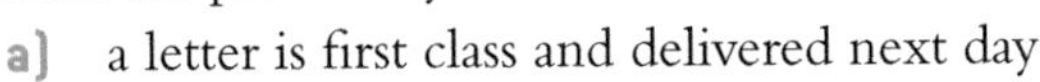

a) a letter is first class and delivered next day

b) the chosen letter is a first-class letter

c) a letter is delivered next working day given that it went first class.

5 Sally and Rashid take turns to tidy up.

When it is Sally's turn, she tidies 75% of the time.

When it is Rashid's turn, he tidies up 90% of the time.

a) Draw a tree diagram to illustrate these probabilities.

b) Calculate the probability that on a random day, the tidying is not done.

c) Given that the tidying is not done, calculate the probability that it was Sally's turn.

Discrete random variables

Year 1

THE LOWDOWN

1. The score you get when you roll the dice is a **discrete random variable** (DRV).
2. Use capital letters for the name of the random variable and the lower-case letters to represent **the values it can take.**
3. The **probability distribution** of the random variable gives the values it can take and the probabilities associated with them, either as a table, or a formula and the set of values.
4. The total probability in a probability distribution is always 1.
5. For a **discrete uniform random variable,** all probabilities are equal.

Discrete means it only takes certain values, **random** means you don't know in advance what it will be and **variable** means it can take different values.

Example The probability distribution for the score for a fair dice

Score (x)	1	2	3	4	5	6
Probability $P(X = x)$	$\frac{1}{6}$	$\frac{1}{6}$	$\frac{1}{6}$	$\frac{1}{6}$	$\frac{1}{6}$	$\frac{1}{6}$

Hint: When a question asks for the **probability distribution** it wants a table showing all the outcomes and their probabilities. It is just a different way of asking the question: 'Find the probability of every possible value.'

Example A six-sided dice is weighted so that the probability of rolling a 6 is 0.25 and all the other scores are equally likely.
Total probability of the other scores is $1 - 0.25 = 0.75$.
So the probability of each is 0.15.
Let Y = the score on the biased dice.
The probability distribution of Y is
$P(Y = 1) = 0.25$
and $P(Y = y) = 0.15$ for $y = 2, 3, 4, 5, 6$.

GET IT RIGHT

A spinner is made so that the probability distribution of the scores is given by the formula $P(X = x) = kx$ for $x = 1, 2, 3, 4, 5$ where k is a constant.

a) Find the value of k.

b) Find the probability that the score is at least 3.

Solution:

a) **Step 1** Create a table of probabilities.

Score (x)	1	2	3	4	5
$P(X = x)$	k	$2k$	$3k$	$4k$	$5k$

Step 2 Put the total probability equal to 1 and solve.

Total $k + 2k + 3k + 4k + 5k = 15k = 1$ giving $k = \frac{1}{15}$.

b) **Step 3** Draw a new table of probabilities without k.

Score (x)	1	2	3	4	5
$P(X = x)$	$\frac{1}{15}$	$\frac{2}{15}$	$\frac{3}{15}$	$\frac{4}{15}$	$\frac{5}{15}$

Step 4 Use the table to answer the follow-up question.

$P(X \geqslant 3) = P(X = 3, 4, 5) = \frac{3}{15} + \frac{4}{15} + \frac{5}{15} = \frac{12}{15} = \frac{4}{5}$

YOU ARE THE EXAMINER

Which one of these solutions is correct? Where are the errors in the other solution?

Wini plays a game with a coin in which she wins:

- 80p prize for throwing a head on the first throw
- 50p prize for the first head on the second thrown
- 10p prize for the first head on the third throw
- and no prize otherwise.

Find the probability distribution of X, the prize money on a game.

NASREEN'S SOLUTION

Prize (x)	80 (H)	50 (T then H)	10 (TT then H)
Probability	0.5	0.25	0.125

PETER'S SOLUTION

Prize (x)	80	50	10	0
P(X = x)	0.5	0.25	0.125	0.125

SKILL BUILDER

1 A discrete random variable, X, has the probability distribution:

$P(X = r) = k(2r^2 - r)$ For $r = 1, 2, 3$

$P(X = r) = 3k$ For $r = 4$

$P(X = r) = 0$ Otherwise

Find the value of k.

A 25 **B** $\frac{1}{22}$ **C** $\frac{1}{54}$ **D** 1 **E** 0.04

Hint: Substitute each value in turn into the relevant formula. The total probability is 1.

2 A discrete random variable, X, has the probability distribution shown in the table:

r	0	1	2	3	4
P(X = r)	0.25	0.2	0.15	0.3	0.1

Find $P(1 \leqslant X < 3)$.

A 0.15 **B** 0.35 **C** 0.45 **D** 0.55 **E** 0.65

3 The table shows the probability distribution of a discrete random variable, X.

x	0	1	2	3
P(X = x)	0.6	0.25	p	0.1

a) Calculate the probability that X is not 2.

b) Find the value of p.

4 The table shows the probability distribution of a discrete random variable, Y.

y	10	20	30	40	50	60
P(Y = y)	k	k	k	$2k$	$3k$	$2k$

a) Find the value of k. b) Calculate $P(Y \leqslant 30)$.

5 A discrete random variable, Y, takes the values 1, 4, 9, 16 and 25.

The probability distribution is given by the formula $P(Y = y) = k\sqrt{y}$.

a) Find the value of k. b) Calculate $P(Y > 9)$.

6 Two coins are thrown together. X is the number of heads.

a) Create a table to show the probability distribution of X.

b) Draw a vertical line graph to show these probabilities.

7 Angelika makes a five-sided fair spinner.

Explain why the score on the spinner can be modelled with a uniform discrete random variable.

Binomial distribution

Year 1

THE LOWDOWN

1. Suppose you throw three dice together and count the sixes scored. The probability of each possible value is shown in the table.

No of sixes (X)	0	1	2	3
Probability	$\left(\frac{5}{6}\right)^3$	$3 \times \left(\frac{1}{6}\right)^1 \times \left(\frac{5}{6}\right)^2$	$3 \times \left(\frac{1}{6}\right)^2 \times \left(\frac{5}{6}\right)^1$	$\left(\frac{1}{6}\right)^3$

2. Suppose you throw n dice. To score x sixes you need $(n - x)$ not sixes and there will be ${}^nC_x$ branches that give x sixes. So $P(X = x) = {}^nC_x \times \left(\frac{1}{6}\right)^x \times \left(\frac{5}{6}\right)^{n-x}$
   Not all exam boards ask you to use the formula.
3. When you do n trials of any experiment where the outcomes are independent and have the same probability p of success each time, the number of successes, X, fits a binomial distribution.
   You write $X \sim B(n, p)$.
4. You can find these probabilities directly from your calculator, by choosing the binomial probability distribution with the correct values for x, n and p.

Example The number of sixes is $X \sim B(10, \frac{1}{6})$
The probability of 4 sixes from 10 dice is $P(X = 4)$.
so $P(X = 4) = 0.0543$.

5. To find the probability of a set of values, you can either add up the probabilities of each value or use the cumulative distribution on your calculator.
   Some calculators find $P(X \leqslant x)$ and you input a value for x, for n and for p.
   Other calculators find $P(\textit{lower} \leqslant X \leqslant \textit{upper})$ instead.
   Find out what your calculator does.
6. If you know the value of $P(X \leqslant x)$ but not x, and you have inverse binomial on your calculator, you can use it to find the value of x. Otherwise use trial and improvement.
7. If you throw n dice you would expect one sixth of them to give a six.
   The expected value for a general binomial distribution is the probability multiplied by the number of trials.
   Write $E(X) = np$.

◀◀ **Probability** on page 94.
▶▶ **Hypothesis testing** on page 102.

Hint: Draw the tree diagram for this. There are three branches that give 1 six.

Hint: Start your solution by writing down the quantity that has a binomial distribution and give the values of n and p as evidence of your working.

Hint: x is a whole number so it may not give the probability exactly.

GET IT RIGHT

10% of people are left-handed. There are 24 students in a class.
Find the probability that there are:

a) 4 left-handed people in the class
b) fewer than 4 left-handed people in the class
c) at least 4 left-handed people in the class.

Solution:

X is the number of left-handed people so $X \sim B(24, 0.1)$

a) $P(X = 4) = 0.1292$ (using binomial probability distribution)

For $P(X \leqslant X)$
b) $P(X < 4) = P(X \leqslant 3) = 0.7857$
c) $P(X \geqslant 4) = 1 - P(X \leqslant 3)$
$= 1 - 0.7857 = 0.2143$

For $P(\textit{lower} \leqslant X \leqslant \textit{upper})$
b) $P(0 \leqslant X \leqslant 3) = 0.7857$
c) $P(X \geqslant 4) = P(4 \leqslant X \leqslant 24)$
$= 0.2143$

There are two methods here – use the one that matches your calculator.

Hint: Round your answer carefully. Rounding to 4 decimal places here is good.

★ YOU ARE THE EXAMINER

Which one of these solutions is correct? Where are the errors in the other solution?

A random variable has a binomial distribution $X \sim B(20, 0.4)$.

Find the largest value of x so that $P(X \leqslant x) < 10\%$.

LILIA'S SOLUTION

Using XInv with $n = 20$ and $p = 0.4$ and area 0.1: XInv = 5

So the value for x that gives a probability nearest to 10% is 5.

MO'S SOLUTION

For $X \sim B(20, 0.4)$, and a list of probabilities,

$P(X \leqslant 5) = 0.125... > 10\%$.

$P(X \leqslant 4) = 0.0509... < 10\%$ so the largest value of x is 4.

✓ SKILL BUILDER

1 For each distribution, work out the expected value of X.

a) $X \sim B(20, 0.4)$ b) $X \sim B(16, 0.5)$ c) $X \sim B(10, 0.35)$ d) $X \sim B(17, 0.25)$

2 X is a binomial random variable. Below are five statements about probabilities involving inequalities. Four of them are true and one is false. Find the one that is false.

A $P(X \geqslant 4) = 1 - P(X \leqslant 4)$

B $P(X < 3) = P(X \leqslant 2)$

C $P(X > 5) = 1 - P(X \leqslant 5)$

D $P(3 < X < 8) = P(X \leqslant 7) - P(X \leqslant 3)$

E $P(2 \leqslant X \leqslant 6) = P(X \leqslant 6) - P(X \leqslant 1)$

3 $X \sim B(40, 0.3)$. Calculate these probabilities.

a) $P(X = 12)$ b) $P(X \leqslant 12)$ c) $P(X > 12)$ d) $P(X \geqslant 12)$

4 $X \sim B(25, \frac{2}{3})$. Calculate these probabilities.

a) $P(0 \leqslant X \leqslant 15)$ b) $P(15 \leqslant X \leqslant 25)$

c) $P(3 \leqslant X \leqslant 10)$ d) $P(3 < X < 12)$

5 X is a binomial random variable. Below are five pairs of inequalities; one in words and one in symbols. Four of the pairs are equivalent and one pair is not equivalent. Find the pair that is not equivalent.

A X is no more than 6; $X \leqslant 6$

B X is at least 5; $X \geqslant 5$

C X is more than 6; $X > 6$

D X is not less than 7; $X \geqslant 7$

E X is at most 7; $X > 7$

6 Sunil enters a quiz where the answer to every question is either 'True' or 'False'.

He guesses for every question so the probability of getting any answer correct is 0.5.

There are 20 questions.

Calculate the probability that he gets:

a) 8 correct answers b) at least 15 correct answers

c) at most 10 correct answers d) more than 12 correct answers.

Hypothesis testing (binomial)

Year 1

THE LOWDOWN

Suppose you thought a dice was biased against sixes but your friend maintained it was fair. To sort out your argument you might roll the dice 60 times and count how many sixes you got. You would expect 10.

But how many sixes would lead you to believe the dice was biased?

You can run a **hypothesis test** to see if there is evidence to suggest the dice is biased.

◀◀ See page 100 for **calculating probabilities with the binomial distribution.**
▶▶ **Hypothesis tests using normal distribution** on page 106.

Hint: p is the probability of success (in this case rolling a six) on each trial.

GET IT RIGHT

Step 1 Write down the null hypothesis, H_0 – this says that the probability given in the question is right even if you think it's not!

$H_0: p = \frac{1}{6}$ where p is the probability of rolling a six.

Step 2 Write down the alternative hypothesis, H_1 – this says that the probability given in the null hypothesis is wrong.

$H_1: p < \frac{1}{6}$ (So the dice is biased against rolling a 6.)

Step 3 Run the experiment n times and count how many 'successes', in this case sixes, there are.

The number of successes X in n trials is called the test statistic.

As there are only two outcomes ('six' and 'not six') you can use the binomial distribution and say that $X \sim B(n, p)$.

Suppose you roll the dice 60 times and you get 4 sixes, then you would write $X \sim B\left(60, \frac{1}{6}\right)$ and $X = 4$.

Step 4 Now calculate the **probability of getting X successes or any more extreme value**, assuming that H_0 is correct.

In this case, work out the probability of getting **4 or fewer sixes** out of 60 rolls – this is called the p-value.

Compare the p-value with an agreed probability, usually **1%**, **5%** or **10%** – called the **significance level.**

$P(X \leqslant 4) = 0.0202 = 2.02\% < 5\%$

Step 5 State your conclusion: There is evidence at the 5% level to reject H_0. There is sufficient evidence that the probability of rolling a six is less than $\frac{1}{6}$.

The **critical region** is the set of values for which H_0 is rejected.

$P(X \leqslant 4) = 0.0202 < 5\%$ and $P(X \leqslant 5) = 0.0512 > 5\%$

So the critical region is $(X \leqslant 4)$, therefore if you get 4 or fewer sixes you should reject H_0.

Watch out! The wording of the question will tell you what H_1 should be. If you thought the dice was biased **towards** sixes you write: $H_1: p > \frac{1}{6}$.

If you just thought the dice was **biased**, you would write $H_1: p \neq \frac{1}{6}$

If the probability has:

- increased $H_1: p > \frac{1}{6}$
- decreased $H_1: p < \frac{1}{6}$
- changed $H_1: p \neq \frac{1}{6}$

If the p-value is less than the significance level then you **reject H_0**.

If the p-value is greater than the significance level – then you **accept H_0**.

YOU ARE THE EXAMINER

Which one of these solutions is correct? Where are the errors in the other solutions?

a) Find the critical region for the test with H_0: $p = 0.75$, H_1: $p > 0.75$ and $n = 30$ at the 5% significance level, where p is the probability of success in each trial.

b) In 30 trials, there are 26 successes. Complete the test.

PETER'S SOLUTION

a) $P(X \leqslant 25) = 0.9021 < 95\%$
$P(X \leqslant 26) = 0.9626 > 95\%$
Critical region {26, 27... 30}

b) 26 is in the critical region so reject H_0. There is proof that the probability p is more than 0.75.

NASREEN'S SOLUTION

a) $P(X \geqslant 26) = 0.0979 > 5\%$
$P(X \geqslant 27) = 0.0374 < 5\%$
Critical region {27, 28, 28, 30}

b) 26 is not in the critical region so accept H_0. There is insufficient evidence at the 5% level that the probability p is more than 0.75.

SKILL BUILDER

1 Copy and complete these statements – stay as close to 5% as you can.

a) $X \sim B(15, 0.6)$, $P(X \leqslant \_) = \_ < 5\%$ $\quad P(X \leqslant \_) = \_ > 5\%$

b) $X \sim B(30, 0.45)$, $P(X \leqslant \_) = \_ < 2.5\%$ $\quad P(X \leqslant \_) = \_ > 2.5\%$

c) $X \sim B(15, 0.6)$, $P(X \geqslant \_) = \_ < 5\%$ $\quad P(X \geqslant \_) = \_ > 5\%$

2 The null hypothesis for a test is $H_0: p = 0.45$ and $n = 40$.
Calculate the p-value of X.

a) H_1: $p < 0.45$ and

i $X = 12$ ii $X = 13$

b) H_1: $p > 0.45$ and

i $X = 23$ ii $X = 24$

3 The hypotheses for a two-tailed test are $H_0: p = 0.5$ and $H_1: p \neq 0.5$ and $n = 30$.
The significance level is 5%, giving two parts to the critical region.

a) Find the largest value of x so that $P(X \leqslant x) < 2.5\%$

b) Find the smallest value of x so that $P(X \geqslant x) < 2.5\%$

c) Use your answers in (a) and (b) to write down the critical region for the test.

d) The value of the test statistic is $X = 8$. Write a conclusion for this test.

For questions 4 and 5, X is the number of successes for a binomial random variable and p is the probability of success on each trial.

4 A sample of size 19 is taken to test the hypotheses H_0: $p = 0.5$ H_1: $p \neq 0.5$ at the 1% level of significance.
Which of the following gives the correct critical region?

A $X \leqslant 3, X \geqslant 15$
B $X \leqslant 3, X \geqslant 16$
C $X \leqslant 4, X \geqslant 15$
D $X \leqslant 5, X \geqslant 14$
E $X \leqslant 5, X \geqslant 16$

5 A sample of size 14 is taken to test the hypotheses H_0: $p = 0.6$ H_1: $p \neq 0.6$ at the 10% level of significance.
Which of the following gives the correct critical region?

A $X \leqslant 4, X \geqslant 10$
B $X \leqslant 4, X \geqslant 11$
C $X \leqslant 4, X \geqslant 12$
D $X \leqslant 5, X \geqslant 12$
E $X \leqslant 3, X \geqslant 13$

Normal distribution

Year 2

THE LOWDOWN

① If you measured the heights of a group of adults, you would typically find that most of them are quite near to the mean height, with fewer tall or short people. The histogram has a distinctive bell shape.

About $\frac{2}{3}$ of the population are less than one standard deviation σ from the **mean** μ. 95% of people are within two standard deviations and almost everyone is less than three away from the **mean**.

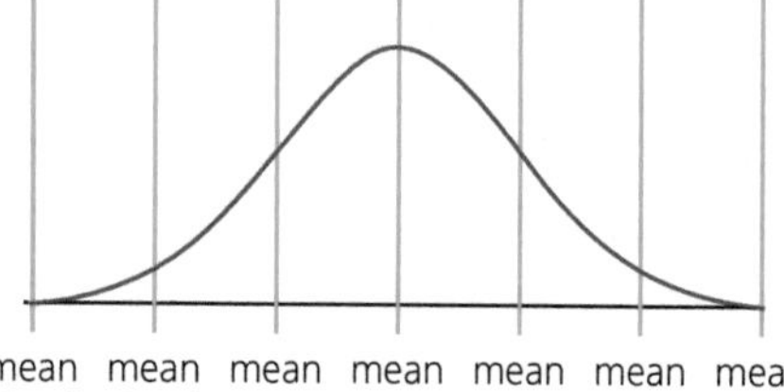

② Write $X \sim N(\mu, \sigma^2)$ for a Normal distribution with mean μ and variance σ^2.

The probability of being between given values is represented by the area under the curve.

③ You can use your calculator to work out the probability that a man's height is between 170 cm and 180 cm (the boundary values), given that the **mean height** is **178 cm** and the standard deviation is 10 cm.

Example

Step 1 Write down: $X \sim N(178, 10^2)$.

Step 2 State the probability you are finding: $P(170 < X < 180)$

Step 3 Use Normal cumulative distribution

lower = 170, upper = 180,

$\sigma = 10$ and $\mu = 178$

So $P(170 < X < 180) = 0.3674$

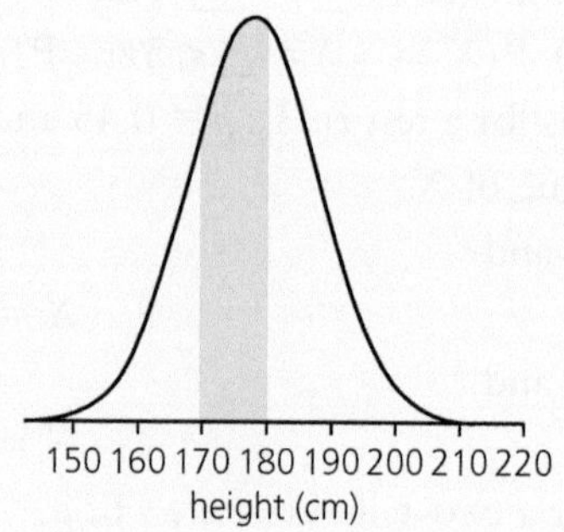

④ If you know the value of $P(X < x)$ (the area of the left tail of the graph) and the **mean** and standard deviation, you can use the inverse Normal function to find the boundary value x.

Example 5% of the adult males lie below the value x.

Using inverse Normal (with left tail),

$P(X < x) = 0.05$

Using area = 0.05, $\sigma = 10$ and

$\mu = 178$ gives xInv = 161.6 cm.

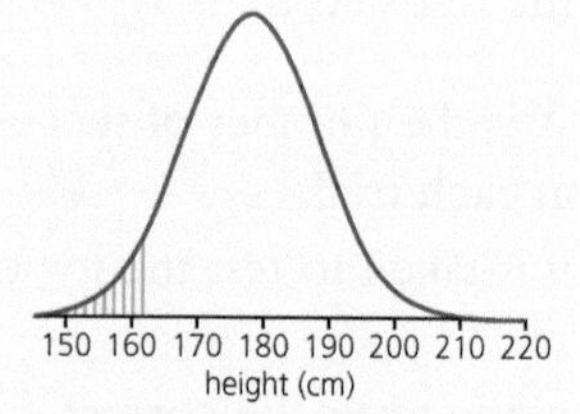

◀◀ See page 88 for **mean** and **standard deviation** and page 86 for **histograms**.
▶▶ **Hypothesis testing** on page 106.

Hint: We usually write a normal distribution with the values of **mean** μ and standard deviation σ or variance σ^2 as evidence of your working.

Your calculator may also give the option of the right tail and the centre.

Hint: A diagram of the distribution can be helpful to check answers.

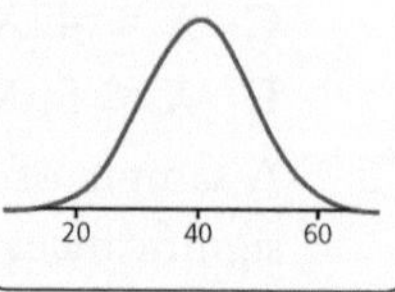

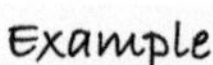

GET IT RIGHT

A survey at a company shows the amounts of time people spend travelling to work fit a Normal distribution with **mean of 40 minutes** and standard deviation of 9 minutes. A person is chosen at random. Find the probability that their travel time is:

a) between 30 and 40 minutes **b)** less than 15 minutes **c)** more than an hour.

Solution:

a) Use cumulative Normal distribution with lower = 30, upper = 40, $\mu = 40$, $\sigma = 9$ and $P(30 < X < 40) = 0.366$.

b) $P(X < 15) = P(-10^{10} < X < 15) = 0.0027$

c) $P(X > 60) = P(60 < X < 10^{10}) = 0.0131$

Hint: For b) and c) choose any impossibly low lower boundary or high upper boundary. For example, -10^{10} and 10^{10}.

YOU ARE THE EXAMINER

Find the correct solution. Where are the errors in the other solutions?

The height of adult females in the UK are normally distributed with mean 165 cm and standard deviation 9 cm.

Find the interval round the mean in which 90% of heights lie.

LILIA'S SOLUTION

$P(X < x) = 90\%$

Inverse Normal area $= 0.9$, $\sigma = 9$, $\mu = 165$

x Inv $= 176.5$. So the region is $X < 176.5$ cm.

NASREEN'S SOLUTION

Inverse Normal with central area 0.9 $\sigma = 9$, $\mu = 165$

x_1 Inv $= 150.2$, x_2 Inv $= 179.8$

So the region is $X < 150.2$ or $X > 179.8$

PETER'S SOLUTION

$P(X < x) = 90\%$

Inverse Normal area $= 0.9$, $\sigma = 9$, $\mu = 165$

x Inv $= 176.5$

$P(X < x) = 10\%$

Inverse Normal area $= 0.1$, $\sigma = 9$, $\mu = 165$

x Inv $= 153.4$

So the region is $153.4 < X < 176.5$ cm

MO'S SOLUTION

$P(X < x) = 5\%$

Inverse Normal area $= 0.05$, $\sigma = 9$, $\mu = 165$

x Inv $= 150.2$

$P(X < x) = 90\%$

Inverse Normal area $= 0.95$, $\sigma = 9$, $\mu = 165$

x Inv $= 179.8$

So the region is $150.2 < X < 179.8$ cm

SKILL BUILDER

1 X is from a Normal distribution with mean 50 and standard deviation 6.5.

Calculate these probabilities.

a) $P(55 < X < 65)$

b) $P(45 < X < 46)$

c) $P(42 < X < 75)$

d) $P(42.5 < X < 43.5)$

2 Y is from a Normal distribution with mean 20 and standard deviation 8.

Calculate these probabilities.

a) $P(Y < 40)$ b) $P(Y < 10)$ c) $P(Y > 25)$ d) $P(Y > 17)$.

3 The heights of a population of adult male elephants are Normally distributed with mean 3.2 m and standard deviation 0.3 m.

Find the proportion of the population with heights:

a) between 3 m and 3.3 m

b) less than 3 m

c) more than 2.8 m

d) more than 4 m.

Hint: This is like asking the probability that an individual chosen at random has at least that height.

4 Records at a hospital show that the masses of new-born baby boys are Normally distributed with mean 3.3 kg and standard deviation 0.4 kg.

The girls' masses have a mean of 3.2 kg and the same standard deviation.

A midwife chooses a boy and a girl at random.

Find the probability that:

a) the boy's mass is over 3.5 kg

b) the girl's mass is over 3.5 kg

c) that both the babies have masses over 3.5 kg. (Assume the choices are independent.)

5 X is from a Normal distribution with mean 50 and standard deviation 6.5.

Calculate the boundaries:

a) $P(X < x) = \frac{1}{4}$

b) $P(X < x) = 0.7$

c) $P(X > x) = 0.1$

d) $P(X > x) = 0.6$.

6 A farmer's hens produce eggs with weights that are Normally distributed with mean 60 g and standard deviation 12 g.

Find the interval round the mean in which 75% of the eggs lie.

Hypothesis testing (normal)

Year 2

THE LOWDOWN

Samir wants to find out if the new roundabout will change his journey time to college.
Before the roundabout is built, his journey time has **mean** μ and standard deviation σ.

Samir records the time for a sample of n journeys using the new roundabout and works out the mean, $\bar{X}$. The standard deviation for $\bar{X}$ is $\frac{\sigma}{\sqrt{n}}$.

Samir runs a **hypothesis test** to see if there is evidence that the journey time has changed. If the sample mean is very different from the **mean** journey time without the roundabout, Samir can conclude that the journey time has changed.

Hint: If n numbers are used for the sample mean, divide the standard deviation by $\sqrt{n}$.

See page 104 for **Normal distribution** and page 102 for **hypothesis test**.

GET IT RIGHT

Samir's journey times to college are Normally distributed with **mean 45 minutes** and standard deviation 6 minutes. With the new roundabout in place, the journey times for a week are 34, 48, 36, 38, 45. Perform a **hypothesis test** at the 5% **significance level** to see if there is evidence that the journey time has changed.

Solution:

Step 1 Write down the null hypothesis, H_0 – this says that the **mean time** has not changed.

H_0: $\mu = 45$ where μ is the population mean.

Step 2 Write down the alternative hypothesis, H_1 – this says that the mean time has changed. (It does not say whether it is larger or smaller!)

H_1: $\mu \neq 45$

Hint: A test is a two-tailed test when both small and large values of sample mean lead you to reject H_0.

Step 3 Use the values to calculate the sample mean

$$\bar{x} = \frac{34+48+36+38+45}{5} = 40.2$$

Step 4 Write down the distribution of the sample mean

Under H_0 $\bar{X} \sim N\left(45, \frac{6^2}{5}\right)$ (standard deviation is $\frac{6}{\sqrt{5}}$)

Step 5 Calculate the z-value for 40.2

$$z = \frac{\bar{x}-45}{\left(6/\sqrt{5}\right)} = -1.789$$

Hint: The z-value tells us how many standard deviations the sample mean is away from the value in H_0.

Step 6 Find the critical region for the significance level.

Under H_0 $Z \sim N(0, 1)$

At the 5% significance level and two-tailed, we need $P(Z < z) = 0.025$ x Inv $= -1.95996$

So the critical region is either $z < -1.95996$ or $z > 1.95996$

You can also use $P(Z < z) = 0.975$ or $P(Z > z) = 0.025$ to get 1.959 96.

Step 7 Check whether the z-value is in the critical region

$z = -1.789 > -1.95996$ so it is not in the critical region.

Step 8 Make a clear conclusion.

Accept H_0 as there is not enough evidence to support the hypothesis that the mean time has changed.

Hint: Draw a diagram to show where the critical region is – there is 2.5% probability in each tail. Show the test statistic.

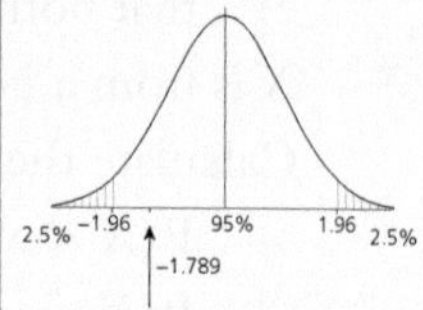

Alternative: The p-value is the probability of the sample mean being more extreme than the actual value. For a two-tailed test double the calculated probability. (you need to have probability at each end.) Compare the p-value with the significance level.

★ YOU ARE THE EXAMINER

Which one of these solutions is correct? Where are the errors in the other solution?

For the hypothesis test on page 106 find the p-value of Samir's journey times.

Complete the hypothesis test at the 5% significance level.

LILIA'S SOLUTION

$$z_{test} = \frac{\bar{x} - 45}{\left(6/\sqrt{5}\right)} = -1.789$$

so $P(Z < -1.789) = 0.0368$

$0.0368 < 5\%$ so reject H_0 as there is evidence at the 5% level that the mean journey time has changed.

NASREEN'S SOLUTION

$$z_{test} = \frac{\bar{x} - 45}{\left(6/\sqrt{5}\right)} = -1.789$$

so $P(Z < -1.789) = 0.0368$

Test is two-tailed so the p-value is $2 \times 0.0368 = 0.0736$

$0.0736 > 5\%$ so accept H_0 as there is insufficient evidence at the 5% level that the mean journey time has changed.

✓ SKILL BUILDER

1 Random samples of size 5 are taken from N(12, 2.5). What is the distribution of the mean of such samples?

A N(2.4, 0.5)
B N(12, 0.5)
C N(12, 0.1)
D N(12, 2.5)
E N(60, 12.5)

Hint: The first value is the mean and the second is variance $\frac{\sigma^2}{n}$.

2 A man has found that the time it takes him to walk to work is Normally distributed with mean 15.5 minutes. After going on a fitness course, he thinks that he may be walking faster on average now. Which are suitable null and alternative hypotheses to use for a test?

A $H_0: \mu = 15.5$
$H_1: \mu < 15.5$

B $H_0: \mu = 15.5$
$H_1: \mu > 15.5$

C $H_0: \mu = 15.5$
$H_1: \mu \neq 15.5$

D $H_0: \mu < 15.5$
$H_1: \mu > 15.5$

E $H_0: \mu > 15.5$
$H_1: \mu < 15.5$

3 Find the critical regions for the z-test statistic for Samir's two-tailed hypothesis test on page 106.

a) at the 1% level
b) at the 10% level.

The critical regions for Z do not change when different values of μ, σ and n are used.

4 Samir repeats the hypothesis test the following week. His values are 38, 27 36, 37, 46. Calculate:

a) the z-value for this data
b) the p-value for this value.

5 Amelia believes that the lengths of a species of fish in a lake have been Normally distributed with mean 21 cm and standard deviation 2.8 cm. After a temporary fishing ban, she believes the mean may have increased.

Amelia measures a sample of fish from the lake and their lengths are 25, 22.5, 20, 25.4,24.5,22.8 cm. She performs a hypothesis test at the 5% level.

a) Write down the null and alternative hypotheses for her test.
b) Calculate the sample mean.
c) Calculate the z-value.
d) Find the critical region for the test. (Remember: this is a one-tailed test.)
e) Finish the hypothesis test, writing your conclusion clearly.
f) Check your answer by calculating the p-value of the test statistic.

Hint: She thinks the length has increased.

Hint: It must be about lengths of fish!

Bivariate data

Year 2

THE LOWDOWN

① Bivariate data is **paired data**, for example, the ages and heights of people.
You can plot a **scatter diagram**. Put the **independent variable** (age) on the **horizontal** axis and the dependent variable (height) on the vertical axis.

② The **shape** reveals further information about the data set.
- Correlation is a linear relationship between the points.
- The correlation coefficient measures the strength of the correlation.
- Perfect correlation gives correlation of ± 1.

③ The regression line shows the linear relationship between age and height.
- The intercept is the value of the height when age = 0.
- The gradient is the rate of increase of height as age increases.

This is the ideal line of best fit.

④ A **hypothesis test** can be used to determine whether a sample of points indicates correlation in the population.

H_0 is always $\rho = 0$ where ρ is the correlation coefficient for the population.

H_1 could be $\rho \neq 0$ for any correlation
$\rho > 0$ for positive correlation
or $\rho < 0$ for negative correlation.

Calculators give the correlation coefficient and the regression line. You do not need to be able to calculate these yourself.

Hint: At A level it's all about interpretation!

Hint: Write what ρ is every time.

◀◀ **Hypothesis testing** on page 102.

GET IT RIGHT

Graph1 shows the ages and heights of people in the playground one afternoon.
Graph 2 shows only the people whose ages are less than 14.

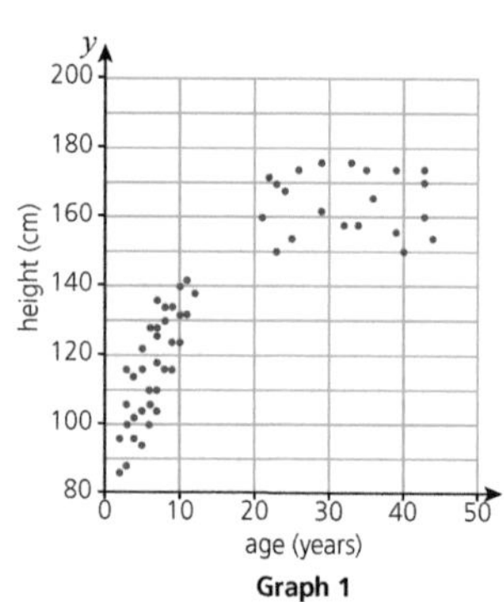

Graph 1

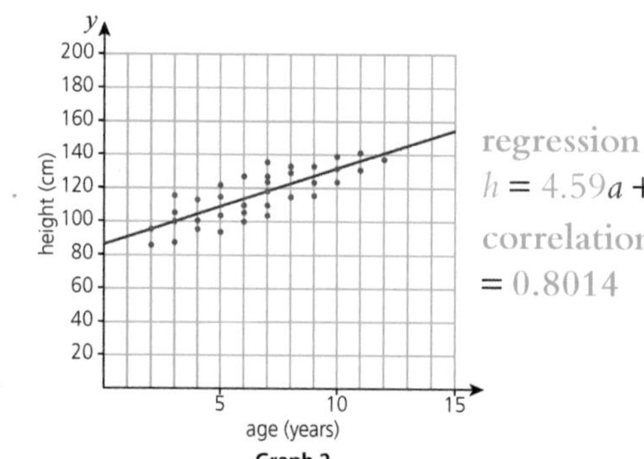

Graph 2

regression line
$h = 4.59a + 86.2$
correlation coefficient
$= 0.8014$

a) Indicate why you should not consider the regression line in graph 1.
b) Comment on the values 4.59 and 86.2 in the equation of the regression line.
c) Why could this data set not be used to estimate the height of a newborn baby?

Solution:

a) The data falls into two distinct groups – the adults and the children. The data as a whole does not fit a linear relationship.

b) 4.59 is the gradient – the rate at which y is increasing - the height increases at a rate of 4.59 cm for every additional year.
86.2 is the intercept – the value of y when $x = 0$ - the height of a child aged 0 years.

c) Using age = 0 is extrapolating beyond the range 2 to 13 years, which is unreliable.

Hint: the gradient of the line indicates how much the height increases as the age increases by 1.

Hint: The intercept gives the height when the age is 0.

YOU ARE THE EXAMINER

Which one of these solutions is correct? Where are the errors in the other solution?

Alan finds information about wind speed on 7 days.

The values for mean wind speed (x) and the maximum gust speed (y) for his sample give a correlation coefficient of 0.6907.

a) Use the table of critical values below to test at the 5% significance level whether there is positive correlation between mean wind speed and maximum gust speed.

b) How could Alan make his test more reliable?

One tailed	5%	2.5%	1%	0.5%
Two tailed	10%	5%	2%	1%
Sample size 7	0.6694	0.7545	0.8329	0.8745

PETER'S SOLUTION

a) $H_0: \rho = 0$
$H_1: \rho \neq 0$
Two-tailed test 5% level, critical value 0.7545 so critical region is $r > 0.7545$ or $r < -0.7545$
Actual value $0.6907 < 0.7545$ so accept H_0 there is no correlation.

b) Alan could measure the wind himself.

MO'S SOLUTION

a) $H_0: \rho = 0$ where ρ is the population correlation coefficient between mean wind speed and gust speed.
$H_1: \rho > 0$
One-tailed test 5% critical value 0.6694 so critical region is $r > 0.6694$
Actual value $0.6907 > 0.6694$ so reject H_0 as there is sufficient evidence at 5% level that there is positive correlation between mean wind speed and maximum gust speed.

b) Use a larger set of values in his test.

SKILL BUILDER

1 Fred exercises for 30 minutes then observes his pulse rate after stopping the exercise. The data are shown in the table and scatter diagram below.

Time after exercise (mins), x	Pulse rate, y
½	133
1	127
1½	98
2	67
2½	68

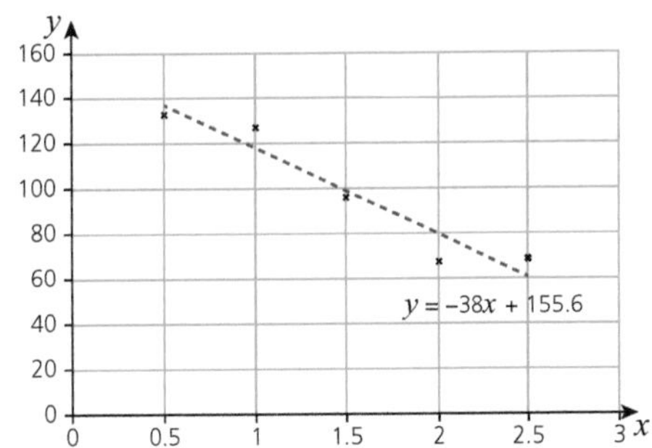

Three of the following statements are false and one is true. Find the one that is true.

A It would be reasonable to use the equation of the regression line to find the pulse rate 4 minutes after exercise as this is not far from the data in the table.

B The equation of the regression line which you have found will apply to the pulse rate of all men who exercise for 30 minutes then stop.

C Time after exercise is the dependent variable and pulse rate is the independent variable.

D 79.6 is the model's prediction of the pulse rate 2 minutes after Fred stops exercise.

2 Use the table in 'You are the examiner' to find the critical region(s) for the hypothesis tests below. In each case, ρ is the population correlation coefficient and the sample size is 7. Complete the test, using the actual value of the correlation coefficient for the sample.

a) 2.5% significance level, $H_0: \rho = 0$, $H_1: \rho > 0$, actual value 0.7012

b) 1% significance level, $H_0: \rho = 0$, $H_1: \rho < 0$, actual value −0.8521

c) 5% significance level, $H_0: \rho = 0$, $H_1: \rho \neq 0$, actual value 0.7243

d) 1% significance level, $H_0: \rho = 0$, $H_1: \rho \neq 0$, actual value −0.8916

Working with graphs

THE LOWDOWN

① **Distance** travelled does not take direction into account.
Displacement is the distance from a fixed point (origin) **in a given direction**.

Example John runs north for 9 s at $4\,\text{m s}^{-1}$ and travels $4 \times 9 = 36$ m.
He runs south for 7 s at $6\,\text{m s}^{-1}$ and travels $6 \times 7 = 42$ m.
John runs a total of $36 + 42 = 78$ m.
Taking north to be the positive direction, John's displacement after 5 s is 36 m, and after 12 seconds is $36 - 42 = -6$ m.

Watch out! Only use the formula distance = speed × time when the **speed is constant.**

This means that he is 6 m south of his starting point.

② **Speed** is the rate of change of distance; the gradient of the speed–time graph.
Velocity is the rate of change of displacement.
Velocity is the **gradient** of a displacement–time graph.

Example John's speeds are $4\,\text{m s}^{-1}$ and $6\,\text{m s}^{-1}$.
John's velocities are $4\,\text{m s}^{-1}$ and $-6\,\text{m s}^{-1}$ (going south).

Watch out! Speed is always positive but velocity is negative when travelling in the negative direction.

③ You can use graphs to represent a journey.

Example

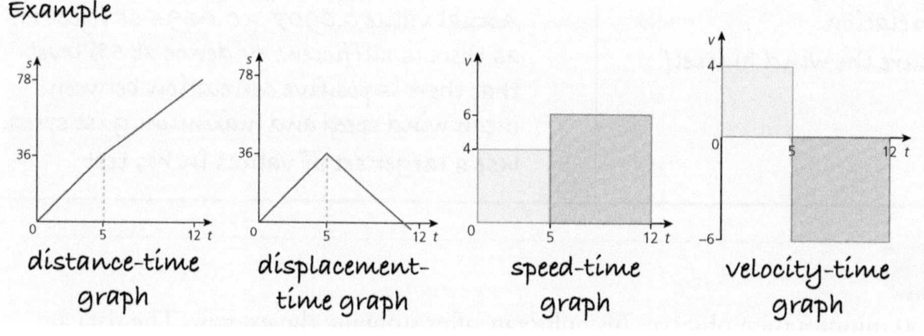

distance–time graph | displacement–time graph | speed–time graph | velocity–time graph

◀◀ See page 34 for **gradient of lines**.
▶▶ See page 112 for *suvat* **equations** and page 120 for **using vectors**.

④ **Distance** is the **area** between the speed–time graph and the horizontal axis.
Displacement is the **area** between the velocity–time graph and the axis.

⑤ **Acceleration** is the rate of change of velocity; the **gradient** of the v–t graph.
It is measured in m s^{-1} per second – written as m s^{-2}.

Watch out! The displacement for the area that lies below the axis is taken to be negative.

GET IT RIGHT

The graph shows Mary's velocity at time t s.

a) Find Mary's acceleration.

b) Find her displacement after 12 s and the average velocity.

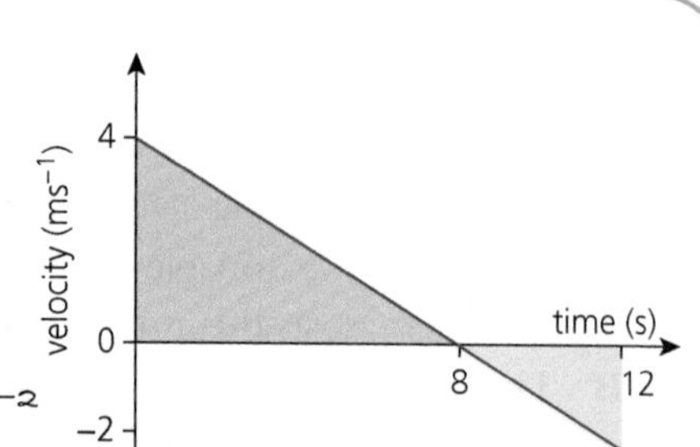

Solution:

a) Acceleration $= \dfrac{\text{change in velocity}}{\text{time}} = \dfrac{-2-4}{12} = -0.5\,\text{m s}^{-2}$

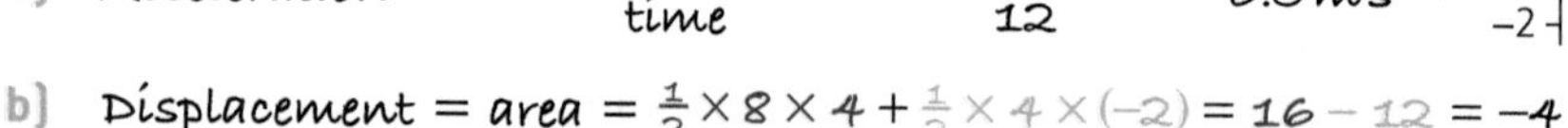

b) Displacement = area $= \frac{1}{2} \times 8 \times 4 + \frac{1}{2} \times 4 \times (-2) = 16 - 12 = -4$ m

Average velocity $= \dfrac{\text{displacement}}{\text{time}} = \dfrac{-4}{12} = -\dfrac{1}{3}\,\text{m s}^{-1}$

★ YOU ARE THE EXAMINER

Sam's solution is muddled. Match the statements to the points on the graph.

In this question, east is the positive direction.

The graph shows the displacement of a child playing on a straight path. Her mother sits at the origin.

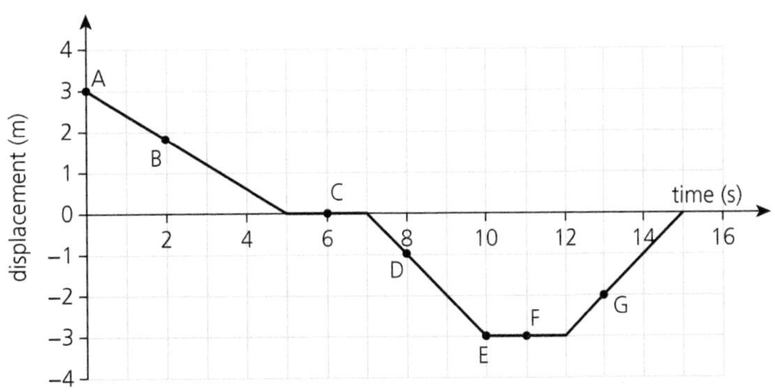

SAM'S SOLUTION

1 The child's velocity is $-0.6\,\mathrm{m\,s^{-1}}$.
2 The child is stationary and with her mother.
3 The child's velocity is $1\,\mathrm{m\,s^{-1}}$.
4 The child stops suddenly.
5 The child is 3 m east of her mother.
6 The child's velocity is $-1\,\mathrm{m\,s^{-1}}$.
7 The child is stationary and 3 m west of her mother.

✓ SKILL BUILDER

1 Which one of the following statements is true?

A The speed of light is $3 \times 10^8\,\mathrm{m\,s^{-1}}$ and the mean distance from the Sun to the Earth is 1.5×10^8 km, so it takes 0.5 s for light to reach the Earth from the Sun.

B The speed of sound is $340\,\mathrm{m\,s^{-1}}$ and it takes 5 s for the sound of thunder to reach me, so I must be 1.7 km away.

C A particle has negative acceleration, so its velocity must also be negative.

D If a particle has zero velocity then its acceleration is also zero.

2 Alan walks 300 m due east in 150 s, and then 150 m due west in 50 s. What is his average velocity?

A $2.25\,\mathrm{m\,s^{-1}}$ **B** $0.75\,\mathrm{m\,s^{-1}}$ **C** $2.25\,\mathrm{m\,s^{-1}}$ east **D** $0.75\,\mathrm{m\,s^{-1}}$ east

In questions 3–6, north is the positive direction.

3 Stewart begins 4 m south of the origin. He travels at $3\,\mathrm{m\,s^{-1}}$ for 12 s.
Draw a displacement–time graph.

4 Mary starts at the origin and travels 10 m north in 5 s. She waits 5 s then returns to the origin in 4 s.
Draw a displacement–time graph, a speed–time graph and a velocity–time graph.

5 Draw a speed–time graph and a velocity–time graph for both of Aleisha's journeys.

a) Aleisha begins by travelling due south at $8\,\mathrm{m\,s^{-1}}$ for 4 s. She then travels north at $5\,\mathrm{m\,s^{-1}}$ for 10 s.

b) Aleisha starts at rest and moves south, with constant acceleration to $5\,\mathrm{m\,s^{-1}}$ in 10 s.
She travels at constant speed for 10 s in the same direction.

6 For each velocity–time graph, find the acceleration for each phase of the motion.
Calculate the distance travelled and the displacement.

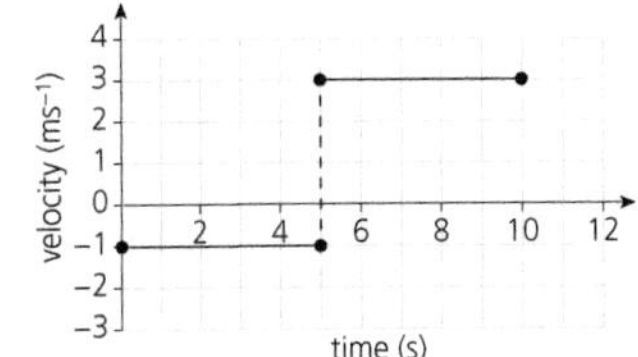

b)

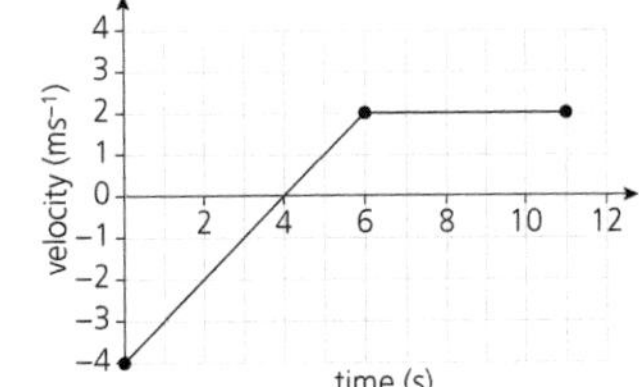

c)

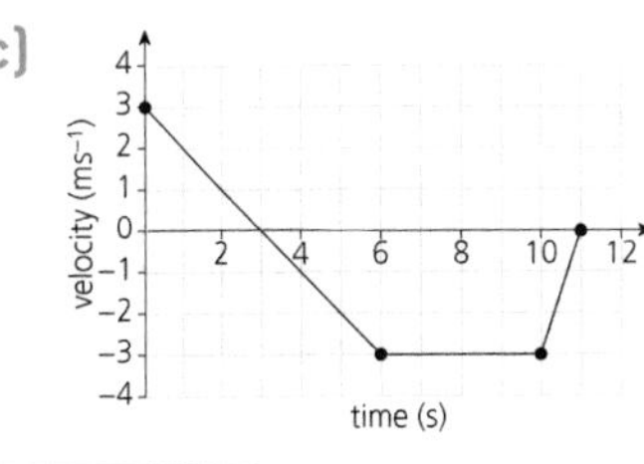

Hint: Find the area for each phase – include the negatives for displacement not distance.

suvat equations

THE LOWDOWN

① Use the *suvat* equations when the acceleration of an object is constant.

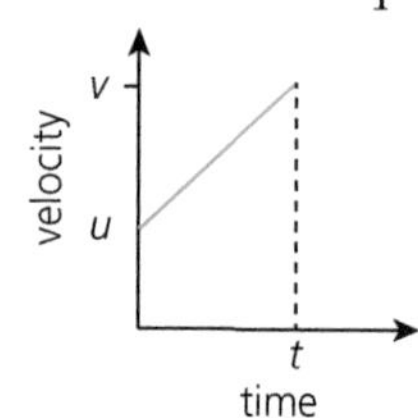

Use s for the displacement,
u for the starting velocity,
v for the final velocity,
a for the acceleration,
t for the time taken.

② The equations are:

- $s = \frac{1}{2}(u + v)t$ (without a)
- $v = u + at$ (without s)
- $s = ut + \frac{1}{2}at^2$ (without v)
- $v^2 = u^2 + 2as$ (without t)
- $s = vt - \frac{1}{2}at^2$ (without u)

③ Many problems can be solved in one step by choosing the correct equation.

Example Velocity after accelerating from rest at $5\,\text{m s}^{-2}$ for 6 s

~~s~~, $u=0$, $v=?$, $a=5$, $t=6$ Use $v=u+at=0+5\times6=30\,\text{m s}^{-1}$

④ Objects travelling vertically are modelled by neglecting air resistance.

- The acceleration is g and is usually taken to be $9.8\,\text{m s}^{-2}$ downwards.
- When an object falls, $u = 0$.
- At the highest point $v = 0$.
- Use the value of s to describe the position of the object.

If an object starts at the origin with up as the positive direction then when:

- the object is above/**below** the starting point, s is positive/**negative**
- the object is moving up/**down** when v is positive/**negative**
- the acceleration, a, is **negative**.

Watch out! These equations can only be used when the **acceleration is constant**.

- The **1st equation** is the **area under the graph**.
- The **2nd equation** is the **equation of the line** with gradient a.
- The other equations are derived from these two.

Hint: Write *suvat* and give each a value or a question mark or cross it out. Choose the equation that **does not** include the letter you crossed out.

▶▶ See page 114 for **variable acceleration** and page 120 for **vectors**.

GET IT RIGHT

A ball is dropped from a point 15 m above the ground.
A stone is thrown upwards at $6\,\text{m s}^{-1}$ from a point 8 m above the ground.
Which hits the ground first if they start at the same time?

Solution:

For the ball:

Step 1 Give values: $s = -15$, $u = 0$, ~~v~~, $a = -g = -9.8$, $t = ?$

Step 2 Choose the equation and substitute values.
Without v so choose $s = ut + \frac{1}{2}at^2$ giving $-15 = 0\times t + \frac{1}{2}(-9.8)t^2$

Step 3 Solve the equation.
$-15 = -4.9t^2$ so $t = \sqrt{\frac{15}{4.9}} = 1.75\,\text{s}$

For the stone:

Step 1 Give values: $s = -8$, $u = 6$, ~~v~~, $a = -g = -9.8$, find t.

Step 2 Choose the equation and substitute values.
Without v so choose $s = ut + \frac{1}{2}at^2$ giving $-8 = 6\times t + \frac{1}{2}(-9.8)t^2$

Step 3 Rearrange and solve the equation.
$4.9t^2 - 6t - 8 = 0$ giving $t = 1.89$ (reject $t = -0.866$)

Step 4 Compare and state the conclusion.
$1.75 < 1.89$ so the ball reaches the ground first.

Hint: Use your calculator to solve the quadratic equation.

Hint: There is no need to consider the upwards and downwards motion separately.

YOU ARE THE EXAMINER

Which one of these solutions is correct? Where are the errors in the other solution?

Fahreed throws a coin upwards, with velocity 7 m s^{-1}, from a point 0.8 m above the ground.

a) Find the height of the coin above the ground at the highest point.

b) Find the total distance the coin has travelled when it hits the ground.

LILIA'S SOLUTION

a) Highest point $u = 7, v = 0, a = 9.8$, find s

Use .. so $0^2 = 7^2 + 2 \times 9.8s$

So $s = \frac{49}{19.6} = 2.5\,\text{m}$

So maximum height is 2.5 m

b) Total distance is $2.5 + 0.8 = 3.3\,\text{m}$

MO'S SOLUTION

a) Highest $s = ?, u = 7, v = 0, a = -9.8$, ~~t~~

Use $v^2 = u^2 + 2as$

so $0^2 = 7^2 + 2(-9.8)s$

So $s = \frac{49}{19.6} = 2.5\,\text{m}$

So total height is $2.5 + 0.8 = 3.3\,\text{m}$

b) Total distance is $2.5 + 2.5 + 0.8 = 5.8\,\text{m}$

SKILL BUILDER

1 Mel jogs along a path, with an initial velocity of 4 m s^{-1}, and has an acceleration of -2 m s^{-2}.
Make a table of values for v and s for $t = 0, 1, 2 \ldots 5$ s.
Describe Mel's jog.

2 For each problem, identify which quantities have which values, and which quantity needs to be found.

a) How long does it take a runner to accelerate from rest to 7 m s^{-1} with an acceleration of 3.5 m s^{-2}?

b) How far does a train travel as it accelerates from 20 m s^{-1} to 25 m s^{-1} in 40 s?

c) A swimmer slows down from 4 m s^{-1} with an acceleration of -0.2 m s^{-2} for 5 s.
Find her displacement.

d) Calculate the acceleration of a bus as it travels 200 m from rest to 18 m s^{-1}.

e) An arrow is fired vertically upwards from ground level and reaches its highest point 3 s later.
The highest point is 10 m above the ground.
Find the initial velocity of the arrow.

3 Complete each of the problems in question 2.

4 A train accelerates from rest at 0.2 m s^{-2} for 4000 m. How long does it take and what's its final speed?

A 200 s, 40 m s^{-1} **B** 200 s, 20 m s^{-1} **C** 20 s, 40 m s^{-1} **D** 20 s, 4 m s^{-1}

5 A car starts at rest and accelerates at 2 m s^{-2} for 4 s. It then travels at constant speed for 10 minutes.
How far has it travelled in this time?

A 76 m **B** 196 m **C** 4800 m **D** 4816 m

6 A ball is thrown upwards from the ground at 8 m s^{-1}.
For how long is the ball visible above a 2 m fence?

Hint: Find the times at which the ball has a displacement of 2 m.

7 A car accelerates from A to B in 10 s and then continues to C, which is 100 m beyond B.
It is at rest at A and has an acceleration of 1.5 m s^{-2} as far as B.
It travels 100 m from B to C with an acceleration of 0.5 m s^{-2}.
Find the velocity of the car at a) B and b) C.

Hint: Use the final velocity of the car at B as the initial velocity for the journey BC.

Variable acceleration

THE LOWDOWN

You can only use the *suvat* equations when the **acceleration** is **constant**. Where the **acceleration** is **variable**, you must use calculus to link these quantities.

You can give displacement s, velocity v and **acceleration** a in terms of t.

① The **gradient** of the displacement–time graph is velocity, so $v = \frac{ds}{dt}$.

Example When $s = (5t - 6t^2)$ m, $v = \frac{ds}{dt} = (5 - 12t)\text{ m s}^{-1}$

② The **gradient** of the velocity–time graph is **acceleration** so $a = \frac{dv}{dt}$.

Example When $v = (5 - 12t)$ m, $a = \frac{dv}{dt} = -12\text{ m s}^{-2}$

③ Displacement is the **area** under the velocity–time graph, so $s = \int v\, dt$.

Example When $v = 3 - 5t^2$ $s = \int(3 - 5t^2)\,dt = 3t - 5 \times \frac{t^3}{3} + c$

Given that when $t = 0$, $s = 8$ then $8 = 3 \times 0 - \frac{5}{3} \times 0^2 + c$

So $s = 3t - \frac{5}{3}t^3 + 8$

④ Velocity is the **area under** the **acceleration–time** graph so $v = \int a\, dt$

Example When $a = 2 - 0.5t$, $v = \int(2 - 0.5t)\,dt = 2t - 0.5 \times \frac{t^2}{2} + c$

Given that initially $v = 4$, then $4 = 2 \times 0 - 0.25 \times 0^2 + c$.

So $v = 2t - 0.25t^2 + 4$.

⑤ **Maximum** and **minimum** values of

- displacement happen when $v = \frac{ds}{dt} = 0$
- velocity happen when $a = \frac{dv}{dt} = 0$.

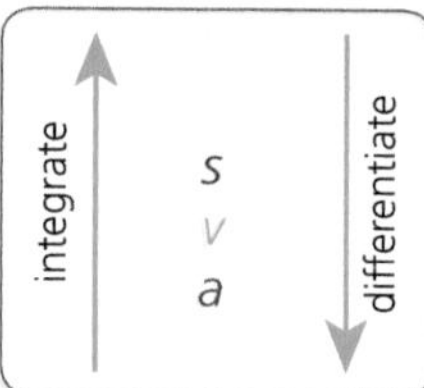

Notice in this case the acceleration is **constant.**

Watch out! Always put $+c$ when you integrate. Use the additional information in the question to find out its value.

Hint: 'Initially' means $t = 0$.

That's why they are called stationary points.

GET IT RIGHT

The velocity of a particle, in m s^{-1}, is given by $v = -0.1t^3 + 0.9t^2 - 1.5t$ for $0 \leqslant t \leqslant 6$ s.

a) Find the acceleration when $t = 0$.

b) Find the maximum and minimum values of velocity for $0 \leqslant t \leqslant 6$ s.

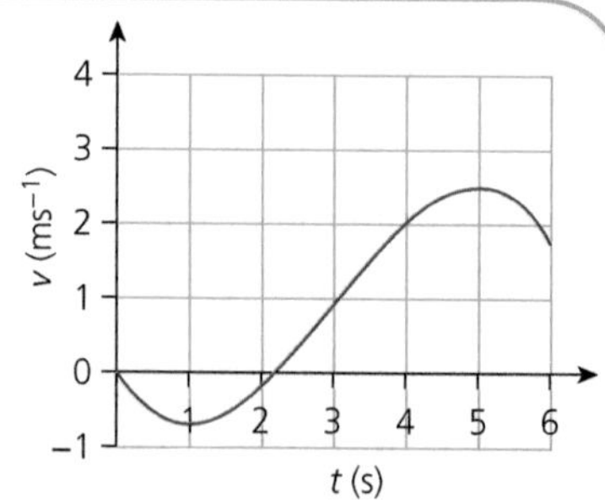

Solution:

a) Step 1 Differentiate: $a = \frac{dv}{dt} = -0.3t^2 + 1.8t - 1.5$

Step 2 Substitute $t = 0$: $a = \frac{dv}{dt} = -0.3 \times 0 + 1.8 \times 0 - 1.5 = -1.5\text{ m s}^{-2}$

b) Step 1 Solve $\frac{dv}{dt} = 0$: $0 = -0.3t^2 + 1.8t - 1.5$ giving $t = 1$ or 5 s

Step 2 Substitute $t = 1$: $v = -0.1 \times 1^3 + 0.9 \times 1^2 - 1.5 \times 1 = -0.7\text{ m s}^{-1}$

Substitute $t = 5$: $v = -0.1 \times 5^3 + 0.9 \times 5^2 - 1.5 \times 5 = 2.5\text{ m s}^{-1}$

◀◀ See page 20 for **polynomials,** page 62 for **differentiation** and page 66 for **integration**.

Watch out! Use the graph to check that the value of v at the end points is not more extreme than these values.

★ YOU ARE THE EXAMINER

Which one of these solutions is correct? Where are the errors in the other solution?

The acceleration of a particle is given by $a = (0.3t^2 - 0.2)\ \text{m s}^{-2}$.

The initial velocity of the particle is $0.5\ \text{m s}^{-1}$.

Find an expression for the velocity at time t.

LILIA'S SOLUTION

$v = u + at$

$v = 0.5 + (0.3t^2 - 0.2)t$

$v = 0.3t^3 - 0.2 + 0.5 = 0.3t^3 + 0.3$

NASREEN'S SOLUTION

$v = \int a\,dt = \int (0.3t^2 - 0.2)\,dt$

$v = 0.1t^3 - 0.2t + c$

When $t = 0$, $v = 0.5$ so $c = 0.5$

$v = 0.1t^3 - 0.2t + 0.5$

✓ SKILL BUILDER

1 A toy is moving in a straight line and its velocity at time t seconds is $v\ \text{m s}^{-1}$, where $v = -4t^2 + t + 5$ for $-1 \leqslant t \leqslant 3$. When is the acceleration of the toy zero?

A $t = 0$ **B** $t = 0.125$ **C** $t = -0.125$ **D** $t = -1$ and $t = 1.25$

2 A particle of grit, G, is stuck to the top of a piece of machinery that is moving up and down a vertical y-axis. The height of G above the ground is y metres at time t seconds where $y = 10t - 2t^2 - 8$. Determine the direction of motion and speed of G when $t = 3$.

A Downwards, $2\ \text{m s}^{-1}$ **B** Downwards, $-2\ \text{m s}^{-1}$ **C** Upwards, $2\ \text{m s}^{-1}$ **D** Upwards, $4\ \text{m s}^{-1}$.

In questions 3 and 4, an insect is moving along an x-axis. At time t seconds, its velocity is $v\ \text{m s}^{-1}$, where $v = 30t - 3t^2 - 63$.

3 Calculate the displacement of the insect from its position when $t = 2$ to its position when $t = 4$.

A -2 m **B** 2 m **C** -150 m **D** -76 m

4 **a)** Calculate the times at which the insect is stationary.
Explain what is happening at these times.

Hint: Solve $v = 0$.

b) Find the displacement of the particle from $t = 2$ to $t = 3$.

c) Find the displacement of the particle from $t = 3$ to $t = 4$.

d) Calculate the total distance travelled by the insect from $t = 2$ to $t = 4$.

5 The displacement of a car at time t s is given by $s = 0.4t^3 - 0.7t^2 + 0.9$.

a) Find an expression for velocity at time t s.

b) Determine whether the acceleration is constant or not.

Hint: Find acceleration – if it is just a number, then the acceleration is constant.

6 An athlete accelerates from rest for 4 s along a straight road.
The acceleration is given by $a = 3t - 0.75t^2$.
The athlete then runs at constant speed until he has completed 100 m.

a) Find an expression for the velocity of the athlete for the first 4 seconds of the motion.

b) Find the velocity of the athlete at $t = 4$.

c) Find an expression for the displacement of the athlete from the start point for the first 4 seconds of the motion.

d) Find the distance the athlete has travelled in the first 4 s.

e) Find the total time he takes to run 100 m.

Hint: Find the distance remaining and use the constant speed from part b).

Understanding forces

THE LOWDOWN

1. **Weight** is a downwards force that the Earth exerts on an object. An object with mass m kg has a weight of mg N.
2. The contact forces when objects touch are:
   - the normal reaction (N or R at 90° to the surface)
   - friction (F along the surface to oppose motion).

   The objects in contact experience equal and opposite forces.
3. When objects are attached by a string there is a tension in the attachment. In a spring or rod there could be tension or compression.

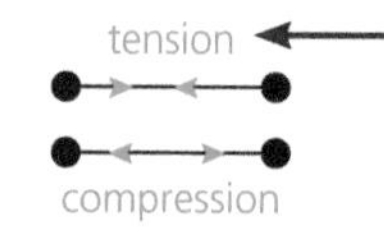

4. Other forces such as air resistance or driving forces may be mentioned in the question. When drawing a force diagram, draw all forces as arrows. Make sure the arrow starts on the right object and points in the correct direction. Label each force. Avoid extra made-up forces.
5. Look for modelling words in the question.
   - **particle** – ignore the size of the object
   - **smooth** – ignore friction
   - **rough** – friction must be included
   - **light** – ignore mass and weight
   - **inextensible** – will not stretch; it means the objects joined together move at the same speed as each other
   - **uniform** – the same all along; it usually means the **weight** goes in the middle of the object in your force diagram.

The acceleration of a falling object is $g = 9.8\,\text{m s}^{-2}$.

Watch out: The normal reaction is often not equal to the **weight**.

Newton's 3rd law.

Hint: The tension will pull both ends in equally.

NOTHING ELSE IS A FORCE! People talk about the force of nature, or centrifugal force or think mass times acceleration is a force – they are not forces in this sense!

▶▶ See page 118 for **calculations involving forces** and page 122 for **resolving forces**.

GET IT RIGHT

Draw a diagram of the forces acting on a tractor pulling a trailer uphill. The tractor experiences a resistance force as it travels.

Solution:

Step 1 Draw the objects first.

Step 2 Put in the **weight,** contact forces and the attachment forces.

Step 3 Add the driving force that tractor provides.

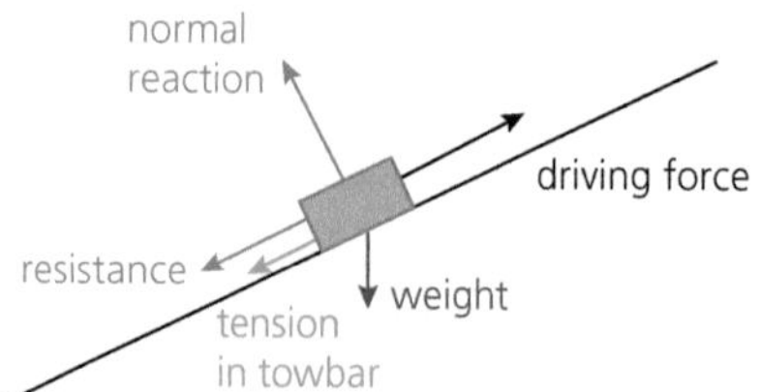

Watch out! Don't put the weight at 90° to the slope.

Watch out! When you draw the forces on the trailer, make sure the **weight,** the normal reaction and the resistance are labelled differently from the ones for the tractor (e.g. N_1 and N_2).

YOU ARE THE EXAMINER

Decide whether each solution is correct. Where have the others gone wrong?

Draw a diagram showing all the forces acting on a car, with mass m kg, in each situation.

LILIA'S SOLUTION

A car driving along a horizontal road.

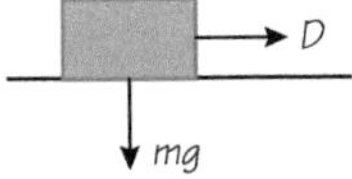

PETER'S SOLUTION

A car sliding down a smooth slope.

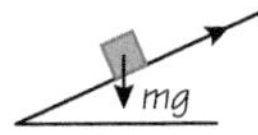

MO'S SOLUTION

A car stopped at traffic lights.

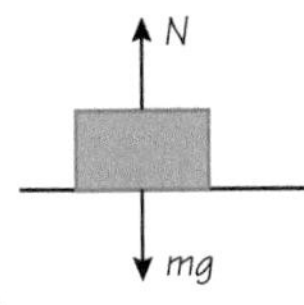

NASREEN'S SOLUTION

A car driving up a slope.

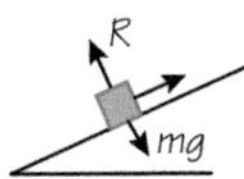

SKILL BUILDER

1 A box is at rest in a rough place. Which of these is the correct diagram showing the forces acting on the box?

A friction normal reaction weight

B

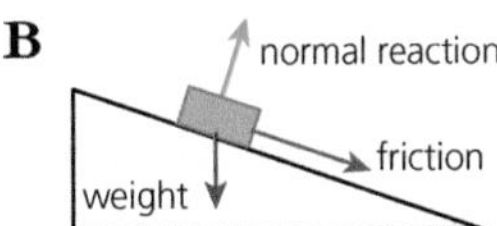

C

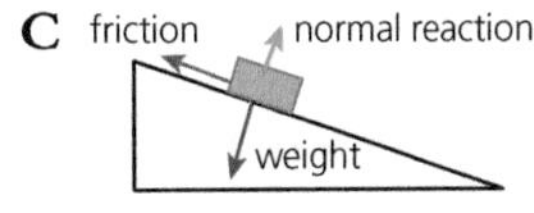

D

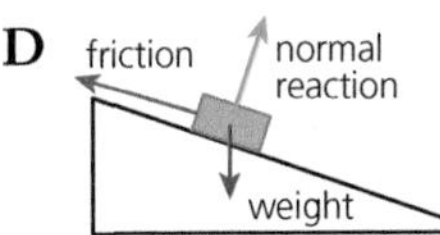

2 A block of mass 5 kg rests on a rough horizontal table. It is attached by a light horizontal string over a smooth pulley to a sphere of mass 1.8 kg. Which of these diagrams correctly shows the forces on the different parts of the system?

A

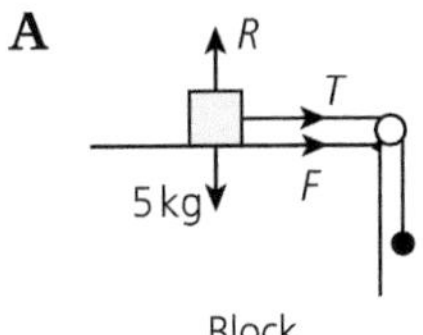

B

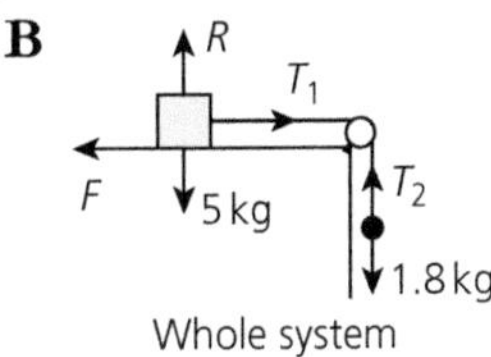

C

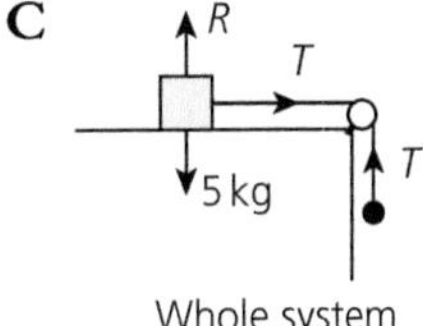

D

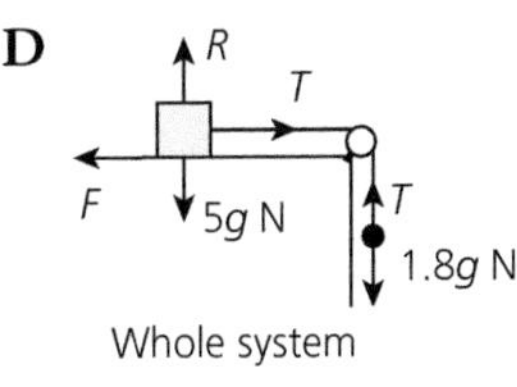

3 Draw force diagrams for the following objects.

a) A cup on a rough table, when a horizontal force P is applied (the cup does not move).

b) A sledge being pulled along horizontal ground by a string at 40° to the vertical.

c) A ruler balancing on two supports, one at each end.

d) A car pulling a trailer along a horizontal road.

4 Draw a force diagram for the trailer described in 'Get it right'.

5 Two bricks are stacked, one on top of the other, on a horizontal surface.

a) Draw separate diagrams showing the forces acting on each brick.

b) Ahmed says the weight of the top brick acts on the bottom brick. Is he right?

Forces in a line

Year 1

THE LOWDOWN

① An object is in **equilibrium** when the forces **balance**.
The object will have no acceleration.
It may be at rest or it could be travelling with constant velocity.

Example The contact force between a 75 kg person standing on the ground is equal to the weight of the person: $75g = 735$ N.

② Newton's 2nd law states the resultant force is equal to mass × acceleration.

Example The resultant force needed for a 3 kg object to accelerate at $5\,\text{m}\,\text{s}^{-2}$ is $3 \times 5 = 15$ N.

③ A unknown force can be calculated when the acceleration in known.

Example The contact force R between a 75 kg person in a lift and the lift floor changes depending on the acceleration of the lift.

Step 1 Find the **resultant force** in the direction of the acceleration.

Step 2 Put the resultant force equal to **mass** × acceleration and solve.

When a lift is accelerating upwards at $0.6\,\text{m}\,\text{s}^{-2}$

$R - 75g = 75 \times 0.6$ so $R = 780$ N

When the lift is slowing down with an acceleration of $-0.8\,\text{m}\,\text{s}^{-2}$

$R - 75g = 75 \times (-0.8)$ so $R = 675$ N

④ When two objects are **connected** they have the **same acceleration**.
You can think of them as a single object and work out the total mass.
You can look at each part separately and find the forces acting on each part.

Watch out! Take care writing $F = ma$ as you may think that F is a separate force or confuse it with the friction.

▶▶ **Force diagrams** on page 116.
▶▶ **Resolving forces** on page 122.

This is called the **equation of motion**.

Hint: Use $g = 9.8\,\text{m}\,\text{s}^{-2}$.

GET IT RIGHT

A car pulls a caravan along a level straight road.
The mass of the car is 1100 kg and mass of the caravan is 900 kg.
Resistances of 400 N and 600 N act on the car and caravan respectively.

a) Draw a diagram of the horizontal forces acting on the car and on the caravan.
b) Find the driving force, D, the car must provide when travelling at constant speed.
c) The driving force increases to 1500 N. Find the acceleration of the car and caravan.
d) Find the tension in the towbar when the car is accelerating.

Solution:

a)

600 N ← caravan (900 kg) →T T← car (1100 kg) → D
400 N ←

b) Constant speed so equilibrium for the combined object.

$D - 400 - 600 = 0$ so $D = 1000$ N

c) Use Newton's second law for the combined object.

Total mass $900 + 1100 = 2000$ kg

$1500 - 400 - 600 = 2000a$ giving .. so $a = 0.25\,\text{m}\,\text{s}^{-2}$

d) Looking just at the caravan

$T - 600 = 900 \times 0.25$ giving $T = 825$ N

(Check by looking at the car $1500 - 400 - T = 1100 \times 0.25$ ✓)

Watch out! The driving force acts only on the car (which changes the tension). Make sure each resistance acts on the correct object.

Hint: The tension pulls the car **backwards** and the caravan **forwards**.

Hint: Use the mass of the caravan. Only include the forces that act on the caravan – not D and not the 400 N.

YOU ARE THE EXAMINER

Which one of these solutions is correct?
Where are the errors in the other solutions?

A box of mass 2 kg is released from hold on a horizontal, smooth table.
It is attached to a sphere of mass 1.5 kg which hangs vertically.
The connecting string is light and inextensible and passes over a smooth pulley.

box
sphere

a) Write down the equation of motion of the box.

b) Find the acceleration of the system.

LILIA'S SOLUTION

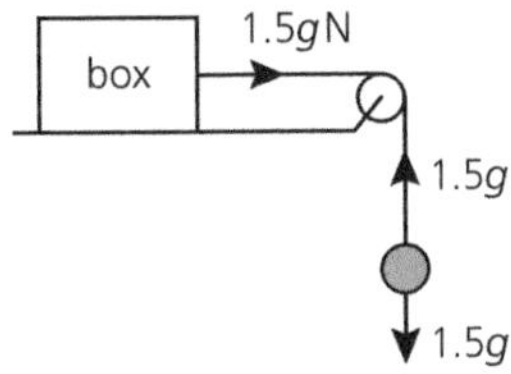

a) Tension is the same throughout the string $1.5g = 2a$

b) $a = 7.35\,\text{ms}^{-1}$

PETER'S SOLUTION

a) $2g - 1.5g = 2a$

b) $a = \frac{0.5g}{2} = 2.45\,\text{ms}^{-2}$

MO'S SOLUTION

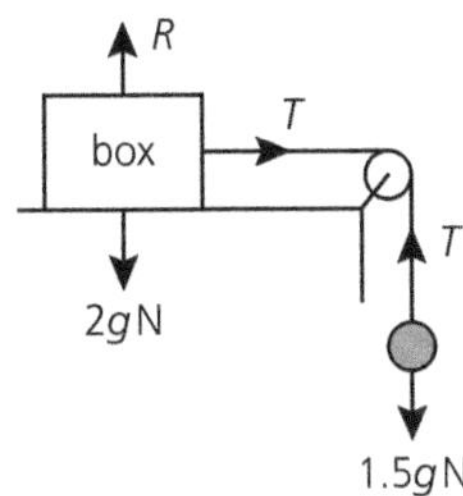

a) Tension is the same throughout the string $T = 2a$

b) For the sphere (down as positive) $1.5g - T = 1.5a$

Adding the two equations gives $1.5g = 3.5a$

so $a = \frac{3}{7}\,\text{ms}^{-2} = 4.2\,\text{ms}^{-2}$

SKILL BUILDER

1 Find the acceleration of the 800 kg car when the following forces act.
Take left to right as the positive direction.

a)

b)

c) 250 N 200 N

2 The mass of a model boat is 4.2 kg.
Find the driving force D needed when the acceleration is

a) $1.2\,\text{m s}^{-2}$ and there is no resistance to motion.

b) $2.5\,\text{m s}^{-2}$ against a resistance of 8 N

c) $-0.5\,\text{m s}^{-2}$ with a resistance of 4.5 N.

Use the following information for questions 3 and 4.

A car of mass 800 kg is pulling a trailer of mass 600 kg along a straight level road. There is a resistance of 90 N on the car and 200 N on the trailer and they are slowing down at a rate of $0.8\,\text{m s}^{-2}$.

3 Calculate the braking force required.

A 550 N **B** 830 N **C** 1120 N **D** 1410 N

4 By considering the motion of the trailer, calculate the force in the tow-bar between the car and trailer.

A Thrust 680 N **B** Thrust 280 N **C** Tension 200 N **D** Tension 280 N

5 Two boxes, A of mass 25 kg and B of mass 35 kg, are stacked onto a fork-lift truck with A on top of B. The truck applies an upward force of 750 N to the bottom of box B and both boxes are lifted. Find the magnitude of the contact force between the boxes correct to 3 significant figures.

A 536 N **B** 312 N **C** 361 N **D** 313 N

> **Hint:** Use the two boxes together to find the acceleration of the boxes. Draw a diagram for box A on its own and use your acceleration to find the contact force.

Using vectors

THE LOWDOWN

1. A vector is a quantity with size (magnitude) and direction.
   Displacement **r**, velocity **v**, acceleration **a** and force **F** are vector quantities.
   Time t and mass m are scalar (non-vector) quantities.
2. Vectors are added by putting the arrows nose-to-tail or adding the components.

Example The resultant of forces
$F_1 = 2i + j$ and $F_2 = 4i - 2j$
so $F_1 + F_2 = 6i - j$

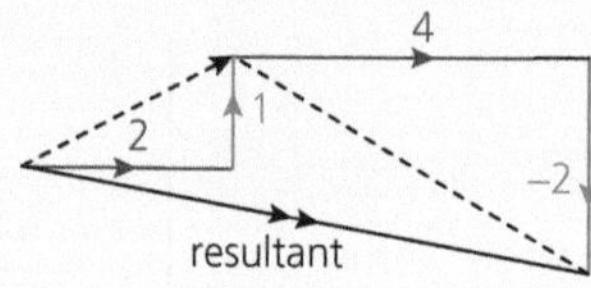

3. **Distance** is the magnitude of the **displacement** vector.
   Speed is the magnitude of the **velocity** vector.
   Pythagoras' theorem is used to find the **magnitude** of a vector.

Example The distance from the origin when displacement $r = (6i - j)$ m:
distance $= |r| = \sqrt{6^2 + (-1)^2} = \sqrt{37}$ m

Example The speed when $v = \begin{pmatrix} -5 \\ 2 \end{pmatrix}$ is $|v| = \sqrt{(-5)^2 + 2^2} = \sqrt{29}$ ms^{-1}

4. **Vector versions** of the ***suvat*** **equations** for constant acceleration **a**.
   - $\mathbf{v} = \mathbf{u} + \mathbf{a}t$
   - $\mathbf{r} = \frac{1}{2}(\mathbf{u} + \mathbf{v})t$
   - $\mathbf{r} = \mathbf{u}t + \frac{1}{2}\mathbf{a}t^2$

Example A particle travelling with velocity $u = 6i - 4j$ ms^{-1} accelerates for 5 s with an acceleration of $a = -i + 2j$ ms^{-2}

$v = u + at = (6i - 4j) + 5(-i + 2j) = 1i + 6j = i + 6j$ ms^{-1}

$r = ut + \frac{1}{2}at^2 = 5(6i - 4j) + \frac{1}{2} \times 5^2(-i + 2j)$

$= (30i - 20j) + (-12.5i + 25j) = 17.5i + 5j$ m

5. **Vectors** and **calculus** can be used together.
   The components are functions of t and can be differentiated and integrated.
6. A stationary object has all components of the velocity vector zero simultaneously.
7. **Newton's second law** can be used with vector notation: $\mathbf{F} = m\mathbf{a}$.
8. An object is in **equilibrium** when the total force is zero.

The vectors **i** and **j** are vectors of length 1 (unit vectors) in the x- and y-directions.

Watch out! If you use your calculator to get $(-1)^2$ make sure you use brackets.

◀◀ See page 112 for ***suvat*** **equations**, page 114 for **using calculus** and page 118 for **forces in a line**.
▶▶ See page 122 for **resolving forces**.

Watch out! You can divide by a scalar but not by a vector.

GET IT RIGHT

In this question **i** is horizontal and **j** is vertically upwards.
An object of mass 3 kg is acted upon by a force **F** N.
The velocity vector at time t is $\mathbf{v} = ((3 + 2t)\mathbf{i} + (\frac{1}{2}t^2)\mathbf{j})$ m s^{-1}.

a) Find an expression for the acceleration.
b) Find force **F** in terms of t.

Solution:

a) $a = \frac{dv}{dt} = \frac{d}{dt}\left((3 + 2t)i + (\frac{1}{2}t^2)j\right) = (2i + tj)$ ms^{-2}

b) Weight $= -3gj$ N

Resultant $F - 3gj = ma = 3(2i + tj)$ giving $F = 6i + (-3g + 3t)j$ N

Hint: The weight has no component in the **i** direction and is in the negative **j** direction.

YOU ARE THE EXAMINER

Which one of these solutions is correct? Where are the errors in the other solutions?

In this question **i** is horizontal and **j** is vertically upwards.

An object of mass 3 kg is acted upon by two external forces. $\mathbf{F}_1 = 2\mathbf{i} + 2\mathbf{j}$ N and $\mathbf{F}_2 = -3\mathbf{i} + 12\mathbf{j}$ N.

Find the magnitude of the acceleration.

LILIA'S SOLUTION

Newton's second law

$F_1 + F_2 = (2i + 2j) + (-3i + 12j) = 3a$

$a = \frac{1}{3}(-i + 14j) = (-\frac{1}{3}i + \frac{14}{3}j)\,\text{ms}^{-2}$

NASREEN'S SOLUTION

Magnitudes $|F_1| = \sqrt{2^2 + 2^2} = \sqrt{8} = 2.828$

$|F_2| = \sqrt{-3^2 + 12^2} = \sqrt{135} = 11.618$

$|a| = \frac{1}{3}(2.828 + 11.619) = 4.816\,\text{ms}^{-2}$

PETER'S SOLUTION

Newton's second law

$F_1 + F_2 + W = (2i + 2j) + (-3i + 12j) - 3gj$

$= 3a$

$a = \frac{1}{3}(-i - 15.4j) = (-0.333i + 5.133j)$

$|a| = \sqrt{0.333^2 + 5.133^2} = 5.14\,\text{ms}^{-2}$

MO'S SOLUTION

Newton's second law

$F_1 + F_2 + W = (2i + 2j) + (-3i + 12j) + 3gj = 3a$

$a = \frac{1}{3}(-i - 43.47j) = (-0.333i + 14.467j)$

$|a| = \sqrt{0.333^2 + 14.467^2} = 14.47\,\text{ms}^{-2}$

SKILL BUILDER

1 Use the vector *suvat* equations in this question.
The velocities are in m s^{-1} and acceleration m s^{-2}.

a) Given $\mathbf{u} = \mathbf{i} - \mathbf{j}$, $\mathbf{v} = -\mathbf{i} + 3\mathbf{j}$, and $t = 2$ s find **r** and **a**.

b) Given $\mathbf{u} = -\mathbf{i} + 3\mathbf{j}$, $\mathbf{v} = -\mathbf{i} + 7\mathbf{j}$ and $t = 4$ s, find **r** and **a**.

c) Given $\mathbf{u} = -\mathbf{i} - \mathbf{j}$, $\mathbf{r} = 5\mathbf{i}$ and $t = 5$ s, find **v** and **a**.

d) Given $\mathbf{u} = 3\mathbf{i} - 7\mathbf{j}$, $\mathbf{v} = \mathbf{i} - 3\mathbf{j}$ and $\mathbf{a} = -\mathbf{i} + 2\mathbf{j}$, find t and **r**.

2 A particle is at the origin and starts with a velocity of $(2\mathbf{i} - 4\mathbf{j})\,\text{m s}^{-1}$ and constant acceleration $3\mathbf{i} + 5\mathbf{j}\,\text{m s}^{-2}$ where **i** is the unit vector East and **j** is the unit vector North.

a) Find the velocity after 5 s.

b) Find the position at time 6 s.

3 Three forces act on a body $\mathbf{F}_1 = (3\mathbf{i} + 2\mathbf{j})$ N, $\mathbf{F}_2 = (6\mathbf{i} - \mathbf{j})$ N, $\mathbf{F}_3 = (-10\mathbf{i} + 2\mathbf{j})$ N.

a) Find the resultant force.

b) The three forces act on an object of mass 4 kg on a smooth horizontal surface. What is the acceleration of the object?

4 Two forces act on a body of mass 7 kg.
Together they produce an acceleration of $(4\mathbf{i} + 8\mathbf{j})\,\text{m s}^{-2}$.
One force is $\mathbf{F}_1 = (25\mathbf{i} + 60\mathbf{j})$ N. What is the other?

5 The velocity of a particle at time t s is given by $\mathbf{v} = ((3t - 3t^2)\mathbf{i} + (2 - 5t)\mathbf{j})\,\text{m s}^{-1}$.

a) Find an expression for the acceleration of the particle.

b) Given that the particle starts 2 m from the origin in the **i** direction, find an expression for the displacement.

6 A particle is in equilibrium under the action of three forces. $\mathbf{F}_1 = 12\mathbf{i} - 5\mathbf{j}$, $\mathbf{F}_2 = 3(\mathbf{i} + \mathbf{j})$ and $\mathbf{F}_3$.
Which of the following is the correct expression for $\mathbf{F}_3$?

A $\mathbf{F}_3 = 15\mathbf{i} - 2\mathbf{j}$ **B** $\mathbf{F}_3 = -15\mathbf{i} + 4\mathbf{j}$ **C** $\mathbf{F}_3 = -13$ **D** $\mathbf{F}_3 = -15\mathbf{i} + 2\mathbf{j}$

Resolving forces

THE LOWDOWN

① A **vector** can be resolved into **two component vectors**, usually at right angles.
You picture the vector being replaced by these two 'parts' added together.
Use trigonometry to find the components.

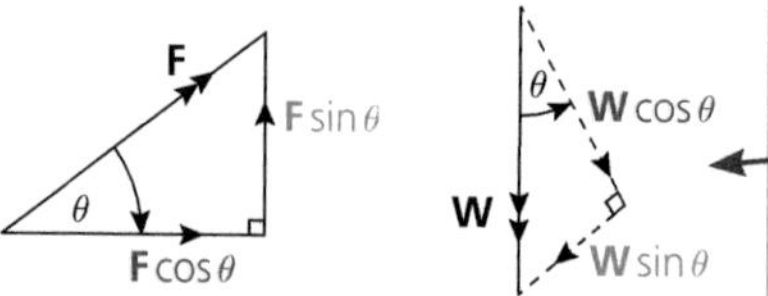

② A particle is in equilibrium when the resultant force in any direction is zero.

Example A lightbulb of weight 30 N hangs on a wire.
A horizontal force F holds the lightbulb so that the wire makes an angle of 20° to the vertical.
Resolve the tension in the wire.

Vertical: $T\cos 20° = 30$

So $T = \dfrac{30}{\cos 20°} = 31.9$ N

Horizontal: $F = T\sin 20°$

So $F = 31.9 \times \sin 20° = 10.9$ N

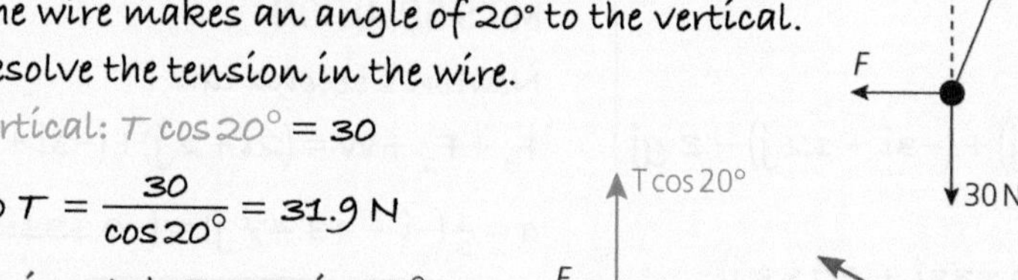

③ To use Newton's second law, resolve the forces in the direction of the acceleration. (Take care with signs – if the component of the force is 'helping' the acceleration it is positive, if it is 'hindering' it is negative.)

Resultant force = mass × acceleration in that direction.

④ For an object on a slope, resolve the weight

- **down the slope**
- towards the slope.

Example A 2 kg box rests on a smooth plane that is at 30° to the horizontal.

Resolve perpendicular to the slope:
(no motion in this direction)t
The normal reaction
$R = 2g\cos 30° = 16.97$ N

Resolve down the slope:
The acceleration of the box
$2g\sin 20° = ma = 2a$
So $a = 3.35\ \text{m s}^{-2}$

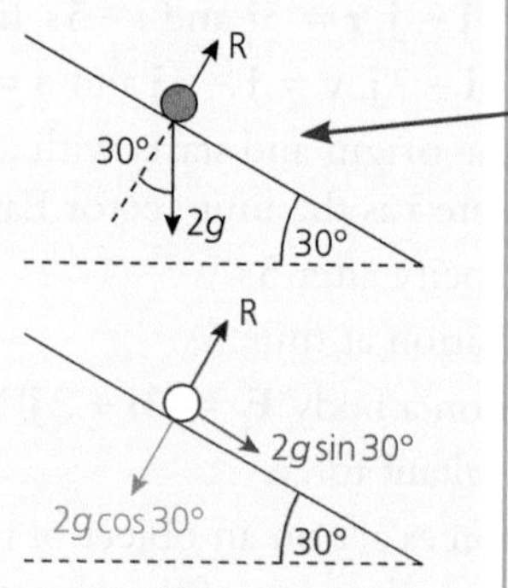

Hint: 'Trapdoor rule': Put your pencil along the vector and turn it **to the direction of the component**, (closing the trapdoor).
cos for close,
sin for open.
Remember:
$\cos(90° - \theta) = \sin\theta$
and cosine is negative when the angle is obtuse.

Watch out!
Do not add the components to your original diagram. Draw the second diagram like this if it helps you.

Hint: The angle the slope makes with horizontal is the angle the weight makes with the perpendicular to the slope.

Hint: Remember $g = 9.8\ \text{m s}^{-2}$.

GET IT RIGHT

An object of weight 100 N hangs from two strings.
Find the tensions T_A and T_B.

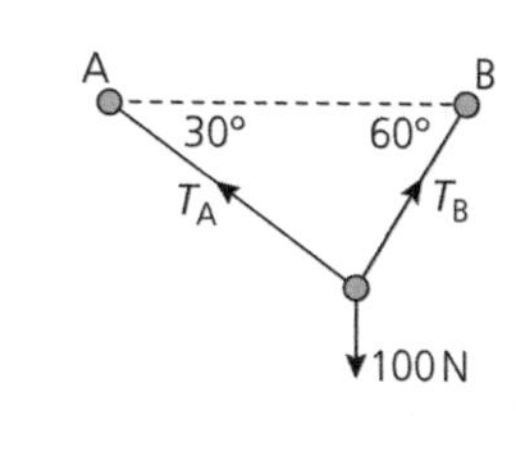

Solution:

Step 1 Resolve vertically: $T_A\sin 30° + T_B\sin 60° = 100$

Step 2 Resolve horizontally: $T_A\cos 30° - T_B\cos 60° = 0$

Step 3 Use calculator to solve simultaneous equations.
$T_A = 50$ N and $T_B = 86.6$ N

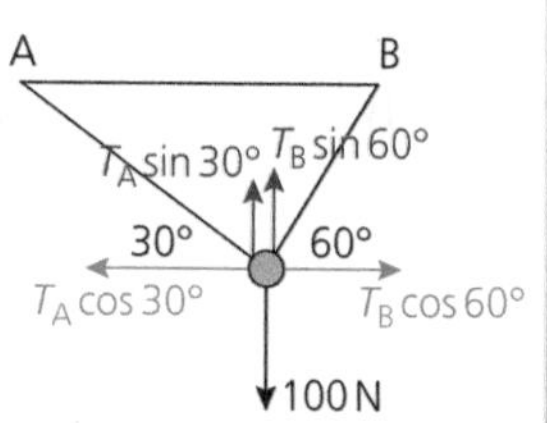

◀◀ **Forces in a line** on page 118.
▶▶ **Friction** on page 126.

Hint: Use total force = 0 here so that the equations are in the same form ready for the calculator.

YOU ARE THE EXAMINER

Which one of these solutions is correct? Where are the errors in the other solution?

Isak pulls a 40 kg sledge by a rope at 25° to the horizontal.

The tension in the rope is 80 N and there is a total resistance to motion of 50 N.

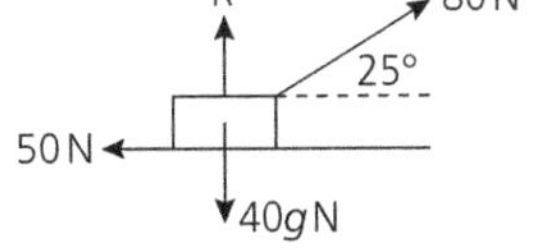

a) Find the normal reaction between the sledge and the ground.

b) Find the acceleration of the sledge.

LILIA'S SOLUTION

a) $R = 40g = 392$ N

b) $80 \sin 25° - 50 + 40g = 40a$

So $a = 9.40$ m s^{-2}

NASREEN'S SOLUTION

a) Resolve vertically

$R + 80 \sin 25° - 40g = 0$ so $R = 358$ N

b) Resolve horizontally

$80 \cos 25° - 50 = 40a$

So $a = 0.563$ m s^{-2}

SKILL BUILDER

1 Find in terms of **i** and **j** the resultant of the forces shown in the diagram.

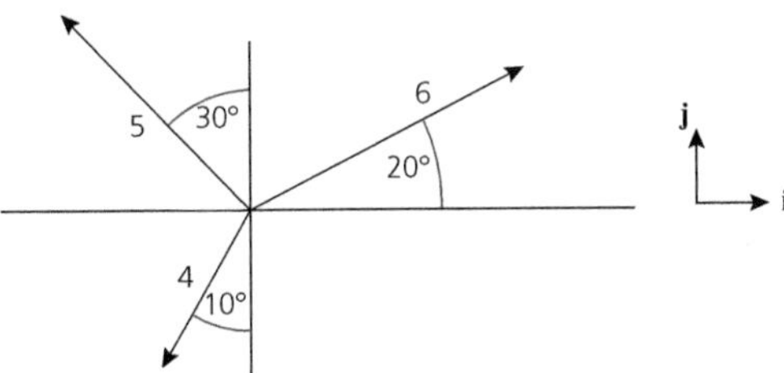

A $-1.14\mathbf{i} + 6.03\mathbf{j}$ **B** $2.44\mathbf{i} + 2.44\mathbf{j}$

C $-2.63\mathbf{i} - 3.86\mathbf{j}$ **D** $9.56\mathbf{i} + 9.61\mathbf{j}$

2 Find the magnitude of the resultant of the vectors shown in the diagram.

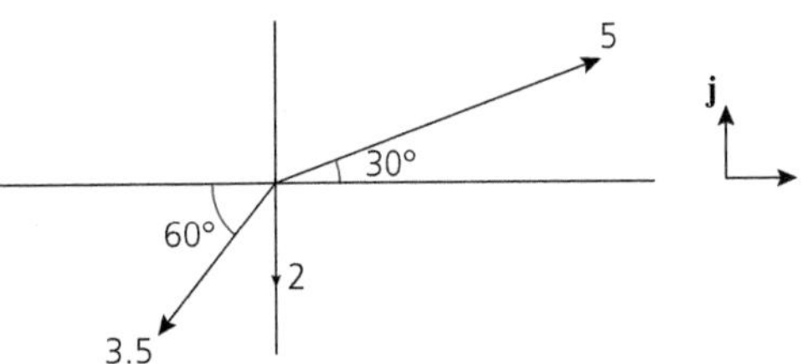

A 0.79 units **B** 3.61 units

C 2.97 units **D** 0.51 units

3 A 5 kg box is put on a smooth plane which is at 35° to the horizontal.

It is held in place by a horizontal force *F*.

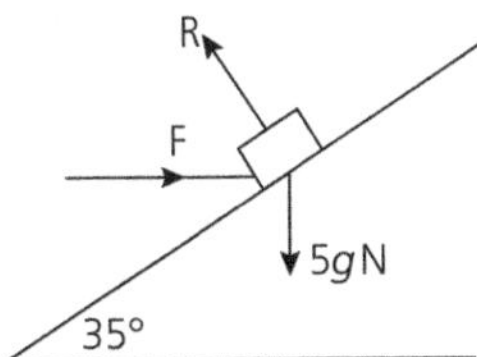

a) Resolve along the slope to find the force *F*.

b) Resolve at right angles to the slope to find the normal reaction *R*.

4 The diagram shows a block on a slope. Three of the following statements are false and one is true. Which one is true?

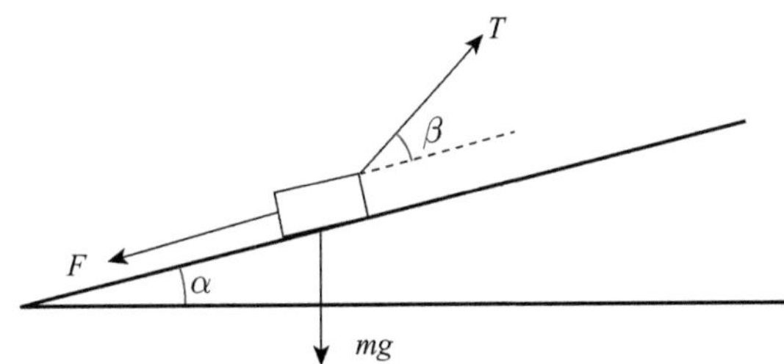

A The vertical component of T is $T \sin \beta$.

B The component of F parallel to the slope is $F \cos \alpha$.

C The component of mg parallel to the slope is $mg \sin \alpha$ down the slope.

D The component of mg perpendicular to the slope is $mg \sin \alpha$.

5 Two people are pushing a 600 kg car as shown. There is a resistance force of 150 N in the line of motion.

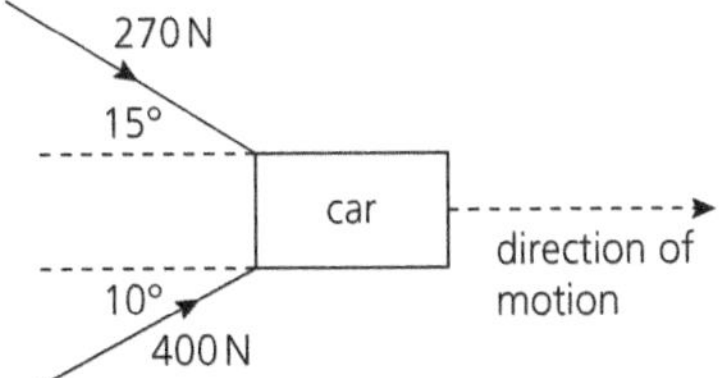

a) Find the acceleration of the car in the forward direction.

b) Show that there is only a very small resultant force perpendicular to the motion.

Projectiles

Year 2

THE LOWDOWN

① A projectile is an object that is moving under gravity.

② You look at the **horizontal** and vertical directions separately.

③ Resolve the initial velocity into **horizontal** and vertical components.

Example When a ball is thrown at $u\,\text{ms}^{-1}$ at $\alpha°$ above the horizontal:
horizontally $u_x = u\cos\alpha°$
vertically $u_y = u\sin\alpha°$

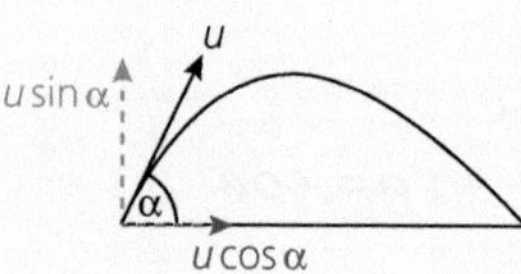

④ Set out all the u and a values in a table.

⑤ The acceleration **horizontally** is zero so u_x **is constant.**

The acceleration vertically upwards is $-g = -9.8\,\text{m s}^{-2}$.

⑥ Use the *suvat* equations in each direction. t links the two components.

$x =$	$y =$
$u_x = u\cos\alpha°$	$u_y = u\sin\alpha°$
$v_x = u\cos\alpha°$	$v_y =$
$a = 0$	$a = -g = -9.8$
$t =$	$t =$

Example A ball is kicked from ground level at $17\,\text{ms}^{-1}$ at 35° above horizontal. Find the height of the ball when hits a wall 20 m away.

Step 1 Complete the table

Step 2 Use x to find t

$x = u_x t + \frac{1}{2}at^2 = (17\cos 35°)\,t = 20$

giving $t = 1.4362$ s

Step 3 Use t to find y

$y = 17\sin 35° \times 1.436 - \frac{1}{2} \times 9.8 \times 1.436^2 = 3.90\,\text{m}$

$x = 20$	$y =$
$u_x = 17\cos 35°$	$u_y = 17\sin 35°$
$v_x = 17\cos 35°$	$v_y =$
$a = 0$	$a = -g = -9.8$
$t = ?$	$t =$

⑦ Key values:

At the **highest point** $v_y = 0$

At **ground level** $y = 0$

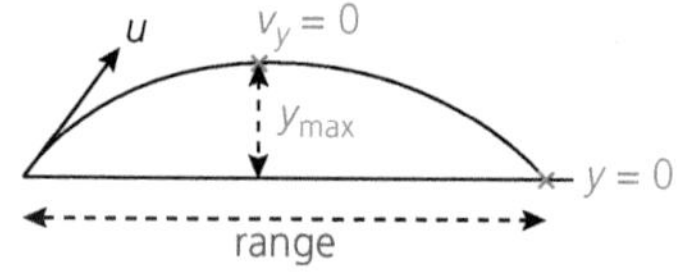

Imagine a person throwing a ball vertically in the air while travelling on a moving train. To the person outside the train, the ball moves like a projectile.

Hint: Use x and y for displacement in the two directions.

When the projectile is thrown **horizontally** $u_x = u \quad u_y = 0$

Some people use vectors for projectiles instead of separating the components.

◀◀ See page 112 for **vertical motion**.

When the projectile is thrown **horizontally** $u_x = u$ $u_y = 0$.

GET IT RIGHT

A projectile is thrown at $15\,\text{m s}^{-1}$ at 40° above horizontal.

a) Find the maximum height the projectile reaches.

b) Find the distance to the point where the projectile lands.

Solution:

a) **Step 1** Complete a table $v_y = 0$

Step 2 Vertical direction $v^2 = u^2 + 2as$

$0^2 = (15\sin 40°)^2 - 2 \times 9.8\,y$

Giving max height 4.74 m

$x = ?$	$y =$
$u_x = 15\cos 40°$	$u_y = 15\sin 40°$
$v_x =$	$v_y = 0$
$a = 0$	$a = -9.8$
$t =$	$t =$

b) **Step 1** Complete a new table $y = 0$.

Step 2 Vertical direction $s = ut + \frac{1}{2}at^2$

$y = 15\sin 40°t - \frac{1}{2}gt^2 = 0$

$t\left(15\sin 40° - \frac{1}{2}gt\right) = 0$

So $t = \dfrac{15\sin 40°}{\frac{1}{2}g} = 1.9677$ s

$x = ?$	$y = 0$
$u_x = 15\cos 40°$	$u_y = 15\sin 40°$
$v_x =$	$v_y =$
$a = 0$	$a = -9.8$
$t =$	$t =$

Step 3 Use t to find x. $x = 15\cos 40° \times 1.9677 = 22.6\,\text{m}$

Watch out! Make sure your calculator is set to degrees.

Hint: Draw a new table when the situation changes.

YOU ARE THE EXAMINER

Which one of these solutions is correct?
Where are the errors in the other solution?

George throws a stone with velocity of $20\,\text{m}\,\text{s}^{-1}$ at 30° above horizontal from the top of a wall 5 m above the ground.
Calculate the distance from the wall where the stone lands.

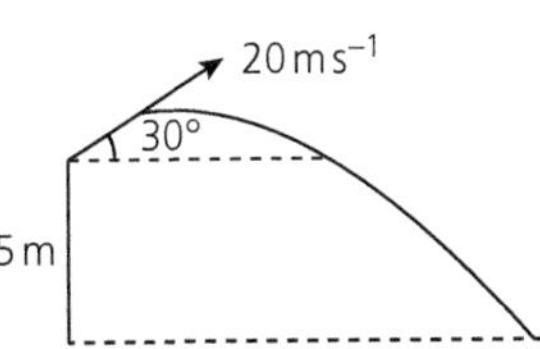

PETER'S SOLUTION

$x = ?$	$y = -5$
$u_x = 20\cos 30°$	$u_y = 20\sin 30°$
$v_x = 20\cos 30°$	$v_y =$
$a = 0$	$a = -9.8$
$t =$	$t =$

Find t when stone lands $y = -5$
$-5 = 20\sin 30°t - 4.9t^2$ so
$4.9t^2 - 20\sin 30°t - 5 = 0$
giving $t = 2.456, -0.415$
$x = 20\cos 30° \times 2.456 = 42.5\,\text{m}$

MO'S SOLUTION

$x = ?$	$y = 5$
$u_x = 30\cos 20°$	$u_y = 30\sin 20°$
$v_x = 30\cos 20°$	$v_y =$
$a = 0$	$a = -9.8$
$t =$	$t =$

Find t when stone lands $y = 5$
$5 = 30\sin 20°t - 4.9t^2$ so
$4.9t^2 - 30\sin 20°t + 5 = 0$
giving $t = 1.32, 0.772$ so stone lands twice
$x = 30\sin 20° \times 1.32 = 13.5\,\text{m}$

SKILL BUILDER

In these questions take the upward direction as positive and use $9.8\,\text{m}\,\text{s}^{-2}$ for g.
All the projectiles start at the origin.

1 For each of these situations, complete a table of values as in the lowdown.
 a) A ball is thrown horizontally with velocity $16\,\text{m}\,\text{s}^{-1}$. It travels 10 m horizontally.
 b) A stone is thrown with a velocity of $16\,\text{m}\,\text{s}^{-1}$ at 45° above horizontal. It reaches its highest point.
 c) A spark is thrown out of a fire with a velocity of $3\,\text{m}\,\text{s}^{-1}$ at 65° above the horizontal. It lands on the ground.
 d) An arrow is fired from a castle wall 28 m above the ground. It has an initial velocity of $35\,\text{m}\,\text{s}^{-1}$ at 25° **below** the horizontal. It lands on the ground.

> **Hint:** Keep upwards as the positive direction. u_y, g and y will all be negative.

2 For the situations described in question 1 complete the calculations.
 a) Find the time taken to travel 10 m horizontally and find the height at that time.
 b) Find the time to reach the highest point and the greatest height above the ground.
 c) Find the time to reach the ground and the horizontal distance it has travelled at that time.
 d) Find the time taken to travel 28 m downwards and the horizontal distance travelled.

3 A particle is projected with a velocity of $30\,\text{m}\,\text{s}^{-1}$ at an angle of 36.9° to the horizontal. What are the x- and y-coordinates of the position of the particle after time t seconds?

A $x = 24.0t;\ y = 18.0t + 4.9t^2$ **B** $x = 18.0t;\ y = 24.0t - 4.9t^2$
C $x = 24.0t;\ y = 18.0t - 4.9t^2$ **D** $x = 24.0;\ y = 18.0 - 9.8t$

4 A ball is kicked with a velocity of $14.7\,\text{m}\,\text{s}^{-1}$ at an angle of 30° above the horizontal. Find the time taken to reach its highest point and its maximum height.

A 1.50 s; 2.76 m **B** 0.75 s; 2.76 m **C** 0.75 s; 9.53 m **D** 1.50 s; 19.1 m

5 Find the horizontal range and time of flight of a particle that has been projected with a velocity of $21\,\text{m}\,\text{s}^{-1}$ at an angle of 60°.

A 16.9 m; 1.86 s **B** 39.0 m; 1.86 s **C** 0 m; 3.71 s **D** 39.0 m; 3.71 s

6 A ball is thrown with a velocity of $30\,\text{m}\,\text{s}^{-1}$ at 60° above the horizontal.
 a) Find an expression for the horizontal distance travelled at time t s.
 b) Find an expression for the vertical height of the ball at time t s.
 c) Make t the subject of your answer in part a).
 d) Substitute for t in your answer to b) to get the equation of the path of the projectile (trajectory).

Friction

▼ THE LOWDOWN

① When two surfaces are in contact, you can think of the force between them as a normal reaction (at 90° to the contact) and **friction** (along the surface).

② **Friction** acts to oppose likely motion.

Example   Sliding down the slope   Being pulled up the slope

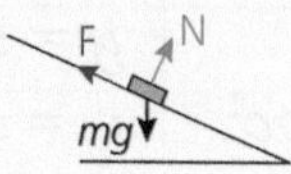

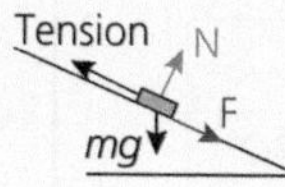

③ **Friction** has a maximum value which is a fraction of the normal reaction. $F \leqslant \mu N$ where μ is a measure of the roughness of the contact, and is called the **coefficient of friction**.

For most surfaces the fraction μ is less than 1 – very sticky surfaces such as rubber may be greater than 1.

Example   When a 3 kg object rests on a horizontal surface the normal reaction is equal to the weight.
When $\mu = 0.4$ the maximum friction is
$F_{max} = 0.4 \times 3g = 11.76\,N$

Watch out! This F is not the same as the F in $F = ma$.

④ When the surface is **smooth**, $\mu = 0$ and there is no friction force.

⑤ The normal reaction is not always equal to the weight of the object.

◀◀ See page 122 for **resolving forces** and page 118 for **motion in a line**.

Example   When an object is on a rough slope
normal reaction $N = mg\cos\alpha$
$F \leqslant \mu N$ giving $F \leqslant \mu mg\cos\alpha$

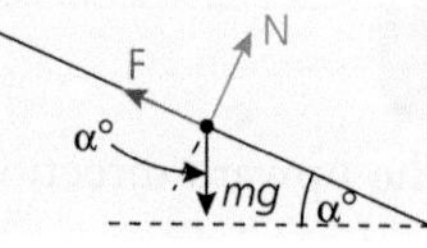

⑥ When an object is on the point of sliding:

- the object is about to slide
- the object is still in equilibrium
- the maximum friction is needed.

⑦ When an object is moving, the friction takes its **maximum value**.

GET IT RIGHT

Mary pulls a 1.2 kg pan across the cooker by its handle. She applies a force of 8 N, as shown in the diagram. The coefficient of friction is 0.6.

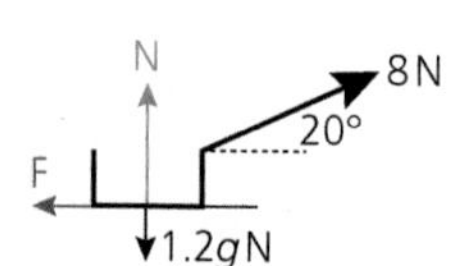

a) Find the maximum friction force.

b) Find the acceleration of the pan.

Solution:

Step 1 Find the normal reaction.
Resolve vertically: $8\sin 20° + N = 1.2g$
giving $N = 1.2 \times 9.8 - 8\sin 20° = 9.024\,N$

Watch out! This is a case where the normal reaction is not equal to the weight.

Step 2 Find the maximum friction
$F_{max} = \mu N = 0.6 \times 9.024 = 5.414\,N$

Step 3 Resolve in the direction of motion (horizontally)
Newton's 2nd law gives $8\cos 20° - 5.414 = ma = 1.2a$
So $a = \dfrac{2.104}{1.2} = 1.75\,ms^{-2}$ (to 3 s.f.)

★ YOU ARE THE EXAMINER

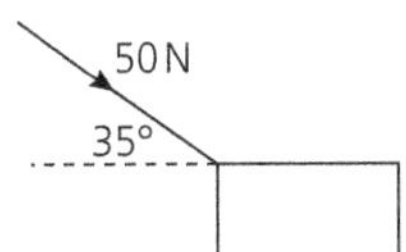

Which one of these solutions is correct? Where are the errors in the other solution?

An 8 kg box stands on horizontal ground.
Gerry pushes the box with a force of 50 N at an angle of 35° to the horizontal as shown.
The coefficient of friction between the box and the ground is 0.75.
Determine whether the box moves.

PETER'S SOLUTION

50 N
N
35°
F
8g N

If the box does not move
Resolve horizontally
$50\cos 35° - F = 0$
F would need to be 40.96 N
Resolve vertically
$N = 8g + 50\sin 35° = 107.08$
Max friction possible $0.75 \times 107.08 = 80.31$ N
Friction needed is less than the maximum so the box does not move.

MO'S SOLUTION

Normal reaction equals weight $= 8g = 78.4$ N
Friction $= 0.75 \times 78.4 = 58.8$ *left*
Resolve horizontally
Newton's second law
$50\cos 35° - 58.8 = -17.84 = ma = 8a$
So $a = -2.23\,\text{m s}^{-2}$
so the box accelerates to the left.

✓ SKILL BUILDER

1 A 5 kg box stands on a surface where the coefficient of friction is 0.4.
Find the normal reaction and the maximum possible value for friction in each case.

a)

N
5g N

b)

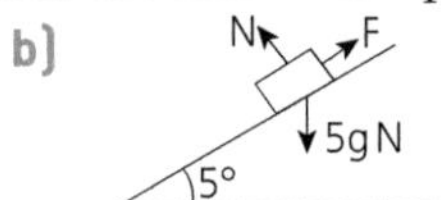

c)

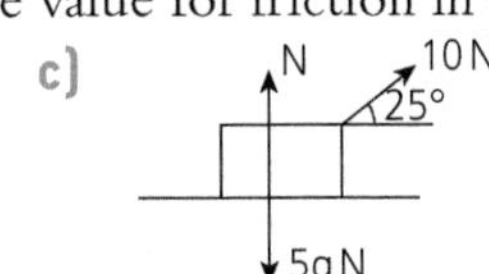

d)

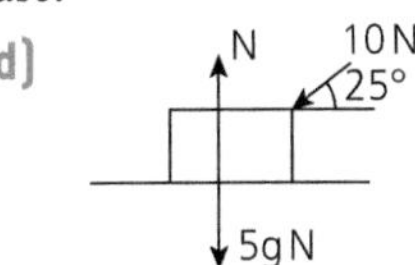

2 Hakim pushes a box of mass 12 kg across a rough horizontal floor. The coefficient of friction between the box and the floor is 0.5. He applies a horizontal force of 75 N. Calculate the acceleration of the box.

A $5.75\,\text{m s}^{-2}$ **B** $1.35\,\text{m s}^{-2}$ **C** $0.138\,\text{m s}^{-2}$ **D** $3.125\,\text{m s}^{-2}$

3 A 3 kg box stands on a rough inclined plane.
The coefficient of friction between the box and the plane is 0.3.

a) The plane makes an angle of 15^0 with the horizontal.
Show that the box does not move.

b) The angle is increased to 25^0 so that the box slides down the slope.
Find the acceleration of the box.

4 A block of mass 2 kg rests on a rough slope which makes an angle of α^0 with the horizontal. It is on the point of sliding down the slope. What is the coefficient of friction between the block and the slope?

A $\sin\alpha$ **B** $\cos\alpha$ **C** $\tan\alpha$ **D** $\dfrac{1}{\cos\alpha}$

5 A block of mass 5 kg is to be pulled along a rough horizontal floor by a rope which is inclined at 40^0 to the horizontal. The coefficient of friction between the block and the floor is 0.6. What is the tension in the rope when the block is on the point of moving?

A 25.5 N **B** 32.4 N **C** 38.4 N **D** 2.6 N

6 Terry pulls a sledge of mass 40 kg up a slope at 30^0 to the horizontal.
He applies a force of 325 N parallel with the slope.
The coefficient of friction between the slope and the sledge is 0.2.

a) Resolve perpendicular to the plane to find the normal reaction.

b) Find the magnitude of the friction force and the direction in which it acts.

c) Resolve along the plane to find the acceleration of the sledge.

Moments

THE LOWDOWN

① The **moment** of a force about a point is its turning effect. O d F
The bigger the force, the bigger the turning effect.
The bigger the distance from the force to the point, the bigger the turning effect.
Moment about O = Fd

Example — The turning effect of a 30 N force at right angles to a door, acting 50 cm from the hinge, has a moment of $30 \times 0.5 = 15\,\text{N m}$.

② The unit for a moment is newton-metre (N m).

③ **Anticlockwise** is usually taken to be the **positive direction**.

④ Moments can be added when several forces act on a body.

⑤ A **rigid body** is an object for which the size and shape are important and fixed.

⑥ The weight appears to act at the **centre of mass** of an object.
When a body is uniform, the weight acts at the centre.

⑦ A rigid body is in **equilibrium** when

- there is no resultant force acting on it in any direction
- there is no resultant moment about any point.

You could write forces up = forces down, forces right = forces left.

Example — When a plank is supported by two supports

R_A 0.7 m 0.5 m R_B 0.2 m 5g N

Use ACM = CM and take moments

About A: $1.2 \times R_B = 0.7 \times 5g$ so $R_B = 28.6\,\text{N}$

About B: $0.5 \times 5g = 1.2 \times R_B$ so $R_A = 20.4\,\text{N}$

Check: Vertical forces in equilibrium so $R_A + R_B = 28.6 + 20.4 = 49 = 5g$

⑧ (Edexcel and OCR only)
Where the given distance is at an angle θ to the force
moment = $Fd\sin\theta$

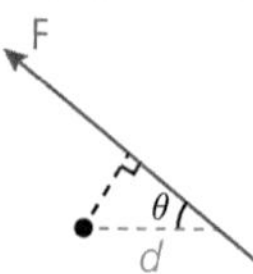

Hint: The distance must be perpendicular to the line of action of the force.

Hint: To help see the direction, imagine that is the only force acting on the body – which way would it turn?

So far, you have only considered particle models – where the size and shape and ability to rotate have been ignored.

You could write anticlockwise moment ACM = clockwise moment CM.

Hint: Take moments about a point where an unknown force acts to make the algebra simpler.

GET IT RIGHT

Three forces act at the corners of an object as shown in the diagram. A force F is applied to the object at right angles to the edge AB.
Find F and the distance from A at which it must be applied.

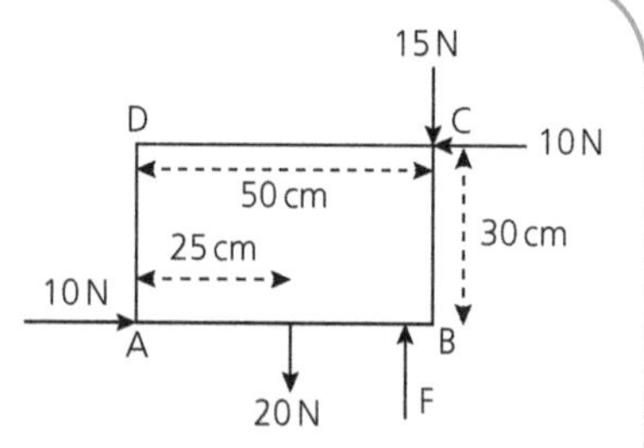

Solution:

Vertical forces $F = 15 + 20 = 35\,\text{N}$ upwards

Clockwise moments about A

$20 \times 0.25 + 15 \times 0.50 = 12.50$

Anticlockwise moments about A

$10 \times 0.3 + Fx = 0.3 + Fx$

Moments are equal so $0.3 + Fx = 12.50$ giving $x = \dfrac{12.50 - 0.3}{35} = 0.271\,\text{m}$

Hint: The 10 N force acting at A has no moment about A.

Watch out! You can use cm for distances as long as **all** the lengths are in cm.

YOU ARE THE EXAMINER

Which one of these solutions is correct? Where are the errors in the other solution?

Anastasia, Billy and their mother want to balance on a see-saw.

Their masses are 40 kg, 50 kg and 75 kg respectively.

Anastasia sits 1.4 m from the centre, Billy 0.9 m from the centre on the same side.

Where should her mother sit so that they balance?

LILIA'S SOLUTION

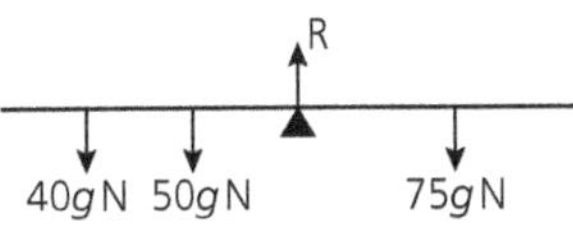

Mum sits d m from the centre

Take moments about the centre

$40g \times 1.4 + 50g \times 0.9 = 75gd$

So $d = \dfrac{101g}{75g} = 1.35$ m from the centre

NASREEN'S SOLUTION

Mum sits d m from the centre

Take moments about the centre

$40g \times 1.4 - 50g \times 0.9 = 75gd$

So $d = \dfrac{11g}{75g} = 0.147$ m

SKILL BUILDER

1 A swing is made by attaching light strings to the ends of a uniform plank of wood AB of mass 1.5 kg and length 50 cm. A girl of mass 30 kg sits in equilibrium on the swing so that her centre of mass is 20 cm from A. Find the tensions in the strings in newtons.

A $T_A = 176.4$, $T_B = 117.6$ **B** $T_A = 154.35$, $T_B = 154.35$

C $T_A = 185.22$, $T_B = 123.48$ **D** $T_A = 183.75$, $T_B = 124.95$

2 Two 50 g particles balance on a 35 g 30 cm ruler. The ruler balances on two pivots so that it is horizontal. One pivot is at the end of the ruler. Where should the other be placed so that the contact forces on the two pivots are equal?

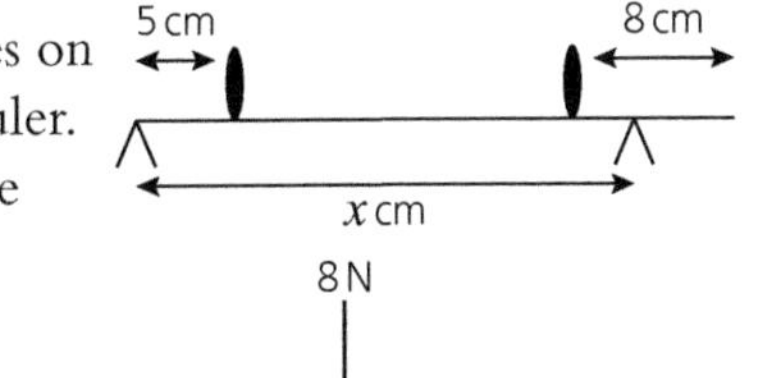

3 The diagram shows three forces acting on an object.
Which of the following is the total moment about the point A, taking anticlockwise as positive?

A −7 N m **B** −1.5 N m

C −1.75 N m **D** −11 N

4 The diagram shows four forces acting on an object.
Which of the following is the total moment about the point P, taking anticlockwise as positive?

A 8 N m **B** −5 N m

C 0.5 N m **D** 4 N m

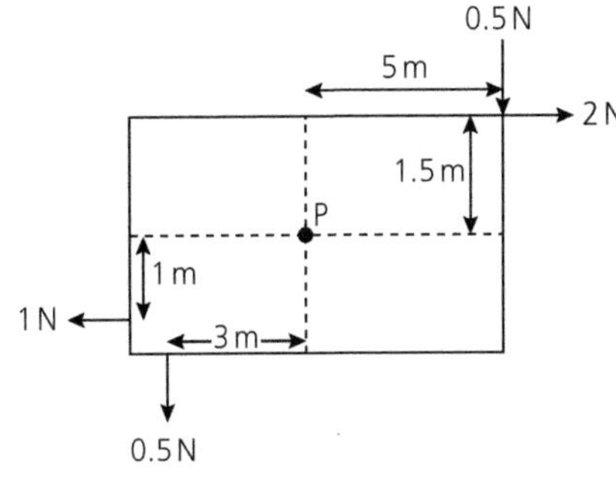

5 A uniform rectangular sign ABCD of mass 3.5 kg is attached to a wall at A.
A horizontal force at D keeps it in place with AB horizontal.

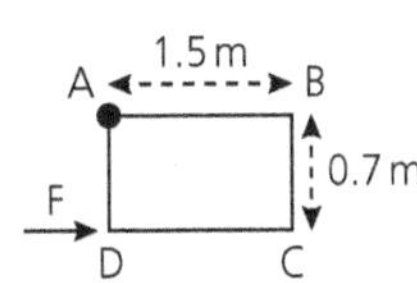

a) Take moments about A to find the size of the horizontal force.

b) Write down the horizontal and vertical components of the attachment force at A.

c) Find the magnitude of the attachment force at A.

Answers

Algebra review 1

You are the examiner

1 Lilia is right; Sam forgot to change the signs when he expanded the 2nd pair of brackets.

2 Sam is right; Lilia has multiplied by 3 twice.

3 Lilia is right; Sam has only 2 terms inside his brackets instead of 3 and his 2nd term is wrong.

4 Sam is right; Lilia has made sign errors when she expanded.

5 Sam is right; Lilia didn't multiply both sides of the equation by 6.

Skill builder

1 a) $5x(4xy - 3) + 3y(2x - y)$
$= 20x^2y - 15x + 6xy - 3y^2$

b) $2x(3x - 3) - (2 - 5x)$
$= 6x^2 - 6x - 2 + 5x = 6x^2 - x - 2$

c) $3x(1 - 2x) - 3x(4x - 2)$
$= 3x - 6x^2 - 12x^2 + 6x = 9x - 18x^2$

2 a) $(x + 5)(x - 4) = x^2 + 5x - 4x - 20$
$= x^2 + x - 20$

b) $(2x + 3)(3x - 5)$
$= 6x^2 - 10x + 9x - 15$
$= 6x^2 - x - 15$

c) $(4x - 2y)(2y - 3x)$
$= 8xy - 12x^2 - 4y^2 + 6xy$
$= 14xy - 12x^2 - 4y^2$

3 a) $12a^2 + 8a = 4a(3a + 2)$

b) $8b^2c - 4b = 4b(2bc - 1)$

c) $3bc - 2b + 6c^2 - 4c$
$= b(3c - 2) + 2c(3c - 2)$
$= (b + 2c)(3c - 2)$

4 a) $5(2x - 4) = 13 \Rightarrow 10x - 20 = 13$
$\Rightarrow x = 3.3$

b) $4(x + 3) = 5(3 - x) \Rightarrow$
$4x + 12 = 15 - 5x \Rightarrow 9x = 3 \Rightarrow x = \frac{1}{3}$

c) $3(2x + 1) - 2(1 - x) = 2(9x + 4)$
$\Rightarrow 6x + 3 - 2 + 2x = 18x + 8$
$\Rightarrow -7 = 10x \Rightarrow x = -0.7$

5 a) $\frac{x}{4} = \frac{2x - 1}{3} \Rightarrow 3x = 8x - 4$
$\Rightarrow 5x = 4 \Rightarrow x = 0.8$

b) $\frac{1}{4x + 3} = \frac{2}{5 - x} \Rightarrow 5 - x = 8x + 6$
$\Rightarrow -1 = 9x \Rightarrow x = -\frac{1}{9}$

c) $\frac{x}{5} - 6 = \frac{2x + 3}{3} \Rightarrow 3x - 90 = 10x + 1$!
$\Rightarrow 7x = -105 \Rightarrow x = -15$

Algebra review 2

You are the examiner

Mo: $\frac{1}{2}at^2 = s - ut$; 2; $at^2 = 2(s - ut)$; t^2;

$a = \frac{2(s - ut)}{t^2}$

Lilia: $6x + 8y = -2$; subtract;

$6x + 8y = -2$
$-\ \ 6x - 5y = -8.5$; $y = \frac{1}{2}$;
$13y = 6.5$

$3x + 4 \times \frac{1}{2} = -1$; $x = -1$

Skill builder

1 C

2 D

3 a) $2(2b - 3) + b = 14 \Rightarrow$
$5b - 6 = 14 \Rightarrow b = 4, a = 5$

b) Subtracting gives:
$-2c = 4 \Rightarrow c = -2,\ d = 6$

c) $10e - 6f = 22$ and $12e + 6f = 54$
$\Rightarrow 22e = 76 \Rightarrow e = \frac{38}{11}, f = \frac{23}{11}$

4 a) False: $y = x + \frac{5}{2} \Rightarrow 2y = 2x + 5$

b) True

c) False: $\frac{2}{3}(x - 4) = 6 \Rightarrow 2(x - 4) = 18$

5 $by = c - ax \Rightarrow y = \frac{c - ax}{b}$

6 $(x + by)y = ax + 1 \Rightarrow$
$xy + by^2 = ax + 1 \Rightarrow xy - ax = 1 - by^2$

$\Rightarrow x(y - a) = 1 - by^2 \Rightarrow x = \frac{1 - by^2}{y - a}$

7 $\sqrt{a^2 - x^2} = \frac{5}{y} \Rightarrow a^2 - x^2 = \frac{25}{y^2} \Rightarrow$

$x^2 = a^2 - \frac{25}{y^2} \Rightarrow x = \pm\sqrt{a^2 - \frac{25}{y^2}}$

Indices and surds

You are the examiner

1 Nasreen is right, as $(-3)^3 = -27$.

2 Sam is right: Nasreen didn't cube the 2 in the denominator and the power of a should be -5.

3 Nasreen is right: $\sqrt{147} - \sqrt{27} \neq \sqrt{120}$.

4 Sam is right: $64^{-\frac{3}{2}} \neq 64^{\frac{2}{3}}$.

Skill builder

1 C

2 D

3 B

4 a) False, $-4^2 = -16$

b) False, square root symbol means the positive square root $\sqrt{9} = 3$

c) False, can't add unlike bases

d) True

e) False, can't simplify further

f) True

5 a) $\frac{4 \times \sqrt{2}}{\sqrt{2} \times \sqrt{2}} = \frac{4 \times \sqrt{2}}{2} = 2\sqrt{2}$

b) $\frac{4(3 - \sqrt{5})}{(3 + \sqrt{5})(3 - \sqrt{5})} = \frac{4(3 - \sqrt{5})}{3^2 - 5}$
$= 3 - \sqrt{5}$

c) $\frac{2(6 + 3\sqrt{2})}{(6 - 3\sqrt{2})(6 + 3\sqrt{2})} = \frac{2(6 + 3\sqrt{2})}{6^2 - (3\sqrt{2})^2}$

$= \frac{6(2 + \sqrt{2})}{36 - 9 \times 2} = \frac{2 + \sqrt{2}}{3}$

6 a) $x^2 - 49 = 15 \Rightarrow x^2 = 64 \Rightarrow x = \pm 8$

b) $5x^3 - 27 = 13 \Rightarrow x^3 = 8 \Rightarrow x = 2$

c) $10x^2 - 40x = 4 - 40x \Rightarrow$
$10x^2 = 4 \Rightarrow x = \pm\frac{2}{\sqrt{10}} = \pm\frac{\sqrt{10}}{5}$

Quadratic equations

You are the examiner

Peter is right. Lilia lost the solution $x = 0$ when she divided by x^2 and she forgot the negative square root.

Skill builder

1 E

2 A

3 E

4 a) $(x + 3)(x + 4)$

b) $(x + 3)(x - 5)$

c) Perfect square: $(x + 3)^2$

d) Perfect square: $(x - 6)^2$

e) Difference of 2 squares:
$(x + 7)(x - 7)$

f) Difference of 2 squares:
$(2x + 10)(2x - 10)$
$= 4(x - 5)(x + 5)$

g) $2x^2 + 6x + x + 3 = (2x + 1)(x + 3)$

h) $6x^2 + 10x - 9x - 15$
$= (3x + 5)(2x - 3)$

i) Perfect square: $(3x - 2)^2$

5 a) $(x + 5)(x - 2) = 0; x = -5$ or $x = 2$

b) $(2x + 1)(x - 1) = 0; x = -\frac{1}{2}$ or $x = 1$

c) $(2x - 3)(x - 4) = 0; x = \frac{3}{2}$ or $x = 4$

d) $x(4x - 5) = 0; x = 0$ or $x = \frac{5}{4}$

e) $(2x - 5)^2 = 0;\ x = \frac{5}{2}$

f) $(3x - 5)(3x + 5) = 0;\ x = \pm\frac{5}{3}$

6 a) i) $(x - 2)(x - 6) = 0; x = 2$ or $x = 6$

ii) $y^2 = 2$ or $y^2 = 6$;
$y = \pm\sqrt{2}$ or $y = \pm\sqrt{6}$

iii) $\sqrt{z} = 2$ or $\sqrt{z} = 6$;
$z = 4$ or $z = 36$

b) i) $(x - 9)(x - 1) = 0; x = 1$ or $x = 9$

ii) $3^y = 1$ or $3^y = 9$; $y = 0$ or $y = 2$

iii) $\sqrt{z} = 1$ or $\sqrt{z} = 9$;
$z = 1$ or $z = 81$

c) i) $(x + 1)(4x - 1) = 0; x = -1$ or $x = \frac{1}{4}$

ii) $y^2 = -1$ ✗ or $y^2 = \frac{1}{4}$; $y = \pm\frac{1}{2}$

iii) $\sqrt{z} = -1$ ✗ or $\sqrt{z} = \frac{1}{4}$; $z = \frac{1}{16}$

7 a) Daisy can't divide by 5 as this changes the function.

b) Yes; the graphs of f(x) and g(x) are different but they have the same roots.

Inequalities

You are the examiner

Sam is right, Lilia didn't rearrange the quadratic to 0 before solving.

Skill builder

1 A

2 D

3 D

4 A

5 E

6 $\{x : -5 \leqslant x \leqslant 3\}$,
$\{x : x < 1\} \cup \{x : x > 5\}$

7 a) $2 + 15t - 5t^2 = 12 \Rightarrow 5t^2 - 15t + 10$
$= 0 \Rightarrow 5(t-1)(t-2) = 0$
$\Rightarrow t = 1$ s or $t = 2$ s

b) i) $1 < t < 2$

ii) $0 \leqslant t < 1$ or $t > 2$

Completing the square

You are the examiner

Mo:

a) $y = 5(x^2 + \boxed{2}x) + 8$
$= 5(x^2 + \boxed{2}x + \boxed{1} - \boxed{1}) + 8$
$= 5\left[\left(x + \boxed{1}\right)^2 - \boxed{1}\right] + 8$
$= 5\left(x + \boxed{1}\right)^2 + \boxed{3}$

b) $\boxed{0}$ real solutions, $b^2 - 4ac \boxed{<} 0$

Lilia:

a) $3(x-4)^2 = \boxed{15}$
$(x-4)^2 = \boxed{5}$
$x - 4 = \pm\sqrt{\boxed{5}}$
$x = \boxed{4} \pm \sqrt{\boxed{5}}$

b) $(\boxed{4}, \boxed{-15})$; $x = \boxed{4}$

Skill builder

1 B

2 E

3 B

4 E

5 $(x+5)^2 = 3 \Rightarrow x = -5 \pm \sqrt{3}$

6 $b^2 - 4ac > 0 \Rightarrow 7^2 - 4 \times 3k > 0$
$\Rightarrow 49 > 12k \Rightarrow k < \dfrac{49}{12}$

7 $b^2 - 4ac < 0 \Rightarrow k^2 - 4 \times 5 \times 5 < 0$
$\Rightarrow k^2 < 100 \Rightarrow -10 < k < 10$

Algebraic fractions

You are the examiner

Lilia is right.

Sam: $\dfrac{x+1}{\dfrac{x^2-1}{x}} = \dfrac{x(x+1)}{(x^2-1)}$

Skill builder

1 D

2 E

3 a) False d) True

b) False e) False

c) True f) True

4 a) $\dfrac{x+3}{2x^2} - \dfrac{5 \times 2x}{2x^2} = \dfrac{x + 3 - 10x}{2x^2}$
$= \dfrac{3 - 9x}{2x^2} = \dfrac{3(1-3x)}{2x^2}$

b) $\dfrac{1}{x} + \dfrac{1}{y} = \dfrac{y}{xy} + \dfrac{x}{xy} = \dfrac{x+y}{xy}$

c) $\dfrac{x^2}{2x} - \dfrac{2(x-3)}{2x} = \dfrac{x^2 - 2(x-3)}{2x}$
$= \dfrac{x^2 - 2x + 6}{2x}$

5 a) $\dfrac{x^2 + y}{2x} \times \dfrac{y^2}{x}$
$= \dfrac{y^2(x^2+y)}{2x^2}$

b) $\dfrac{(x+5)(x-5)}{2x} \times \dfrac{x^2}{(x+5)(x+1)}$
$= \dfrac{x(x-5)}{2(x+1)}$

c) $\dfrac{x^2 \times x}{\left(3 + \dfrac{2}{x}\right) \times x} = \dfrac{x^3}{3x+2}$

Proof

You are the examiner

Sam: D, E, B, C, A

Lilia: $2 = \dfrac{a^2}{b^2}$; $2b^2 = a^2$, a^2 and a are even.

$2b^2 = 4n^2$, $b^2 = 2n^2$, b^2 and b are even.

But if a and b are both even then $\frac{a}{b}$ can't be in its lowest terms.

This contradicts my original assumption so $\sqrt{2}$ must be irrational.

Skill builder

1 D

2 D

3 E

Functions

You are the examiner

a) Sam is right, exclude $x = -\frac{3}{2}$ as you can't divide by 0.

b) Lilia is right. Sam has found the reciprocal, not the inverse.

Skill builder

1 C

2 D

3 C

4 C

5 C

6 a) $x \xrightarrow{\text{q}} 2x - 1 \xrightarrow{\text{p}} \dfrac{1}{2x-1}$ so
$\text{pq}(x) = \dfrac{1}{2x-1}$

b) $\dfrac{1}{2x-1} = x \Rightarrow 1 = 2x^2 - x$
$\Rightarrow 2x^2 - x - 1 = 0$
$(2x+1)(x-1) = 0 \Rightarrow$
$x = -\frac{1}{2}$ or $x = 1$

Polynomials

You are the examiner

a) Sam is right. Lilia hasn't shown her working.

b) Sam is right. Lilia has given roots, not factors.

Skill builder

1 C

2 D

3 B

4 B

5 C

6 a) $(2x-1)$ is a factor $\Rightarrow \text{f}\left(\frac{1}{2}\right) = 0$
$\Rightarrow 2 \times \left(\frac{1}{2}\right)^3 - 3 \times \left(\frac{1}{2}\right)^2 + k \times \left(\frac{1}{2}\right) + 6 = 0$
$\frac{2}{8} - \frac{3}{4} + \frac{1}{2}k + 6 = 0 \Rightarrow \frac{1}{2}k = -5.5$
$\Rightarrow k = -11$

b) $2x^3 - 3x^2 - 11x + 6$
$= (2x-1) \times$ (a quadratic)

×	x^2		-6
$2x$	$2x^3$	$-2x^2$	$-12x$
-1	$-x^2$		$+6$

×	x^2	$-x$	-6
$2x$	$2x^3$	$-2x^2$	$-12x$
-1	$-x^2$	$+x$	$+6$

$2x^3 - 3x^2 - 11x + 6$
$= (2x-1)(x^2 - x - 6)$
$= (2x-1)(x-3)(x+2)$

c) When $x = 0$, $y = 6$
When $y = 0$, $x = \frac{1}{2}$, $x = 3$, $x = -2$

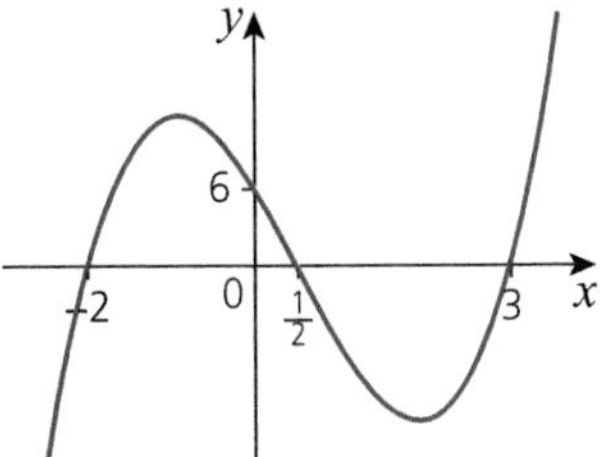

Exponentials and logs

You are the examiner

Lilia is wrong, $\dfrac{\ln\left(\frac{1}{2}\right)}{\ln e}$ is not the same as $\ln\frac{1}{2} - \ln e$ since $\ln x - \ln y = \ln\frac{x}{y}$, not $\frac{\ln x}{\ln y}$.

Skill builder

1 B

2 C

3 B

4 A

5 B

6 E

7 a) $\log_3 \dfrac{x}{(2x-1)} = \log_3 4 \Rightarrow \dfrac{x}{(2x-1)} = 4$
$\Rightarrow x = 8x - 4 \Rightarrow x = \dfrac{4}{7}$

b) $\ln x + 2\ln x = 6 \Rightarrow 3\ln x = 6$
$\Rightarrow \ln x = 2 \Rightarrow x = e^2$

c) $3\log_{10} x - \dfrac{1}{2}\log_{10} x = 5$
$\Rightarrow \dfrac{5}{2}\log_{10} x = 5 \Rightarrow \log_{10} x = 2$
$\Rightarrow x = 100$

8 a) $(x+3)(2x-1) \Rightarrow x = -3$ or $x = \frac{1}{2}$

b) $3^y = -3$ has no solutions
$3^y = \frac{1}{2} \Rightarrow y\log 3 = \log\left(\frac{1}{2}\right)$
$\Rightarrow y = \dfrac{\log\left(\frac{1}{2}\right)}{\log 3} = -0.631$ (3 s.f.)

Straight lines and logs

You are the examiner

Sam is correct. Lilia should have written $\ln y = 1.6095x + 1.386$.

Skill builder

1 a) is right
$y = 5 \times x^3 \Rightarrow \ln y = \ln(5 \times x^3)$
$\Rightarrow \ln y = \ln 5 + \ln x^3 \Rightarrow \ln y = 3\ln x + \ln 5$

2 c) is right
$y = 2 \times 7^x \Rightarrow \log y = \log(2 \times 7^x)$
$\Rightarrow \log y = \log 2 + \log 7^x \Rightarrow \log y$
$= \log 2 + x\log 7$

3 A

4 B

Sequences and series

You are the examiner

a) Lilia is right. Sam has made a mistake with his signs when he expanded the brackets.

b) Sam is right. Lilia should have written $100n - \frac{3}{2}n^2 + \frac{3}{2}n = n$ on line 3.

Skill builder

1 a) i) $a = 2, d = 3$ so $a_n = 2 + 3(n-1) = 3n - 1$

ii) $a_{50} = 3 \times 50 - 1 = 149$

iii) $S_{100} = \frac{100}{2}[2 \times 2 + (100-1) \times 3] = 15\,050$

b) i) $a = 40, d = -3$ so $a_n = 40 - 3(n-1) = 43 - 3n$

ii) $a_{50} = 43 - 3 \times 50 = -107$

iii) $S_{100} = \frac{100}{2}[2 \times 40 + (100-1) \times (-3)] = -10\,850$

c) i) $a = -\frac{7}{4}, d = \frac{1}{4}$ so $a_n = -\frac{7}{4} + \frac{1}{4}(n-1) = \frac{n}{4} - 2$

ii) $a_{50} = \frac{50}{4} - 2 = 10.5$

iii) $S_{100} = \frac{100}{2}[2 \times \left(\frac{-7}{4}\right) + (100-1) \times \frac{1}{4}] = 1062.5$

2 a) i) $a = 2, r = -3$ so $a_n = 2 \times (-3)^{n-1}$

ii) $a_{10} = 2 \times (-3)^9 = -39\,366$

iii) $S_{10} = \frac{2(1-(-3)^{10})}{(1-(-3))} = -29\,524$

iv) No sum to infinity as $r > 1$

b) i) $a = 256, r = \frac{1}{4}$ so $a_n = 256 \times \left(\frac{1}{4}\right)^{n-1}$

ii) $a_{10} = 256 \times \left(\frac{1}{4}\right)^9 = \frac{1}{1024}$

iii) $S_{10} = \frac{256\left(1-\left(\frac{1}{4}\right)^{10}\right)}{\left(1-\left(\frac{1}{4}\right)\right)} = 341.333$ (3 d.p.)

iv) $S = \frac{256}{1-\frac{1}{4}} = 341\frac{1}{3}$

c) i) $a = 1, r = \frac{1}{2}$ so $a_n = \left(\frac{1}{2}\right)^{n-1}$

ii) $a_{10} = 1 \times \left(\frac{1}{2}\right)^9 = \frac{1}{512}$

iii) $S_{10} = \frac{\left(1-\left(\frac{1}{2}\right)^{10}\right)}{\left(1-\left(\frac{1}{2}\right)\right)} = \frac{1023}{512} = 1.998$ (3 d.p.)

iv) $S = \frac{1}{1-\frac{1}{2}} = 2$

3 a) $\sum_{r=1}^{5} r^2 = 1^2 + 2^2 + 3^2 + 4^2 + 5^2 = 55$

b) $\sum_{k=1}^{3} a_k = (2 \times 1 + 3) + (2 \times 2 + 3) + (2 \times 3 + 3) = 21$

c) $\sum_{r=1}^{20}(3r-7) - \sum_{r=1}^{18}(3r-7) = \sum_{r=19}^{20}(3r-7) = (3 \times 19 - 7) + (3 \times 20 - 7) = 103$

4 C

5 B

6 C

Binomial expansions

You are the examiner

Mo is right. Nasreen forgot the binomial coefficients and she has a sign error as she didn't use brackets.

Skill builder

1 a) $(1+x)^4 = {}^4C_0 \times 1^4 + {}^4C_1 \times 1^3 \times x + {}^4C_2 \times 1^2 \times x^2 + {}^4C_3 \times 1 \times x^3 + {}^4C_4 \times x^4$
$= 1 + 4x + 6x^2 + 4x^3 + x^4$

b) $(t-2)^5 = {}^5C_0 \times t^5 + {}^5C_1 \times t^4 \times (-2) + {}^5C_2 \times t^3 \times (-2)^2 + {}^5C_3 \times t^2 \times (-2)^3 + {}^5C_4 \times t \times (-2)^4 + {}^5C_5 \times (-2)^5$
$= t^5 - 10t^4 + 40t^3 - 80t^2 + 80t - 32$

c) $(1+2y)^3 = {}^3C_0 \times 1^3 + {}^3C_1 \times 1^2 \times (2y) + {}^3C_2 \times 1 \times (2y)^2 + {}^3C_3 \times (2y)^3$
$= 1 + 6y + 12y^2 + 8y^3$

d) $(5-2p)^3 = {}^3C_0 \times 5^3 + {}^3C_1 \times 5^2 \times (-2p) + {}^3C_2 \times 5 \times (-2p)^2 + {}^3C_3 \times (-2p)^3$
$= 125 - 150p + 60p^2 - 8p^3$

2 B

3 A

4 D

5 A

6 B

7 D

Partial fractions

You are the examiner

Lilia is right. Mo has cancelled incorrectly. His final answer shouldn't have decimals in the numerator.

Skill builder

1 a) $A(x+2) + B(x-3)$

b) $A(x+5) + B(2x-1)$

2 a) $x = -3: -6 = A(-5) \Rightarrow A = \frac{6}{5}$
$x = 2: 4 = 5B \Rightarrow B = \frac{4}{5}$

b) $4x + 1 \equiv A(x-3) + B(2x-1)$
$x = 3: 13 = 5B \Rightarrow B = \frac{13}{5}$
$x = \frac{1}{2}: 3 = -\frac{5}{2}A \Rightarrow A = -\frac{6}{5}$

3 $3x + 1 = A(x+2) + B(x-3)$
$x = 3: 10 = 5A \Rightarrow A = 2$
$x = -2: -5 = -5B \Rightarrow B = 1$
$\frac{2}{(x-3)} + \frac{1}{(x+2)}$

4 $2x - 7 = A(x-2) + B(x+1)$
$x = -1: -9 = -3A \Rightarrow A = 3$
$x = 2: -3 = 3B \Rightarrow B = -1$
$\frac{3}{(x+1)} - \frac{1}{(x-2)}$

5 $4x^2 + 9x - 1 = A(x+1)(x+2) + B(x-1)(x+2) + C(x-1)(x+1)$
$x = 1: 12 = 6A \Rightarrow A = 2$
$x = -1: -6 = -2B \Rightarrow B = 3$
$x = -2: -3 = 3C \Rightarrow C = -1$
$\frac{2}{(x-1)} + \frac{3}{(x+1)} - \frac{1}{(x+2)}$

6 $4x^2 - 4x + 3 = A(x-1)^2 + Bx(x-1) + Cx$
$x = 0: 3 = A$
$x = 1: 3 = C \Rightarrow C = 3$
Compare coefficients of x^2:
$4 = A + B \Rightarrow B = 1$
$\frac{3}{x} + \frac{1}{(x-1)} + \frac{3}{(x-1)^2}$

7 a) $(2x+3)(x-2)$

b) $\frac{x-9}{(2x+3)(x-2)} = \frac{A}{(2x+3)} + \frac{B}{(x-2)}$
$x - 9 = A(x-2) + B(2x+3)$
$x = -\frac{3}{2}: -\frac{21}{2} = -\frac{7}{2}A \Rightarrow A = 3$
$x = 2: -7 = 7B \Rightarrow B = -1$
$\frac{3}{(2x+3)} - \frac{1}{(x-2)}$

8 D

9 B

Using coordinates

You are the examiner

Mo is right.

In part a) Peter has not used a long enough square root symbol for AB – it must cover the full calculation. For CD, he has subtracted rather than added. He's also made sign errors.

In part b) Peter has made a sign error and then used the result instead of showing it.

Skill builder

1 D

2 A

3 B

4 B

5 A

6 a) Q is 1.5 right and 2 up from midpoint, so P is 1.5 left and 2 down from midpoint. P(−2, −1).

b) $m_{PQ} = \frac{3-(-1)}{1-(-2)} = \frac{4}{3}$ so $m_{PR} = -\frac{3}{4}$
$m_{PR} = \frac{y-(-1)}{2-(-2)} = \frac{y+1}{4} = -\frac{3}{4}$
$\Rightarrow y + 1 = -3 \Rightarrow y = -4$
So R(2, −4)

c) $PR = \sqrt{(2-(-2))^2 + (-4-(-1))^2} = \sqrt{4^2 + (-3)^2} = \sqrt{16+9} = 5$
$PQ = \sqrt{(1-(-2))^2 + (3-(-1)^2)} = \sqrt{3^2 + 4^2} = \sqrt{9+16} = 5$
Two equal sides ⇒ isosceles (3rd side can't be equal as there is a right angle at P)

d) $\frac{1}{2} \times 5 \times 5 = 12.5$ square units.

Straight line graphs

You are the examiner

Peter is right. Mo has made two sign errors: he should have written $y - (-5) = -\frac{9}{5}(x-2)$ on line 3, then he makes a mistake expanding the brackets on the RHS.

Skill builder

1 a) $y = 2x + 3$

b) $y - 3 = 3(x-1) \Rightarrow y = 3x$

c) $y = 5x - 7$

d) $y - 2 = \frac{1}{2}(x - (-2)) \Rightarrow 2y - 4 = x + 2 \Rightarrow 2y = x + 6$

e) $3y - 2x = 1 \Rightarrow y = \frac{2}{3}x + \frac{1}{3}$ so gradient is $\frac{2}{3}$
$y + 2 = \frac{2}{3}(x - 1) \Rightarrow 3y + 6$
$= 2x - 2 \Rightarrow 2x - 3y - 8 = 0$

2 a) $m = -3 \Rightarrow m_{\perp} = \frac{1}{3}$ so $y = \frac{1}{3}x - 1$
b) $m = \frac{1}{2} \Rightarrow m_{\perp} = -2$ so
$y + 2 = -2(x + 3) \Rightarrow 2x + y + 8 = 0$
c) $2y + 3x + 7 = 0 \Rightarrow y = -\frac{3}{2}x - \frac{7}{2}$ so
$m = -\frac{3}{2} \Rightarrow m_{\perp} = \frac{2}{3}$
so $y - 4 = \frac{2}{3}(x + 1) \Rightarrow 3y - 12$
$= 2(x + 1) \Rightarrow 3y = 2x + 14$

3 C
4 C
5 A
6 C
7 B
8 B

Circles

You are the examiner

Mo is right. Peter subtracted when he found the centre. He halved the square of the diameter, which is not the same as the square of the radius. He should have found the diameter and then halved the result.

Skill builder

1 a) $x^2 + y^2 = 9$
b) $(x - 2)^2 + (y - 5)^2 = 16$
c) $(x + 1)^2 + (y + 3)^2 = 36$
d) $(x + 4)^2 + y^2 = 1$

2 Centre is $\left(\frac{-2+4}{2}, \frac{4+0}{2}\right) = (1, 2)$
Radius $= \sqrt{(1-4)^2 + (2-0)^2}$
$= \sqrt{(-3)^2 + 2^2} = \sqrt{13}$
So $(x - 1)^2 + (y - 2)^2 = 13$

3 Gradient of radius $= \frac{3}{1} = 3$
Gradient of tangent $= -\frac{1}{3}$
Equation of tangent is $y - 3 = -\frac{1}{3}(x - 1)$
$\Rightarrow 3y - 9 = -x + 1 \Rightarrow x + 3y = 10$

4 B
5 A
6 C
7 B
8 E

Intersections

You are the examiner

Peter is right. Mo has made a mistake with his signs when he expanded the brackets. He should have written $x^2 - 6x + 9 + y^2 - 2y + 1 = 5$. Mo also should have found the y-values and given his answer as coordinates.

Skill builder

1 a) $x^2 = x + 6 \Rightarrow x^2 - x - 6 = 0$
$\Rightarrow (x - 3)(x + 2) = 0 \Rightarrow x = 3, -2$
When $x = 3, y = 3 + 6 = 9 \Rightarrow (3, 9)$
When $x = -2, y = -2 + 6 = 4 \Rightarrow (-2, 4)$
b) $x^2 + (x + 3)^2 = 9 \Rightarrow 2x^2 + 6x + 9 = 9$
$\Rightarrow 2x^2 + 6x = 0 \Rightarrow 2x(x + 3) = 0$
$\Rightarrow x = 0, x = -3$
When $x = 0, y = 0 + 3 = 3 \Rightarrow (0, 3)$
When $x = -3, y = -3 + 3 = 0$
$\Rightarrow (-3, 0)$

2 E
3 E
4 A
5 $x^2 + 3x - 4 = x - k \Rightarrow$
$x^2 + 2x + k - 4 = 0$
$b^2 - 4ac = 2^2 - 4 \times 1 \times (k - 4)$
$= 4 - 4k + 16$
$= 20 - 4k$
At a tangent $b^2 - 4ac = 0 \Rightarrow 20 - 4k = 0$
$\Rightarrow k = 5$
When $k = 5$,
$x^2 + 2x + k - 4 = x^2 + 2x + 1 = 0$
$(x + 1)^2 = 0 \Rightarrow x = -1$
$y = x - 5 \Rightarrow y = -1 - 5 = -6$
P$(-1, -6)$

Transformations

You are the examiner

a) Sam is right. The order doesn't matter because one transformation affects the x-coordinates and the other affects the y-coordinates.
b) Sam is right. Peter has used the wrong sign inside the brackets.

Skill builder

1 D
2 C
3 C
4 E
5 A
6 D

Modulus functions

You are the examiner

Sam is right. Peter should have drawn a graph as he has found an extra invalid solution.

Skill builder

1 a) $|2x| = 6 \Rightarrow 2x = \pm 6 \Rightarrow x = \pm 3$
b) $|x + 1| = 5 \Rightarrow x + 1 = 5 \Rightarrow x = 4$
or $x + 1 = -5 \Rightarrow x = -6$
c) $|x - 3| = 8 \Rightarrow x - 3 = 8 \Rightarrow x = 11$
or $x - 3 = -8 \Rightarrow x = -5$
d) $|2x - 5| = 7 \Rightarrow 2x - 5 = 7 \Rightarrow x = 6$
or $2x - 5 = -7 \Rightarrow x = -1$

2 a)

b) i) $|x - 2| = 4 \Rightarrow x - 2 = 4 \Rightarrow x = 6$
or $x - 2 = -4 \Rightarrow x = -2$
ii) Using the graph: $-2 < x < 6$
iii) $x - 2 = -(2x + 1)$
$x - 2 = -2x - 1$
$3x = 1$
$x = \frac{1}{3}$
From graph the blue line is above the red line when $x > \frac{1}{3}$

3 D
4 C
5 A
6 D
7 B

Vectors

You are the examiner

Mo is right. Lilia has worked out $\overrightarrow{QP}$ and has not used brackets when she worked out the magnitude.

Skill builder

1 C
2 D
3 A
4 a) $|\overrightarrow{OA}| = \sqrt{3^2 + 2^2} = \sqrt{13}$
b) $|\overrightarrow{OB}| = \sqrt{5^2 + (-12)^2} = \sqrt{169} = 13$
c) $|\overrightarrow{OC}| = \sqrt{4^2 + (-2)^2 + (-4)^2} = \sqrt{36} = 6$
d) $\overrightarrow{AB} = \overrightarrow{OB} - \overrightarrow{OA} = \begin{pmatrix} 5 \\ -12 \end{pmatrix} - \begin{pmatrix} 3 \\ 2 \end{pmatrix}$
$= \begin{pmatrix} 2 \\ -14 \end{pmatrix}$
$|\overrightarrow{AB}| = \sqrt{2^2 + (-14)^2} = \sqrt{200} = 10\sqrt{2}$

5 a) $\overrightarrow{AB} = \overrightarrow{OB} - \overrightarrow{OA} = \begin{pmatrix} -3 \\ -2 \end{pmatrix} - \begin{pmatrix} -5 \\ 1 \end{pmatrix}$
$= \begin{pmatrix} 2 \\ -3 \end{pmatrix}$
b) $2\overrightarrow{AB} = 2\begin{pmatrix} 2 \\ -3 \end{pmatrix} = \begin{pmatrix} 4 \\ -6 \end{pmatrix} = \overrightarrow{OC}$ so parallel
c) $\overrightarrow{AC} = \overrightarrow{OC} - \overrightarrow{OA} = \begin{pmatrix} 4 \\ -6 \end{pmatrix} - \begin{pmatrix} -5 \\ 1 \end{pmatrix}$
$= \begin{pmatrix} 9 \\ -7 \end{pmatrix}$
$\overrightarrow{BC} = \overrightarrow{OC} - \overrightarrow{OB} = \begin{pmatrix} 4 \\ -6 \end{pmatrix} - \begin{pmatrix} -3 \\ -2 \end{pmatrix}$
$= \begin{pmatrix} 7 \\ -4 \end{pmatrix}$
Perimeter $= |\overrightarrow{AB}| + |\overrightarrow{AC}| + |\overrightarrow{BC}|$
$= \sqrt{2^2 + (-3)^2} + \sqrt{9^2 + (-7)^2}$
$+ \sqrt{7^2 + (-4)^2}$
$= \sqrt{13} + \sqrt{130} + \sqrt{65} = 23.1$ (to 3 s.f.)
d) $\overrightarrow{OM} = \overrightarrow{OA} + \frac{1}{2}\overrightarrow{AB} = \begin{pmatrix} -5 \\ 1 \end{pmatrix} + \frac{1}{2}\begin{pmatrix} 2 \\ -3 \end{pmatrix}$
$= \begin{pmatrix} -4 \\ -0.5 \end{pmatrix}$

Numerical methods

You are the examiner

Lilia is right. Sam's iterative formula doesn't converge – the values increase with each iteration (diverges). Sam is incorrect to conclude that there is no root, he just couldn't find it as his formula diverged.

Skill builder

1 B
2 D
3 C
4 A
5 a) $x = 1$: $1^4 - 3 \times 1 + 1 = -1$
$x = 2$: $2^4 - 3 \times 2 + 1 = 11$
Change of sign so there is a root.

b) $f(x) = x^4 - 3x + 1 \Rightarrow f'(x) = 4x^3 - 3$

$$x_{n+1} = x_n - \frac{x_n^{\,4} - 3x_n + 1}{4x_n^{\,3} - 3}$$

$$x_1 = 2 \Rightarrow x_2 = 2 - \frac{2^4 - 3\times 2 + 1}{4\times 2^3 - 3}$$

$= 1.620...$

$x_3 = 1.404...,\ x_4 = 1.320...,$
$x_5 = 1.307...$

$x_6 = 1.307...\ x_7 = 1.307...$

Root is 1.31 to 2 d.p.

6 a) $x = 2$: $e^2 - 5\times 2 - 2 = -4.61...$
$x = 3$: $e^3 - 5\times 3 - 2 = 3.08...$
Change of sign so there is a root.

b) $e^x - 5x - 2 = 0 \Rightarrow e^x = 5x + 2$
$\Rightarrow x = \ln(5x + 2)$

c) $x_{n+1} = \ln(5x_n + 2)$

$x_1 = 1 \Rightarrow x_2 = \ln(5\times 1 + 2) = 1.945...$

$x_3 = 2.462...,\ x_4 = 2.660...,\ x_5 = 2.728...$

$x_6 = 2.749...,\ x_7 = 2.756...,\ x_8 = 2.758...,$

$x_9 = 2.759...,\ x_{10} = 2.759...,\ x_{11} = 2.759...,$

$x_{12} = 2.760...$

Root is 2.76 to 2 d.p.

The trapezium rule

You are the examiner

Peter is right. Sam has used the x-values for the heights of the trapeziums. Peter's comment is correct.

Skill builder

1 D

2 C

3 a) You are asked for 1 d.p,, so you need to work with at least 2 d.p.

x	1	3	5	7	9	11	13
y	100	73.21	61.80	54.86	50	46.33	43.43

b)

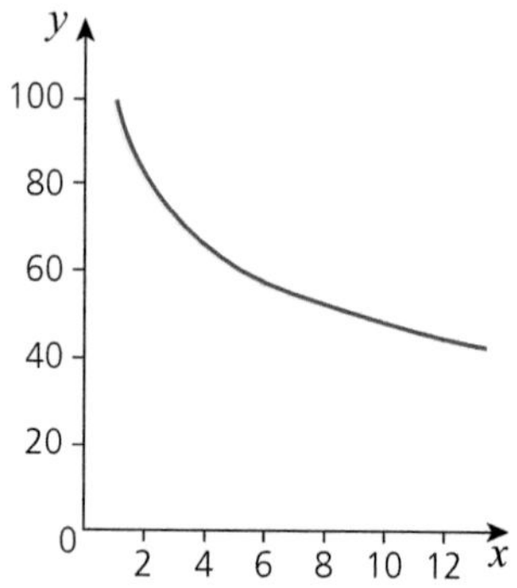

c) Area $= \frac{1}{2}\times 2\times[100 + 43.43 + 2\times (73.21 + 61.80 + 54.86 + 50 + 46.33)]$
$= 715.8$ (to 1 d.p.)

d) Overestimate as trapeziums lie above the curve.

Using more strips will decrease the answer as it will become closer to the true area.

Trigonometry review

You are the examiner

Peter is right, Sam should have found all the possible values for $2x$ first before he halved them.

Skill builder

1 a) $a = 60°$; $b = 60° \div 2 = 30°$;
$c = 90° \div 2 = 45°$

b) $d^2 = 2^2 - 1^2 = 3 \Rightarrow d = \sqrt{3}$ cm
$e^2 = 1^2 + 1^2 = 2 \Rightarrow e = \sqrt{2}$ cm

c)

θ	0°	30°	45°	60°	90°
$\sin\theta$	0	$\frac{1}{2}$	$\frac{1}{\sqrt{2}} = \frac{\sqrt{2}}{2}$	$\frac{\sqrt{3}}{2}$	1
$\cos\theta$	1	$\frac{\sqrt{3}}{2}$	$\frac{1}{\sqrt{2}} = \frac{\sqrt{2}}{2}$	$\frac{1}{2}$	0
$\tan\theta$	0	$\frac{1}{\sqrt{3}} = \frac{\sqrt{3}}{3}$	1	$\sqrt{3}$	undefined

2 Missing answers in order are:

a) 360; 2; origin; 1; −1; $-1 \leqslant \sin\theta \leqslant 1$

b) 360; y; 1; −1; $-1 \leqslant \cos\theta \leqslant 1$

c) 180; 2; origin; 90; 270

3 a) i) From calculator: $x = 30°$; 2nd value is $360° - 30° = 330°$

ii) From part i:
$x + 10° = 30°$ or $330°$
$\Rightarrow x = 20°$or $320°$

iii) From part i: $2x = 30°, 330°$
or $30° + 360° = 390°$
or $330° + 360° = 690°$
$\Rightarrow x = 15°, 165°, 195°$ or $345°$

b) i) From calculator: $x = 30°$; 2nd value is $180° - 30° = 150°$

ii) From calculator: $x = -30°$; 2nd value is $180° - (-30°) = 210°$
So $-30° + 360° = 330°$

iii) $\sin^2 x = \frac{1}{4} \Rightarrow \sin x = \pm\frac{1}{2}$
From parts i and ii:
$x = 30°, 150°, 210°$ or $330°$

c) i) $5\tan x = 1 \Rightarrow \tan x = \frac{1}{5}$
From calculator: $x = 11.3°$; other values $= 11.3° + 180° = 191.3°$

ii) From part i:
$x - 30° = 11.3°$ or $191.3°$
$\Rightarrow x = 41.3°$or $221.3°$

iii) $5\tan\left(\frac{x}{2}\right) - 1 = 0 \Rightarrow \tan\left(\frac{x}{2}\right) = \frac{1}{5}$
From part i: $\frac{x}{2} = 11.3°$ or $191.3°$
$\Rightarrow x = 22.6°$or $382.6°$
$382.6°$ is out of range so answer is $x = 22.6°$

4 B

Triangles without right angles

You are the examiner

Mo is right. Sam has got the sides 12 and 15 the wrong way around. For Sam's value of θ there is no ambiguous case as $145.6° + 45° > 180°$.

Skill builder

1 Area $=$
$\frac{1}{2}\times 5\times 8\sin 125° = 16.4\ \text{cm}^2$ (3 s.f.)
Let x = longest side:
$x^2 = 5^2 + 8^2 - 2\times 5\times 8\cos 125° = 134.89...$
$\Rightarrow x = 11.6$
Perimeter = 5 cm + 8 cm + 11.6 cm = 24.6 cm (to 3 s.f.)

2 Let x = side length:
$\frac{1}{2}\times x^2 \sin 60° = 444 \Rightarrow x^2 = \frac{2\times 444}{\sin 60°}$
$= 1025.37...$
So $x = 32.02$ cm

3 Smallest angle is opposite the shortest side.
$\cos\theta = \frac{7^2 + 9^2 - 5^2}{2\times 7\times 9} = \frac{5}{6} \Rightarrow \theta = 33.6°$

4 C

5 D

6 B

7 C

Working with radians

You are the examiner

Lilia is right. Sam has used the formula for arc length, not sector area. His calculator was in 'degrees' mode so the area of the triangle is wrong.

Skill builder

1 a) $30°\times\frac{\pi}{180°} = \frac{\pi}{6}$

b) $90°\times\frac{\pi}{180°} = \frac{\pi}{2}$

c) $45°\times\frac{\pi}{180°} = \frac{\pi}{4}$

d) $210°\times\frac{\pi}{180°} = \frac{7\pi}{6}$

e) $75°\times\frac{\pi}{180°} = \frac{5\pi}{12}$

f) $18°\times\frac{\pi}{180°} = \frac{\pi}{10}$

2 a) $\pi\times\frac{180°}{\pi} = 180°$

b) $4\pi\times\frac{180°}{\pi} = 720°$

c) $\frac{\pi}{5}\times\frac{180°}{\pi} = 36°$

d) $\frac{2\pi}{3}\times\frac{180°}{\pi} = 120°$

e) $1\times\frac{180°}{\pi} = 57.3°$ (to 1 d.p.)

f) $2\times\frac{180°}{\pi} = 114.6°$ (to 1 d.p.)

3 a) $\cos\pi = \cos 180° = -1$

b) $\sin\left(\frac{\pi}{2}\right) = \sin 90° = 1$

c) $\tan\left(\frac{\pi}{6}\right) = \tan 30° = \frac{\sqrt{3}}{3}$

d) $\sin\left(\frac{\pi}{4}\right) = \sin 45° = \frac{\sqrt{2}}{2}$

e) $\cos\left(\frac{\pi}{6}\right) = \cos 30° = \frac{\sqrt{3}}{2}$

f) $\tan\left(\frac{3\pi}{4}\right) = \tan 135° = -1$

4 Area $= \frac{1}{2}\times 12^2\times\frac{2\pi}{3} = 48\pi\ \text{cm}^2$
Perimeter $=$
$2\times 12 + 12\times\frac{2\pi}{3} = (24 + 8\pi)$cm

5 Arc length $= \frac{\pi}{3}\times 8 = 8.377...$ cm
Straight side:
$a^2 = 8^2 + 8^2 - 2\times 8\times 8\times\cos\left(\frac{\pi}{3}\right)$
$= 128 - 128\times\frac{1}{2} = 64$
So $a = 8$ cm (triangle is equilateral as $\frac{\pi}{3} = 60°$)
Perimeter is $8 + 8.377... = 16.4$ cm (3 s.f.)

6 a) i) $\cos x = \frac{1}{2} \Rightarrow x = \frac{\pi}{3}$
2nd value is $2\pi - \frac{\pi}{3} = \frac{5\pi}{3}$

ii) From i: $(x + \frac{\pi}{4}) = \frac{\pi}{3}$ or $\frac{5\pi}{3}$
So
$x = \frac{\pi}{3} - \frac{\pi}{4} = \frac{\pi}{12}$ or $\frac{5\pi}{3} - \frac{\pi}{4} = \frac{17\pi}{12}$

iii) From i:
$2x = \frac{\pi}{3}$ or $\frac{5\pi}{3}$ or $\frac{\pi}{3} + 2\pi$ or $\frac{5\pi}{3} + 2\pi$

So $2x = \frac{\pi}{3}, \frac{5\pi}{3}, \frac{7\pi}{3}$ or $\frac{11\pi}{3}$

So $x = \frac{\pi}{6}, \frac{5\pi}{6}, \frac{7\pi}{6}$ or $\frac{11\pi}{6}$

b) i) $\tan x = \frac{\sqrt{3}}{3} \Rightarrow x = \frac{\pi}{6}$

2nd value is $\frac{\pi}{6} + \pi = \frac{7\pi}{6}$

ii) $\tan x = -\frac{\sqrt{3}}{3} \Rightarrow x = -\frac{\pi}{6}$

Answers in range: $-\frac{\pi}{6} + \pi = \frac{5\pi}{6}$

and $\frac{5\pi}{6} + \pi = \frac{11\pi}{6}$

iii) $\tan^2 x = \frac{1}{3} \Rightarrow \tan x = \pm\frac{\sqrt{3}}{3}$

From parts i and ii:

$x = \frac{\pi}{6}, \frac{5\pi}{6}, \frac{7\pi}{6}$ or $\frac{11\pi}{6}$

c) i) $\sin x = \frac{3}{4} \Rightarrow x = 0.848$ rads

2nd value is

$\pi - 0.848$ rads $= 2.294$ rads

ii) $4\sin(2x) - 3 = 0 \Rightarrow \sin(2x) = \frac{3}{4}$

From part i:

$2x = 0.848$ rads, 2.293 rads

or $(0.848 + 2\pi)$ rads

or $(2.293 + 2\pi)$ rads

So $x = 0.424$ rads, 1.147 rads

or 3.566 rads or 4.288 rads

iii) $\sin^2 x = \frac{3}{4} \Rightarrow \sin x = \pm\frac{\sqrt{3}}{2}$

$\sin x = \frac{\sqrt{3}}{2} \Rightarrow x = \frac{\pi}{3}$ or $\pi - \frac{\pi}{3} = \frac{2\pi}{3}$

$\sin x = -\frac{\sqrt{3}}{2} \Rightarrow x = -\frac{\pi}{3}$ or

$\pi - (-\frac{\pi}{3}) = \frac{4\pi}{3}$ or $2\pi + (-\frac{\pi}{3}) = \frac{5\pi}{3}$

So $x = \frac{\pi}{3}, \frac{2\pi}{3}, \frac{4\pi}{3}$ or $\frac{5\pi}{3}$

7 b) $\frac{\sin\theta}{\tan\theta} \approx \frac{\theta}{\theta} = 1$

c) $\frac{\tan^2\theta}{\theta} \approx \frac{\theta^2}{\theta} = \theta$

d) $4\sin\theta\tan\theta + 4\cos^2\theta$

$\approx 4\theta \times \theta + 4\left(1 - \frac{\theta^2}{2}\right)^2$

$= 4\theta^2 + 4\left(1 - \theta^2 + \frac{\theta^4}{4}\right)$

$= 4\theta^2 + 4 - 4\theta^2 + \theta^4$

$= 4 + \theta^4$

(Note: often a question will tell you ignore θ^3 or higher, in which case the answer would be just 4.)

8 B

Trigonometric identities

You are the examiner

D, F, B, E, A, C, H, G

Skill builder

1 a) i) $\sin\theta = \frac{a}{c}$

ii) $\cos\theta = \frac{b}{c}$

iii) $\tan\theta = \frac{a}{b}$

iv) $\text{cosec}\,\theta = \frac{c}{a}$

v) $\sec\theta = \frac{c}{b}$

vi) $\cot\theta = \frac{b}{a}$

b) $a = c\sin\theta$ and $b = c\cos\theta$

$a^2 + b^2 = c^2$

$\Rightarrow c^2\sin^2\theta + c^2\cos^2\theta = c^2$

$\Rightarrow \sin^2\theta + \cos^2\theta = 1$

2 Draw a right-angled triangle with sides 5, 12 and 13.

a) $\sin\theta = \frac{12}{13}$

b) $\sec\theta = \frac{13}{5}$

c) $\text{cosec}\,\theta = \frac{13}{12}$

d) $\cot\theta = \frac{5}{12}$

3 B

4 a) i) $\sec x = 2 \Rightarrow \cos x = \frac{1}{2}$

$x = 60°, 300°$

ii) Use i: $2x = 60°, 300°, 420°, 660°$

$x = 30°, 150°, 210°, 330°$

iii) Use i: $x - 30° = 60°, 300°$

$x = 90°, 330°$

b) i) $\text{cosec}\,x = \frac{2}{\sqrt{3}} \Rightarrow \sin x = \frac{\sqrt{3}}{2}$

$x = 60°, 120°$

ii) Use i: $\frac{1}{2}x = 60°, 120°$;

$x = 120°, 240°$

iii) Use i: $\frac{1}{2}x + 60° = 60°, 120°$

$\frac{1}{2}x = 0°, 60°$ $x = 0°, 120°$

c) i) $\cot x = \frac{1}{2} \Rightarrow$ so $\tan x = 2$

$x = 63.4°, 243.4°$

ii) $\cot 2x = \frac{1}{2} \Rightarrow$

$2x = 63.4°, 243.4°, 423.4°, 603.4°$

$x = 31.7°, 121.7°, 211.7°, 301.7°$

iii) $\cot x = \frac{1}{2} \Rightarrow x = 63.4°, 243.4°$

$\cot x = -\frac{1}{2} \Rightarrow$

$x = -63.4°, 116.6°, 296.6°$

So $x = 63.4°, 116.6°, 243.4°, 296.6°$

5 a) $(1 - \cos\theta)(1 + \cos\theta) \equiv 1 - \cos^2\theta$

$\equiv \sin^2\theta$

b) $\cos^2\theta(1 + \tan^2\theta)$

$\equiv \cos^2\theta + \cos^2\theta \times \frac{\sin^2\theta}{\cos^2\theta}$

$\equiv \cos^2\theta + \sin^2\theta \equiv 1$

c) $\tan\theta + \cot\theta = \frac{\sin\theta}{\cos\theta} + \frac{\cos\theta}{\sin\theta}$

$= \frac{\sin^2\theta + \cos^2\theta}{\cos\theta\sin\theta}$

$= \frac{1}{\cos\theta\sin\theta}$

$= \sec\theta\,\text{cosec}\,\theta$

d) $(\sin\theta + \cos\theta)^2$

$\equiv \sin^2\theta + 2\sin\theta\cos\theta + \cos^2\theta - 1$

$\equiv 2\sin\theta\cos\theta + \sin^2\theta + \cos^2\theta - 1$

$\equiv 2\sin\theta\cos\theta + 1 - 1$

$\equiv 2\sin\theta\cos\theta$

6 a) $\sin\theta\tan\theta + \cos\theta \equiv \frac{\sin^2\theta}{\cos\theta} + \cos\theta$

$\equiv \frac{\sin^2\theta}{\cos\theta} + \frac{\cos^2\theta}{\cos\theta}$

$\equiv \frac{\sin^2 + \cos^2\theta}{\cos\theta}$

$\equiv \frac{1}{\cos\theta} \equiv \sec\theta$

b) $\sec\theta = \sqrt{2} \Rightarrow \cos\theta = \frac{1}{\sqrt{2}}$

$\Rightarrow \theta = \frac{\pi}{4}, \frac{7\pi}{4}$

Compound angles

You are the examiner

Sam and Mo are both correct.

Rationalising the denominator gives

Sam: $\frac{(1+\sqrt{3})(1+\sqrt{3})}{(1-\sqrt{3})(1+\sqrt{3})}$

$= \frac{1+2\sqrt{3}+3}{1-3} = \frac{4+2\sqrt{3}}{-2} = -2-\sqrt{3}$

Mo: $\frac{(3+\sqrt{3})(\sqrt{3}+3)}{(\sqrt{3}-3)(\sqrt{3}+3)}$

$= \frac{9+6\sqrt{3}+3}{3-9} = \frac{12+6\sqrt{3}}{-6} = -2-\sqrt{3}$

Skill builder

1 Isobel is wrong.

$\cos(x + 60°) = \cos x\cos 60° - \sin x\sin 60°$

$= \frac{1}{2}\cos x - \frac{\sqrt{3}}{2}\sin x$

2 a) $\cos(x - 45°) = \cos x\cos 45° + \sin x\sin 45°$

$= \frac{\sqrt{2}}{2}\cos x + \frac{\sqrt{2}}{2}\sin x$

b) $\sin(x + 60°) = \sin x\cos 60° + \cos x\sin 60°$

$= \frac{1}{2}\sin x + \frac{\sqrt{3}}{2}\cos x$

c) $\tan(x + 45°) = \frac{\tan x + \tan 45°}{1 - \tan x\tan 45°}$

$= \frac{\tan x + 1}{1 - \tan x}$

d) $\tan(x - 45°) = \frac{\tan x - \tan 45°}{1 + \tan x\tan 45°}$

$= \frac{\tan x - 1}{1 + \tan x}$

3 a) $\cos(x - \frac{\pi}{2}) = \cos x\cos\frac{\pi}{2} + \sin x\sin\frac{\pi}{2}$

$= \cos x \times 0 + \sin x \times 1$

$= \sin x$

b) $\sin(x + \pi) = \sin x\cos\pi + \cos x\sin\pi$

$= \sin x \times -1 + \cos x \times 0$

$= -\sin x$

c) $\sin(x - \frac{\pi}{3}) = \sin x\cos\frac{\pi}{3} - \cos x\sin\frac{\pi}{3}$

$= \frac{1}{2}\sin x - \frac{\sqrt{3}}{2}\cos x$

d) $\tan(x + \frac{\pi}{3}) = \frac{\tan x + \tan\frac{\pi}{3}}{1 - \tan x\tan\frac{\pi}{3}}$

$= \frac{\tan x + \sqrt{3}}{1 - \sqrt{3}\tan x}$

4 a) $\cos(2x) = \cos(x + x)$

$= \cos x\cos x - \sin x\sin x$

$= \cos^2 x - \sin^2 x$

b) $\cos(2x) = \cos^2 x - (1 - \cos^2 x)$

$= 2\cos^2 x - 1$

$\cos(2x) = (1 - \sin^2 x) - \sin^2 x$

$= 1 - 2\sin^2 x$

5 a) i) $\cos 75° = \cos(45° + 30°)$

$= \cos 45°\cos 30° - \sin 45°\sin 30°$

$= \frac{\sqrt{2}}{2} \times \frac{\sqrt{3}}{2} - \frac{\sqrt{2}}{2} \times \frac{1}{2} = \frac{\sqrt{6}-\sqrt{2}}{4}$

ii) $\sin 75° = \sin(45° + 30°)$

$= \sin 45°\cos 30° + \cos 45°\sin 30°$

$= \frac{\sqrt{2}}{2} \times \frac{\sqrt{3}}{2} + \frac{\sqrt{2}}{2} \times \frac{1}{2} = \frac{\sqrt{6}+\sqrt{2}}{4}$

iii) $\tan 75° = \tan(45° + 30°)$

$= \frac{\tan 45° + \tan 30°}{1 - \tan 45°\tan 30°}$

$= \frac{1 + \frac{\sqrt{3}}{3}}{1 - \frac{\sqrt{3}}{3}} = \frac{3+\sqrt{3}}{3-\sqrt{3}}$

$= \frac{(3+\sqrt{3})(3+\sqrt{3})}{(3-\sqrt{3})(3+\sqrt{3})}$

$= \frac{9+6\sqrt{3}+3}{9-3}$

$= \frac{12+6\sqrt{3}}{6} = 2+\sqrt{3}$

b) i) $\cos 15° = \cos(45° - 30°)$

$= \cos 45° \cos 30° + \sin 45° \sin 30°$

$= \frac{\sqrt{2}}{2} \times \frac{\sqrt{3}}{2} + \frac{\sqrt{2}}{2} \times \frac{1}{2} = \frac{\sqrt{6}+\sqrt{2}}{4}$

ii) $\sin 15° = \sin(45° - 30°)$

$= \sin 45° \cos 30° - \cos 45° \sin 30°$

$= \frac{\sqrt{2}}{2} \times \frac{\sqrt{3}}{2} - \frac{\sqrt{2}}{2} \times \frac{1}{2} = \frac{\sqrt{6}-\sqrt{2}}{4}$

iii) $\tan 15° = \tan(45° - 30°)$

$= \frac{\tan 45° - \tan 30°}{1 + \tan 45° \tan 30°}$

$= \frac{1 - \frac{\sqrt{3}}{3}}{1 + \frac{\sqrt{3}}{3}} = \frac{3-\sqrt{3}}{3+\sqrt{3}}$

$= \frac{(3-\sqrt{3})(3-\sqrt{3})}{(3+\sqrt{3})(3-\sqrt{3})}$

$= \frac{9 - 6\sqrt{3} + 3}{9 - 3}$

$= \frac{12 - 6\sqrt{3}}{6} = 2 - \sqrt{3}$

6 a) $\sin(x + x) = \sin x \cos x + \cos x \sin x$

$= 2 \sin x \cos x$

b) $2 \sin x \cos x = \cos x$

$\Rightarrow 2 \sin x \cos x - \cos x = 0$

$\Rightarrow \cos x(2 \sin x - 1) = 0$

$\Rightarrow \cos x = 0$ so $x = 90°, 270°$

or $\sin x = \frac{1}{2}$ so $x = 30°, 150°$

$x = 30°, 90°, 150°$ or $270°$

7 A

8 D

The form $r\sin(\theta + \alpha)$

You are the examiner

Lilia is right. r is always positive so you should ignore the negative square root.

Peter has used degrees and not radians and has got the angle wrong anyway – in degrees it is 35.3°.

Skill builder

1 a) i) one way stretch, scale factor 2, parallel to y-axis and translation by the vector $\begin{pmatrix}-15°\\0\end{pmatrix}$ (in either order)

ii) one way stretch, scale factor $\sqrt{3}$, parallel to y-axis and translation by the vector $\begin{pmatrix}50°\\0\end{pmatrix}$ (in either order)

b) $y = \sin\theta$ has a maximum at (90°, 1) and a minimum at (270°, −1).

i) $y = 2\sin(\theta + 15°)$ has a maximum at (75°, 2) and a minimum at (255°, −2).

ii) $y = \sqrt{3}\sin(\theta - 50°)$ has a maximum at (140°, $\sqrt{3}$) and a minimum at (320°, $-\sqrt{3}$).

2 a) $r\sin(\theta + \alpha) = r\sin\theta\cos\alpha + r\cos\theta\sin\alpha$

b) Compare $r\sin\theta\cos\alpha + r\cos\theta\sin\alpha$ with $3\sqrt{3}\sin\theta + 3\cos\theta$ gives

$r\cos\alpha = 3\sqrt{3}$ and $r\sin\alpha = 3$

$\frac{r\sin\alpha}{r\cos\alpha} = \frac{3}{3\sqrt{3}} \Rightarrow \tan\alpha = \frac{1}{\sqrt{3}}$

$\Rightarrow \alpha = 30°$

$r = \sqrt{3^2 + (3\sqrt{3})^2}$

$= \sqrt{9 + 27} = \sqrt{36} = 6$

So $3\sqrt{3}\sin\theta + 3\cos\theta = 6\sin(\theta + 30°)$

c) $6\sin(\theta + 30°) = 3\sqrt{2}$

$\Rightarrow \sin(\theta + 30°) = \frac{3\sqrt{2}}{6} = \frac{\sqrt{2}}{2}$

So $\theta + 30° = 45°$ or $180° - 45° = 135$

So $\theta = 15°$ or $105°$

3 a) $r\cos(\theta + \alpha) = r\cos\theta\cos\alpha$

$- r\sin\theta\sin\alpha$

b) Compare $r\cos\theta\cos\alpha - r\sin\theta\sin\alpha$ with $3\cos\theta - 4\sin\theta$ gives

$r\cos\alpha = 3$ and $r\sin\alpha = 4$

$\frac{r\sin\alpha}{r\cos\alpha} = \frac{4}{3} \Rightarrow \tan\alpha = \frac{4}{3} \Rightarrow \alpha = 53.1°$

$r = \sqrt{4^2 + 3^2} = \sqrt{16 + 9} = \sqrt{25} = 5$

So $3\cos\theta - 4\sin\theta = 5\cos(\theta + 53.1°)$

c) $5\cos(\theta + 53.1°) = 1$

$\Rightarrow \cos(\theta + 53.1°) = \frac{1}{5}$

So $\theta + 53.1° = 78.5°$

or $360° - 78.5° = 281.5°$

So $\theta = 25.3°$ or $228.4°$

4 B

5 C

6 D

Differentiation

You are the examiner

Sam is right. Lilia should have simplified before she differentiated.

Skill builder

1 a) $\frac{dy}{dx} = 7 \times 3x^2 - 4 \times 2x^1 + 5$

$= 21x^2 - 8x + 5$

b) $\frac{dy}{dx} = 4 \times 10x^9 - 3 \times 9x^8$

$= 40x^9 - 27x^8$

c) $\frac{dA}{dr} = \pi \times 2r^1 = 2\pi r$

d) $f'(x) = 4 \times 5x^4 - 3 \times 4x^3 + 2 \times 8x^2 - 2x^1$

$= 20x^4 - 12x^3 + 6x^2 - 2x$

2 a) $\frac{d^2y}{dx^2} = 21 \times 2x^1 - 8 = 42x - 8$

b) $\frac{d^2y}{dx^2} = 40 \times 9x^8 - 27 \times 8x^7$

$= 360x^8 - 216x^7$

c) $\frac{d^2A}{dr^2} = 2\pi$

d) $f''(x) = 20 \times 4x^3 - 12 \times 3x^2 + 6 \times 2x^1 - 2$

$= 80x^3 - 36x^2 + 12x - 2$

3 a) $y = x^3 - 2x^2 + 3 \Rightarrow \frac{dy}{dx} = 3x^2 - 4x$

When $x = 1$ then

$\frac{dy}{dx} = 3 \times 1^2 - 4 \times 1 = -1$

b) i) Gradient of tangent $= -1$ so at $(1, 2)$:

$y - 2 = -1(x - 1) \Rightarrow y = 3 - x$

ii) Gradient of normal $= 1$ so at $(1, 2)$:

$y - 2 = 1(x - 1) \Rightarrow y = x + 1$

4 E

5 B

6 B

7 A

8 C

9 B

10 C

11 E

Stationary points

You are the examiner

Sam is right. Mo should have checked the gradient either side of the stationary point.

Skill builder

1 C

2 D

3 E

4 $y = x^3 - 6x^2 + 12x$

$\Rightarrow \frac{dy}{dx} = 3x^2 - 12x + 12$

At a stationary point

$\frac{dy}{dx} = 0 \Rightarrow 3x^2 - 12x + 12 = 0 \Rightarrow x = 2$

When $x = 2$ then

$y = 2^3 - 6 \times 2^2 + 12 \times 2 = 8$

Check gradient either side of $x = 2$

At $x = 1$: $\frac{dy}{dx} = 3 \times 1^2 - 12 \times 1 + 12 = 3 > 0$

At $x = 3$: $\frac{dy}{dx} = 3 \times 3^2 - 12 \times 3 + 12 = 3 > 0$

Same sign both sides $\Rightarrow$ (2,8) is a point of inflection.

5 C

6 A

Integration

You are the examiner

Nasreen is right. Mo should have found the area of the two regions separately.

Skill builder

1 a) $\int (x^3 - 2)\,dx = \frac{1}{4}x^4 - 2x + c$

b) $\int (t^4 - 3t^3 + 5)\,dt = \frac{1}{5}t^5 - \frac{3}{4}t^4 + 5t + c$

c) $\int (5y^4 - 2y^2 - 1)\,dy = y^5 - \frac{2}{3}y^3 - y + c$

2 a) $\int_1^3 (2x - 1)\,dx = [x^2 - x]_1^3$

$= (3^2 - 3) - (1^2 - 1) = 6$

b) $I := \int_{-1}^{2} (x^2 - x)\,dx$

$\Rightarrow I = \left[\frac{1}{3}x^3 - \frac{1}{2}x^2\right]_{-1}^{2}$

$\Rightarrow I = \frac{1}{3}(2)^3 - \frac{1}{2}(2)^2 - \left[\frac{1}{3}(-1)^3 - \frac{1}{2}(-1)^2\right]$

$\Rightarrow I = \frac{8}{3} - 2 - \left(-\frac{1}{3} - \frac{1}{2}\right)$

$\Rightarrow I = \frac{8}{3} + \frac{1}{3} - 2 + \frac{1}{2}$

$\Rightarrow I = \frac{9}{3} - \frac{3}{2}$

$\Rightarrow I = \frac{3}{2}$

c) $\int_{-3}^{0} (4x^3 + 2x + 3)\,dx = [x^4 + x^2 + 3x]_{-3}^{0}$

$= (0^4 + 0^2 + 3 \times 0) -$

$((-3)^4 + (-3)^2 + 3 \times (-3))$

$= (0) - (81) = -81$

3 B

4 D

5 D

6 A

7 D

8 a) $A := \int_0^3 x^2\,dx$

$\Rightarrow A = \frac{1}{3}x^3|_0^3$

$\Rightarrow A = \frac{3^3}{3} = 9$

b) When $y = 0, 4x - x^2 = 0$

$\Rightarrow x(4 - x) = 0 \Rightarrow x = 0$ or $x = 4$

$\text{Area} = \int_0^4 (4x - x^2)\,dx = [2x^2 - \frac{1}{3}x^3]_0^4$

$= (2 \times 4^2 - \frac{1}{3} \times 4^3) - (2 \times 0^2 - \frac{1}{3} \times 0^3)$

$= 10\frac{2}{3}$ sq. units

c) Region 1:

$\text{Area} = \int_{-2}^0 (x^3 - 4x)\,dx = [\frac{1}{4}x^4 - 2x^2]_{-2}^0$

$= (\frac{1}{4} \times 0^4 - 2 \times 0^2) - (\frac{1}{4} \times (-2)^4 - 2 \times (-2)^2) = 4$ sq. units

Region 2:

$\text{Area} = \int_0^2 (x^3 - 4x)\,dx = [\frac{1}{4}x^4 - 2x^2]_0^2$

$= (\frac{1}{4} \times 2^4 - 2 \times 2^2) - (\frac{1}{4} \times 0^4 - 2 \times 0^2)$

$= -4$

Total area $= 4 + 4 = 8$ sq. units

Extending the rules

You are the examiner

Mo is right. Peter has made a mistake with the 2nd term when he re-wrote the function $\frac{1}{3x^2} = \frac{1}{3}x^{-2}$ not $\frac{1}{3x^2} = 3x^{-2}$ and he also made a mistake with the powers after he differentiated $x^{-\frac{2}{3}} = \frac{1}{\sqrt[3]{x^2}}$ and not $\frac{1}{\sqrt{x^3}}$.

Skill builder

1 D

2 A

3 A

4 E

5 B

6 A

7 B

8 D

9 A

Calculus with other functions

You are the examiner

Mo is right. Lilia worked in degrees – you must use radians when carrying out calculus with trig functions.

Skill builder

1 a) $f'(x) = 2\cos 2x - 3\sin 3x$

b) $f'(x) = 2\sec^2 2x$

c) $f'(x) = 4 \times (\frac{1}{2})\cos(\frac{x}{2}) - 3 \times (-\frac{1}{6})\sin(\frac{x}{6})$

$= 2\cos(\frac{x}{2}) + \frac{1}{2}\sin(\frac{x}{6})$

2 a) $\frac{dy}{dx} = 5e^{5x} + e^{-x} + e^x$

b) $y = e^{-3x} + e \Rightarrow \frac{dy}{dx} = -3e^{-3x}$

c) $y = \ln(4x) + \ln 2 = \ln 4 + \ln x + \ln 2$

$\Rightarrow \frac{dy}{dx} = \frac{1}{x}$

3 a) $-\frac{1}{2}\cos 2x + 4\sin(\frac{x}{4}) + c$

b) $\frac{1}{5}\tan(5x) + c$

c) $[3\sin x + 2\cos x]_{\frac{\pi}{3}}^{\frac{\pi}{6}} = (3\sin(\frac{\pi}{6}) + 2\cos(\frac{\pi}{6})) - (3\sin(\frac{\pi}{3}) + 2\cos(\frac{\pi}{3}))$

$= (3 \times \frac{1}{2} + 2 \times \frac{\sqrt{3}}{2}) - (3 \times \frac{\sqrt{3}}{2} + 2 \times \frac{1}{2})$

$= \frac{1-\sqrt{3}}{2}$

4 a) $\frac{1}{2}e^{2x} + 3e^x + c$

b) $5\ln|x| + c$

c) $\int_1^2 \frac{6e^{3x}}{e^x}\,dx = \int_1^2 6e^{2x}\,dx = [3e^{2x}]_1^2$

$= 3e^4 - 3e^2 = 142$ (to 3.s.f)

5 B

6 D

7 B

8 A

The chain rule

You are the examiner

Nasreen is right. Peter made a mistake differentiating u^{-5} – he wrongly added 1 to the power. Peter has also incorrectly moved the 5 to the denominator.

Skill builder

1 Let $u = 3x - 2$ in parts a)– f)

a) $\frac{dy}{dx} = 3 \times 4u^3 = 12(3x - 2)^3$

b) $y = (3x - 2)^{-7} \Rightarrow \frac{dy}{dx} = 3 \times (-7)u^{-8}$

$= \frac{-21}{(3x - 2)^8}$

c) $y = \sqrt{3x - 2} \Rightarrow (3x - 2)^{\frac{1}{2}}$

$\Rightarrow \frac{dy}{dx} = \frac{3}{2}(3x - 2)^{-\frac{1}{2}} = \frac{3}{2\sqrt{3x - 2}}$

d) $\frac{dy}{dx} = 3\cos u = 3\cos(3x - 2)$

e) $\frac{dy}{dx} = 3 \times \frac{1}{u} = \frac{3}{3x - 2}$

f) $\frac{dy}{dx} = 3 \times e^u = 3e^{3x-2}$

2 a) $u = 4x^2 - 5x + 1$ and $y = u^{10}$

$\frac{du}{dx} = 8x - 5$ and $\frac{dy}{du} = 10u^9$

$\frac{dy}{dx} = (8x - 5) \times 10u^9$

$= 10(8x - 5)(4x^2 - 5x + 1)^9$

b) $u = 3x^2$ and $y = e^u$

$\frac{du}{dx} = 6x$ and $\frac{dy}{du} = e^u$

$\frac{dy}{dx} = 6xe^u = 6xe^{3x^2}$

c) $u = x^4 - 3x^2$ and $y = \sqrt{u}$

$\frac{du}{dx} = 4x^3 - 6x$ and $\frac{dy}{du} = \frac{1}{2}u^{-\frac{1}{2}}$

$\frac{dy}{dx} = (4x^3 - 6x) \times \frac{1}{2}u^{-\frac{1}{2}}$

$= \frac{1}{2}(4x^3 - 6x)(x^4 - 3x^2)^{-\frac{1}{2}}$

$= \frac{2x^3 - 3x}{\sqrt{x^4 - 3x^2}}$

d) $u = e^{4x} - 1$ and $y = u^5$

$\frac{du}{dx} = 4e^{4x}$ and $\frac{dy}{du} = 5u^4$

$\frac{dy}{dx} = 4e^{4x} \times 5u^4 = 20e^{4x}(e^{4x} - 1)^4$

e) $u = \cos x$ and $y = u^2$

$\frac{du}{dx} = -\sin x$ and $\frac{dy}{du} = 2u$

$\frac{dy}{dx} = -\sin x \times 2u = -2\sin x\cos x$

f) $u = \cos x$ and $y = \ln u$

$\frac{du}{dx} = -\sin x$ and $\frac{dy}{du} = \frac{1}{u}$

$\frac{dy}{dx} = -\sin x \times \frac{1}{u} = -\frac{\sin x}{\cos x} = -\tan x$

3 a) $V = x^3 \Rightarrow \frac{dV}{dx} = 3x^2$

b) $\frac{dx}{dV} = \frac{1}{3x^2}$ and $\frac{dV}{dt} = 0.02$

So $\frac{dx}{dt} = \frac{dx}{dV} \times \frac{dV}{dt} = \frac{1}{3x^2} \times 0.02$

When $x = 4$:

$\frac{dx}{dt} = \frac{1}{3 \times 4^2} \times 0.02 = \frac{1}{2400}$

4 C

5 B

6 C

7 C

Product and quotient rules

You are the examiner

Peter is right – Nasreen has made mistakes when differentiating both u and v.

Skill builder

1 a) $u = \sqrt{x} \Rightarrow \frac{du}{dx} = \frac{1}{2}x^{-\frac{1}{2}}$

$v = (x + 4)^3 \Rightarrow \frac{dv}{dx} = 3(x + 4)^2$

So $\frac{dy}{dx} = (x + 4)^3 \times \frac{1}{2}x^{-\frac{1}{2}} + \sqrt{x} \times 3(x + 4)^2$

Tidy up to give

$\frac{dy}{dx} = \frac{(x + 4)^3}{2\sqrt{x}} + 3\sqrt{x}(x + 4)^2$

b) $u = 3x + 2 \Rightarrow \frac{du}{dx} = 3$

$v = 2x^2 - 1 \Rightarrow \frac{dv}{dx} = 4x$

So $\frac{dy}{dx} = \frac{(2x^2 - 1) \times 3 - (3x + 2) \times 4x}{(2x^2 - 1)^2}$

Tidy up to give $\frac{dy}{dx} = \frac{-6x^2 - 8x - 3}{(2x^2 - 1)^2}$

2 a) $u = x$ $\quad \frac{du}{dx} = 1$

$v = e^x$ $\quad \frac{dv}{dx} = e^x$

$\frac{dy}{dx} = xe^x + e^x \times 1$

$= e^x(x + 1)$

b) $u = x^2$ $\quad \frac{du}{dx} = 2x$

$v = \sin x$ $\quad \frac{dv}{dx} = \cos x$

$\frac{dy}{dx} = x^2\cos x + 2x\sin x$

c) $u = e^{2x}$ $\quad \frac{du}{dx} = 2e^{2x}$

$v = \ln x$ $\quad \frac{dv}{dx} = \frac{1}{x}$

$\frac{dy}{dx} = e^{2x} \times \frac{1}{x} + (\ln x) \times 2e^{2x}$

$= e^{2x}\left(\frac{1}{x} + 2\ln x\right)$

3 a) $u = 2x - 3$ $\quad \frac{du}{dx} = 2$

$v = 4x + 1$ $\quad \frac{dv}{dx} = 4$

$\frac{dy}{dx} = \frac{(4x + 1) \times 2 - (2x - 3) \times 4}{(4x + 1)^2}$

$= \frac{14}{(4x + 1)^2}$

b) $u = \sin x \quad \frac{du}{dx} = \cos x$

$v = x^2 + 1 \quad \frac{dv}{dx} = 2x$

$$\frac{dy}{dx} = \frac{(x^2+1)\times\cos x - (\sin x)\times 2x}{(x^2+1)^2}$$

$$= \frac{(x^2+1)\cos x - 2x\sin x}{(x^2+1)^2}$$

c) $u = \cos x \quad \frac{du}{dx} = -\sin x$

$v = e^x + x \quad \frac{dv}{dx} = e^x + 1$

$$\frac{dy}{dx} = \frac{(e^x+x)\times(-\sin x) - (\cos x)\times(e^x+1)}{(e^x+x)^2}$$

$$= \frac{-\sin x(e^x+x) - \cos x(e^x+1)}{(e^x+x)^2}$$

4 a) $u = \cos x \quad \frac{du}{dx} = -\sin x$

$v = \sin x \quad \frac{dv}{dx} = \cos x$

$f'(x) = -\sin x\sin x + \cos x\cos x$

$f'(\pi) = \cos^2\pi - \sin^2\pi = 1$

b) Given that $f(x) = \frac{\cos x}{\sin x}$ find $f'(\frac{\pi}{6})$

$u = \cos x \quad \frac{du}{dx} = -\sin x$

$v = \sin x \quad \frac{dv}{dx} = \cos x$

$$f'(x) = \frac{(\sin x)\times(-\sin x) - (\cos x)\times(\cos x)}{(\sin x)^2}$$

$$= \frac{-\sin^2 x - \cos^2 x}{\sin^2 x}$$

$$= \frac{-1}{\sin^2 x}$$

$$f'(\tfrac{\pi}{6}) = \frac{-1}{\sin^2(\frac{\pi}{6})} = \frac{-1}{(\frac{1}{2})^2} = -4$$

5 C

6 A

7 B

Integration by substitution

You are the examiner

Nasreen is correct. Peter hasn't adjusted the top of the first fraction properly – he needs to multiply by 2 and then divide by 2. $\frac{1}{x^2}$ is not in the form the 'top is the derivative of the bottom' so he should have written it as a power of x first as Nasreen did.

Skill builder

1 a) i) $\int \frac{2x-6}{x^2-6x-1}dx = \ln|x^2-6x-1| + c$

ii) $2\int \frac{2x-6}{x^2-6x-1}dx$

$= 2\ln|x^2-6x-1| + c$

iii) $\frac{1}{2}\int \frac{2x-6}{x^2-6x-1}dx$

$= \frac{1}{2}\ln|x^2-6x-1| + c$

b) i) $\int \frac{2e^{2x}}{e^{2x}+5}dx = \ln|e^{2x}+5| + c$

ii) $\frac{1}{2}\int \frac{2e^{2x}}{e^{2x}+5}dx = \frac{1}{2}\ln|e^{2x}+5| + c$

iii) $\frac{5}{2}\int \frac{2e^{2x}}{e^{2x}+5}dx = \frac{5}{2}\ln|e^{2x}+5| + c$

c) i) $\int \frac{\cos x}{\sin x}dx = \ln|\sin x| + c$

ii) $\frac{1}{3}\int \frac{\cos x}{\sin x}dx = \frac{1}{3}\ln|\sin x| + c$

iii) $-\int \frac{-\sin x}{\cos x}dx = -\ln|\cos x| + c$

2 a) Let $u = 2x + 1 \Rightarrow \frac{du}{dx}$

$= 2 \Rightarrow \frac{1}{2}du = dx$

$\int (2x+1)^3 dx = \int \frac{1}{2}u^3 du = \frac{1}{8}u^4 + c$

$= \frac{1}{8}(2x+1)^4 + c$

b) Let $u = 4x + 3 \Rightarrow \frac{du}{dx} = 4$

$\Rightarrow \frac{1}{4}du = dx$

$\int \sqrt{4x+3}\,dx = \int \sqrt{u}\,\frac{1}{4}du$

$= \int \frac{1}{4}u^{\frac{1}{2}} du = \frac{1}{4} \times \frac{2}{3}u^{\frac{3}{2}} + c$

$= \frac{1}{6}(4x+3)^{\frac{3}{2}} + c$

3 A, E

4 B

5 D

6 C

Integration by parts

You are the examiner

Lilia is right. You can't integrate $\ln x$ directly but you can differentiate it to get $\frac{1}{x}$ (which is simpler) so Mo was wrong to choose $\frac{dv}{dx}$ as $\ln x$.

Skill builder

1 a) $\int xe^x\,dx$

$u = x \xrightarrow{\text{differentiate}} \frac{du}{dx} = 1$

$\frac{dv}{dx} = e^x \xrightarrow{\text{integrate}} v = e^x$

$\int xe^x\,dx = xe^x - \int e^x\,dx$

$= xe^x - e^x + c$

b) $\int x\ln x\,dx$

$u = \ln x \xrightarrow{\text{differentiate}} \frac{du}{dx} = \frac{1}{x}$

$\frac{dv}{dx} = x \xrightarrow{\text{integrate}} v = \frac{1}{2}x^2$

$\int x\ln x\,dx = \frac{1}{2}x^2\ln x - \int \frac{1}{2}x^2 \times \frac{1}{x}dx$

$= \frac{1}{2}x^2\ln x - \int \frac{1}{2}x\,dx$

$= \frac{1}{2}x^2\ln x - \frac{1}{4}x^2 + c$

2 a) $\int x\cos x\,dx$

$u = x \xrightarrow{\text{differentiate}} \frac{du}{dx} = 1$

$\frac{dv}{dx} = \cos x \xrightarrow{\text{integrate}} v = \sin x$

$\int x\cos x\,dx = x\sin x - \int \sin x\,dx$

$= x\sin x + \cos x + c$

b) $\int 9xe^{3x}\,dx$

$u = 9x \xrightarrow{\text{differentiate}} \frac{du}{dx} = 9$

$\frac{dv}{dx} = e^{3x} \xrightarrow{\text{integrate}} v = \frac{1}{3}e^{3x}$

$\int 9xe^{3x}\,dx = 9x \times (\frac{1}{3}e^{3x}) - \int \frac{1}{3}e^{3x} \times 9\,dx$

$= 3xe^{3x} - \int 3e^{3x}\,dx$

$= 3xe^{3x} - e^{3x} + c$

c) $\int 2x\sin 2x\,dx$

$u = 2x \xrightarrow{\text{differentiate}} \frac{du}{dx} = 2$

$\frac{dv}{dx} = \sin 2x \xrightarrow{\text{integrate}}$

$v = -\frac{1}{2}\cos 2x$

$\int 2x\sin 2x\,dx = 2x \times (-\frac{1}{2}\cos 2x) - \int 2 \times (-\frac{1}{2}\cos 2x)dx$

$= -x\cos 2x + \int \cos 2x\,dx$

$= -x\cos 2x + \frac{1}{2}\sin 2x + c$

3 C

4 D

5 C

6 A

Further calculus

You are the examiner

Mo is right, Sam has forgotten to multiply by $\frac{dy}{dx}$ when he differentiated y^2 and he has differentiated an expression so shouldn't have written $\frac{dy}{dx} =$ on the left hand side. Sam would have been correct to write $\frac{dy}{dx} =$ if he had been asked to differentiate $y = \ldots$

Skill builder

1 a) i) $5x^4$

ii) $5y^4\frac{dy}{dx}$

iii) 0

iv) $5x^4 + 5y^4\frac{dy}{dx}$

b) i) $3e^{3x}$

ii) $2e^{2y}\frac{dy}{dx}$

iii) $3\frac{dy}{dx}$

iv) $3e^{3x} + 2e^{2y}\frac{dy}{dx} + 3\frac{dy}{dx}$

c) i) $-2\sin 2x$

ii) $3\cos 3y\frac{dy}{dx}$

iii) $\sec^2 y\frac{dy}{dx}$

iv) $-2\sin 2x + 3\cos 3y\frac{dy}{dx} + \sec^2 y\frac{dy}{dx}$

d) i) $12x^5$

ii) $3y^2\frac{dy}{dx}$

iii) $\frac{dy}{dx}[3x^2y^3]$

$= 6xy^3 + (3x^2)(3y^2) \times \frac{dy}{dx}$

$= 6xy^3 + 9x^2y^2\frac{dy}{dx}$

e) i) $2e^{2x}$

ii) $\cos y \frac{dy}{dx}$

iii) $e^{2x}\cos y \frac{dy}{dx} + 2e^{2x}\sin y$

2 A

3 A

4 C

5 a) $\frac{dy}{dx} = \frac{6x^2}{y^3} \Rightarrow \int y^3\,dy = \int 6x^2\,dx$

$\frac{1}{4}y^4 = \frac{6}{3}x^3 + c \Rightarrow y^4 = 8x^3 + c_1$

b) When $x = 2, y = 3$

so $c_1 = 3^4 - 8 \times 2^3 = 17$

so $y^4 = 8x^3 + 17$

6 $\frac{dy}{dx} = 3x^2y \Rightarrow \int \frac{1}{y}dy = \int 3x^2\,dx$

$\ln y = x^3 + c \Rightarrow y = e^{x^3+c} \Rightarrow y = Ae^{x^3}$

When $x = 0$ then $y = 10$, so

$10 = Ae^0 \Rightarrow A = 10$

So $y = 10e^{x^3}$

7 a) $\frac{dy}{dx} = \frac{4\cos 2x}{e^y} \Rightarrow \int e^y\,dy$

$= \int 4\cos 2x\,dx$

$e^y = 2\sin 2x + c \Rightarrow y = \ln|2\sin 2x + c|$

b) $0 = \ln|2\sin(2\frac{\pi}{2}) + c| \Rightarrow$

$0 = \ln|2 \times 0 + c| \Rightarrow 0 + c = 1 \Rightarrow c = 1$

$y = \ln|2\sin 2x + 1|$

Parametric equations

You are the examiner

Sam is right. Peter has made a mistake when he differentiated $y = 3\ln t$ and when he divided by $\frac{dx}{dt}$. When you have to divide by a fraction it is easier to turn it upside down and multiply.

Skill builder

1 a) When $t = 3$: $x = 4 \times 3 - 1 = 11$ and $y = 3^2 + 3 \times 3 = 18$ so point is $(11, 18)$

b) $\frac{dy}{dt} = 2t + 3$ and $\frac{dx}{dt} = 4$ so

$\frac{dy}{dx} = \frac{2t+3}{4}$

c) $\frac{dy}{dx} = 0 \Rightarrow \frac{2t+3}{4} = 0 \Rightarrow t = -\frac{3}{2}$

When $t = -\frac{3}{2}$ then

$x = 4 \times (-\frac{3}{2}) - 1 = -7$ and

$y = (-\frac{3}{2})^2 + 3 \times (-\frac{3}{2}) = -2.25$

So stationary point is at $(-7, -2.25)$

d) $x = 4t - 1 \Rightarrow t = \frac{x+1}{4}$

So $y = \left(\frac{x+1}{4}\right)^2 + 3\left(\frac{x+1}{4}\right)$

2 a) $t = \frac{\pi}{6}$: $x = 4\cos(\frac{\pi}{6}) = 4 \times (\frac{\sqrt{3}}{2}) = 2\sqrt{3}$

and $y = 4\sin(\frac{\pi}{6}) = 4 \times \frac{1}{2} = 2$

So the point is $(2\sqrt{3}, 2)$

b) $\frac{dy}{dt} = 4\cos t$ and $\frac{dx}{dt} = -4\sin t$

So $\frac{dy}{dx} = \frac{4\cos t}{-4\sin t} = -\frac{\cos t}{\sin t} = -\cot t$

c) $\frac{dy}{dx} = -\frac{\cos(\frac{\pi}{6})}{\sin(\frac{\pi}{6})} = -\frac{\frac{\sqrt{3}}{2}}{\frac{1}{2}} = (-\frac{\sqrt{3}}{2}) \div \frac{1}{2}$

$= (-\frac{\sqrt{3}}{2}) \times 2 = -\sqrt{3}$

d) i) Tangent is line through $(2\sqrt{3}, 2)$ with gradient $-\sqrt{3}$ is

$y - 2 = -\sqrt{3}(x - 2\sqrt{3})$

$y - 2 = -\sqrt{3}x + 6 \Rightarrow y = 8 - \sqrt{3}x$

ii) Gradient of normal is $\frac{1}{\sqrt{3}}$ so equation of normal

$y - 2 = \frac{1}{\sqrt{3}}(x - 2\sqrt{3})$

$\sqrt{3}y - 2\sqrt{3} = x - 2\sqrt{3}$

$\Rightarrow \sqrt{3}y = x$

e) circle of radius 4, centre $(0, 0)$ is $x^2 + y^2 = 16$

3 B

4 D

5 A

Sampling and displaying data 1

You are the examiner

Lilia is correct. Mo has not used the stem as part of the number at all.

Skill builder

1 D

2 C

3
- Simple random sample: easy to understand, not biased, difficult to find the people to include in the survey.
- Systematic sample: may be biased if the list is ordered in some way, difficult to find the people to include in the survey.
- Cluster sample: simpler to collect the data, easy to introduce bias.
- Stratified sample: representative of a population which is already in groups, difficult to find the people to include in the survey.
- Opportunity sample: easy to collect but likely to be biased.
- Self-selecting sample: easy to collect if sufficient responses are received but most likely to be biased as only people with strong feelings will be included.

4 a) In the order they arise

3	5 8 5
4	1 2 5
5	2 8 2 7 4 8
6	7 5 8 4

Ordered

3	5 5 8
4	1 2 5
5	2 2 4 7 8 8
6	4 5 7 8

Key: 3|5 means 35

b) Modes 35, 52 and 58; median 53

5 Modes Africa 61, 66, 72 and 76. Mode Asia 76, which is equal to one of the Africa modes, but higher than the others. Median Africa 8th value 66. Median Asia 10th value 73, which is higher also.

Displaying data 2

You are the examiner

Peter is correct. Lilia has used the vertical scale as frequency. She should have added the areas.

Skill builder

1

Age	Frequency	fr density
$0 \leqslant x < 5$	9	1.8
$5 \leqslant x < 9$	12	3
$9 \leqslant x < 11$	10	5
$11 \leqslant x < 14$	14	4.667
$14 \leqslant x < 19$	2	0.4

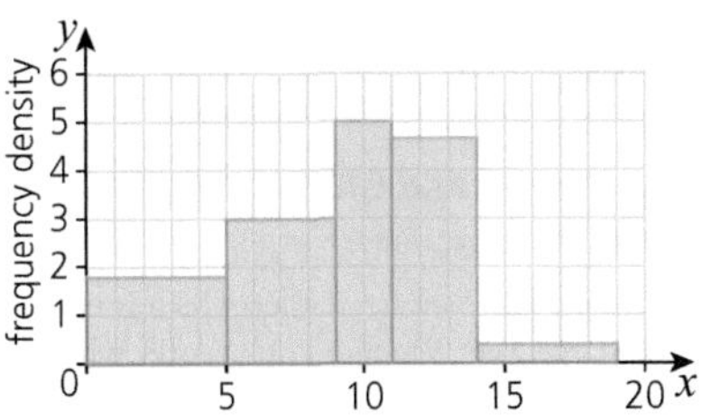

Negative skew.

2 a)

Speed (mph)	Lower boundary	Upper boundary	Class width	Frequency	Frequency density
20–29	19.5	29.5	10	70	7
30–34	29.5	34.5	5	40	8
35–44	34.5	44.5	10	27	2.7
45–60	44.5	60.5	16	3	0.1875

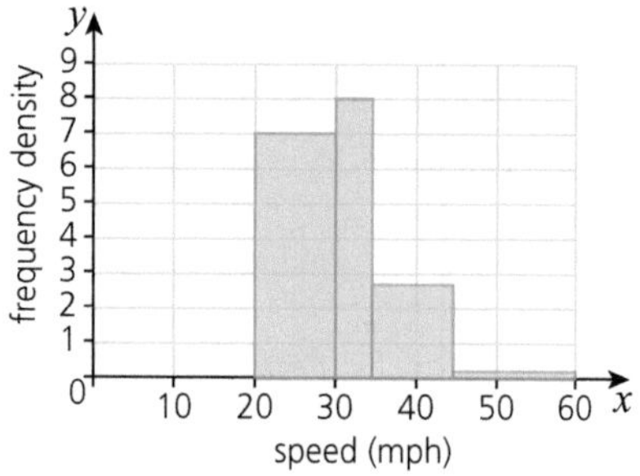

b) positive skew

3

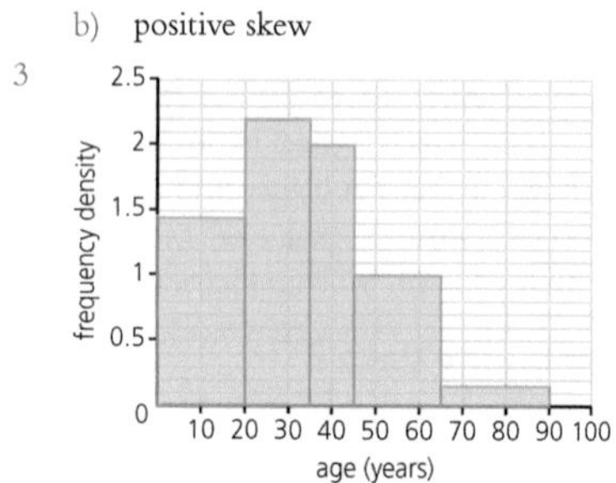

Age (years)	Frequency	Frequency density
$0 \leqslant x < 20$	29	1.45
$20 \leqslant x < 35$	33	2.2
$35 \leqslant x < 45$	20	2
$45 \leqslant x < 65$	20	1
$65 \leqslant x < 90$	4	0.16

Averages and measures of spread

You are the examiner

Lilia is right. Sam has a value which is much too large. Peter's method only works when there are equal numbers of men and women.

Skill builder

1 a) Mean = 562.9, mode = 568, median = 565

b) Range 590 − 529 = 61, IQR = 570 − 554 = 16, standard deviation $\sigma_x = 15.78, s_x = 16.64$

c) $562.9 \pm 2 \times 15.78$ – outliers below 531.34 or above 594.46

$562.9 \pm 2 \times 16.64$ – outliers below 529.62 or above 596.18

529 is an outlier.

d) $LQ - 1.5 \times IQR = 554 - 1.5 \times 16 = 530$ – so below 530

$UQ + 1.5 \times IQR = 570 + 1.5 \times 16 = 594$ – so above 594

529 is an outlier.

2 Mode £5, median £7, mean £8.59 So they are both right and both wrong.

3 Gary is wrong: the mean does not have to be a whole number.

Henry is right: the total will be approximately 1.7 × 100.

4 Colin: mean = 13.7, s.d. is 1.49 or 1.58.

Ahmed: mean = 14.53, s.d. is 1.14 or 1.20.

So on average, Colin is quicker (lower mean time).

Ahmed is more consistent (smaller s.d.).

5 a) Total £275

b) total for other 9: 275 − 120 = 155.

Mean: $\frac{155}{9}$ = £17.22

c) Average for 11 to be £30 total needs to be £330

Person must give £330 − 275 = £55

Cumulative frequency graphs and boxplots

You are the examiner

Nasreen and Sam are correct. Mo is not correct – there are other countries in the world. Lilia is not correct as we know about the rate and not the numbers of infant deaths.

Skill builder

1 B

2 Median 55, LQ 36, UQ 66

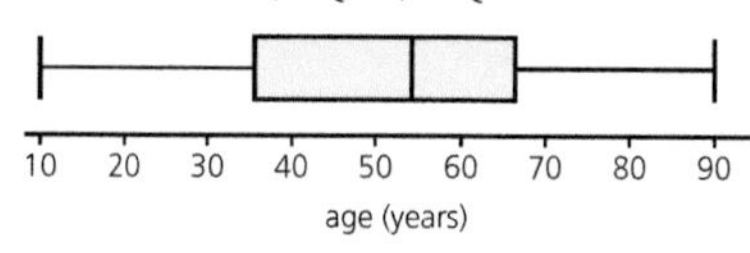

3 a) Cum Fr 3,13,41,89,144,204

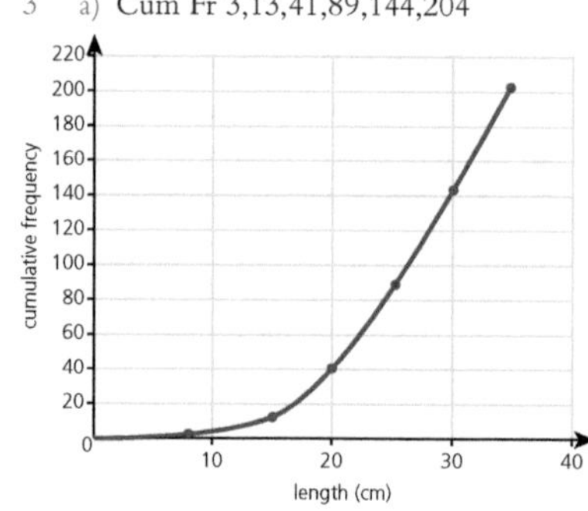

b) Median = 26, LQ = 21, UQ = 31

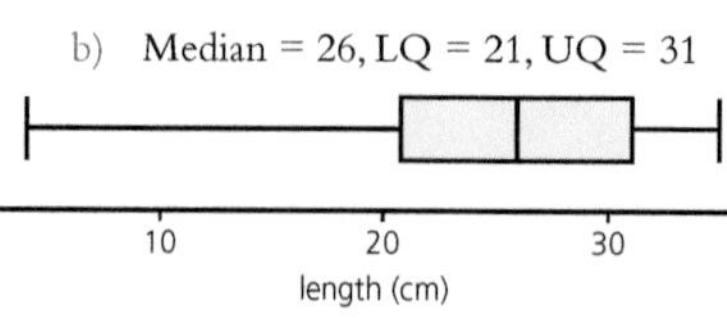

Grouped frequency calculations

You are the examiner

Nasreen is correct. Lilia has used the largest value in each class and Peter the smallest value. Mo has found the average frequency in each class.

Skill builder

1 a) Mean = 4.65, mode = 3, median = 4

Standard deviation

$\sigma_x = 2.46, s_x = 2.49$ and range $10 - 2 = 8$

b) On average the words are shorter in the children's book (mean 3.6 compared with 4.63). There is more variation in the A level textbook (s.d 2.46 compared with 1.34)

2 A

3 E

4 White-faced

Midpoint	Frequency
2.7	0
2.9	1
3.1	6
3.35	34
3.65	14

Mean = 3.39 kg $\sigma_n = 0.178$ kg;

Kaapori

Midpoint	Frequency
2.7	7
2.9	42
3.1	42
3.35	8
3.65	0

Mean = 3.01 kg $\sigma_n = 0.157$ kg;

5 White-faced 27.5th value in 3.2–3.5 class

Median = $3.2 + \frac{20.5}{34} \times 0.3 = 3.38$

Kaapori 49.5th value in 3.0–3.2 class

Median = $3.0 + \frac{0.5}{42} \times 0.2 = 3.002$

6 White-faced capuchins are heavier on average than kaapori capuchins. There is more variation in white-faced capuchins than kaapori capuchins.

Probability

You are the examiner

Peter is correct – the total of all the numbers is 80. Mo has used 32 and 15 for hearts only and spots only, not subtracting the 11 that are in the intersection. His total is correct (he has 22 not 44 in the border). Mo also used $H' \cup S'$ when he means the intersection $H' \cap S'$.

Skill builder

1 C

2 C

3

A		B
38	21	18
23 (outside)		

4 a)

	Fiction	Not fiction
Hardback	4	21
Not hardback	17	18

b) Hardback fiction books $\frac{4}{60} = \frac{1}{15}$

5 a) $P(A) = \frac{35 + 29}{120} = \frac{8}{15}$

$P(B) = \frac{35 + 47}{120} = \frac{41}{60}$

b) $P(A \cap B) = \frac{35}{120} = \frac{7}{24}$

c) $P(A) \times P(B) = \frac{8}{15} \times \frac{41}{60}$

$= \frac{82}{225} \neq P(A \cap B)$

So not independent.

Conditional probability

You are the examiner

Peter is correct. Lilia has two errors – she should have P(*B*) in the formula. The 0.3 is not the probability of A.

Skill builder

1 B

2 E

3 a) $P(H) = \frac{58 + 55}{300} = \frac{113}{300}$

b) $P(M \cap H) = \frac{58}{300} = \frac{29}{150}$

c) $P(H|M) = \frac{58}{58 + 120} = \frac{29}{89}$

d) $P(M|H) = \frac{58}{58 + 55} = \frac{58}{113}$

4 a) $P(F \cap N) = 0.45$

b) $P(F) = 0.05 + 0.45 = 0.5$

c) $P(N|F) = \frac{P(N \cap F)}{P(F)} = \frac{0.45}{0.5} = 0.9$

5 a)

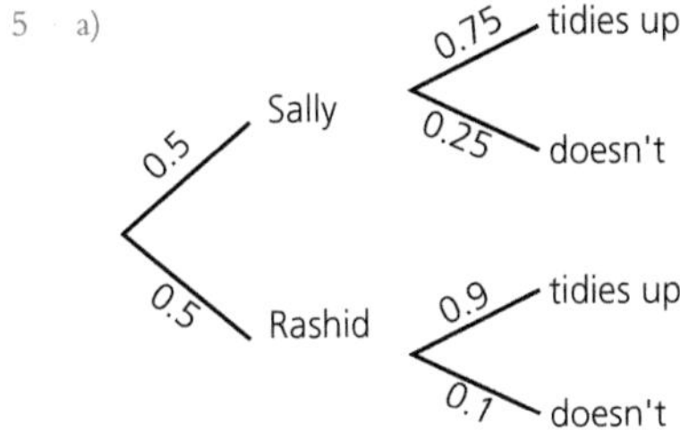

b) $P(\text{no tidying}) = 0.5 \times 0.25 + 0.5 \times 0.1$
$= 0.175$

$P(\text{Sally}|\text{no tidying}) = \frac{0.5 \times 0.25}{0.175} = \frac{5}{7}$

Discrete random variables

You are the examiner

Peter is correct. Nasreen has omitted the possibility of not winning so her probabilities do not add to 1.

Skill builder

1 E

2 B

3 a) $P(\text{not } 2) = 0.6 + 0.25 + 0.1 = 0.95$

b) $p = 1 - 0.95 = 0.05$

4 a) $k + k + k + 2k + 3k + 2k = 1$ so $k = \frac{1}{10}$

b) $P(Y \leq 30) = k + k + k = 0.3$

5 a)

Y	1	4	9	16	25
$P(Y = y)$	k	2k	3k	4k	5k

So $15k = 1$ giving $k = \frac{1}{15}$

b) $P(Y = 16, 25) = 4k + 5k = \frac{9}{15} = \frac{3}{5}$

6 a)

No. of heads x	0	1	2
$P(X = x)$	0.25	0.5	0.25

b)

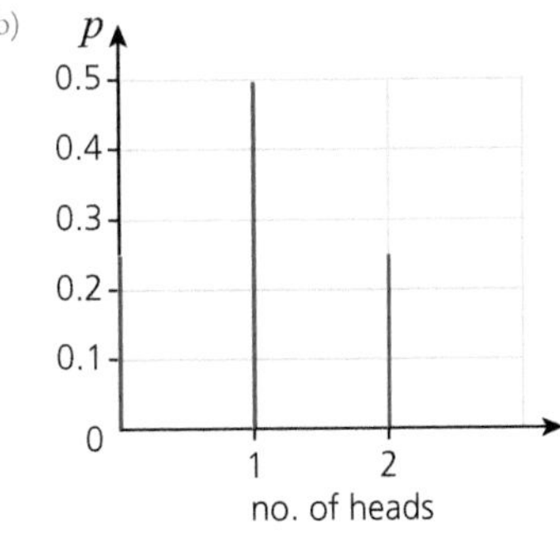

7 The score only takes the values 1, 2, 3, 4 and 5 so discrete. If the spinner is fair the probability of each score is equal so the distribution is uniform.

Binomial distribution

You are the examiner

Mo is correct. Lilia has correctly used the inverse binomial function and found the value with the probability nearest to 10% but it is over, not under, 10%.

Skill builder

1 a) $np = 20 \times 0.4 = 8$

b) $np = 16 \times 0.5 = 8$

c) $np = 10 \times 0.35 = 3.5$

d) $np = 17 \times 0.25 = 4.25$

2 A

3 a) 0.1366

b) 0.5772

c) 0.4228

d) 0.5594

4 a) 0.3044

b) 0.8220

c) 0.0056

d) 0.0164

5 E

6 Number correct $X \sim B(20, 0.5)$

a) $P(X = 8) = 0.1201$

b) $P(X \geq 15) = 0.0207$

c) $P(X \leq 10) = 0.5881$

d) $P(X > 12) = 0.1316$

Hypothesis testing (binomial)

You are the examiner

Nasreen is correct. Peter's solution has correct probabilities, so $P(X > 26) = 1 - 0.9626 < 5\%$ so 26 is not in the critical region. His conclusion follows from his working but it is the wrong conclusion. He uses the word 'proof' which is never correct for a hypothesis test.

Skill builder

1 a) $P(X \leq 5) = 0.0338 < 5\%$

$P(X \leq 6) = 0.0950 > 5\%$

b) $P(X \leq 7) = 0.0121 < 2.5\%$

$P(X \leq 8) = 0.0312 > 2.5\%$

c) $P(X \geq 13) = 0.0271 < 5\%$

$P(X \geq 12) = 0.0905 > 5\%$

2 a) i) $P(X \leq 12) = 0.0386$

ii) $P(X \leq 13) = 0.0751$

b) i) $P(X \geq 23) = 0.0767$

ii) $P(X \geq 24) = 0.0405$

3 a) $P(X \leq 9) = 0.0214 < 2.5\%$

b) $P(X \geq 21) = 0.0214 < 2.5\%$

c) Critical region $X \leq 9$ or $X \geq 21$

d) $X = 8$ is in the critical region, so reject H_0 there is enough evidence at the 5% level that $p \neq 0.5$

4 B

5 C

Normal distribution

You are the examiner

Mo is correct. Nasreen has the correct boundary values but uses the extremes and not the central area. Lilia does not have a central region but finds the value below which 90% of the population lies. Peter has 10% in each tail, so only 80% of the population lie in his interval.

Skill builder

1 a) 0.2104

b) 0.0483

c) 0.8907

d) 0.0344

2 a) 0.9938

b) 0.1056

c) 0.2660

d) 0.6462

3 a) 0.3781

b) 0.2525

c) 0.9088

d) 0.0038

4 a) 0.3085

b) 0.2266

c) 0.0699 (4 d.p.)

5 a) 45.62

b) 53.41

c) 58.33

d) 48.35

6 Using either Central with area 0.75 or left tail with area 0.125 and left tail with area 0.875 giving $46.20 < X < 73.80$

Hypothesis test (normal)

You are the examiner

Nasreen is correct – there are equally extreme values for *X* at the top and bottom end. Lilia has only considered the bottom tail.

Skill builder

1 B

2 A

3 a) Using either central 99% for N(0, 1) or left tail for 0.5% gives $Z < -2.576$ or $Z > 2.576$

b) Using either central 90% for N(0, 1) or left tail for 5% gives $Z < -1.645$ or $Z > 1.645$

4 a) $\bar{x} = \frac{184}{5} = 36.8$ $\bar{X} \sim N\left(45, \frac{6^2}{5}\right)$

So $z_{test} = \frac{36.8 - 45}{\left(6/\sqrt{5}\right)} = -3.056$

b) $P(Z < -3.056) = 0.0011$

2-tailed so p-value $= 2 \times 0.0011$
$= 0.0022$

5 a) $H_0 : \mu = 21 \; H_1 : \mu > 21$

b) $\overline{X} = \frac{140.2}{6} = 23.37$

c) $\overline{X} \sim N\left(21, \frac{2.8^2}{6}\right)$ so

$z_{test} = \frac{23.37 - 21}{\left(2.8/\sqrt{6}\right)} = 2.07$

d) Critical region left tail $\overline{X} > 22.88$ or for 95% N(0, 1) $Z > 1.645$

e) $z_{test} = 2.07$ is in the critical region so reject H_0. There is enough evidence at the 5% level that the mean length of fish has increased from 21 cm.

f) p-value = $P(Z > 2.070) = 0.0192 < 5\%$

Bivariate data

You are the examiner

Mo is correct. Peter does not define ρ. He uses the wrong alternative hypothesis for positive correlation. His test is otherwise correct. There is no reason to think that Alan could measure the wind more accurately than the data he has used.

Skill builder

1 D

2 a) Critical region $X > 0.7545$. Actual value $0.7012 < 0.7545$ so accept H_0. There is insufficient evidence of positive correlation in the population.

b) Critical region $X < -0.8329$. Actual value $-0.8521 < -0.8329$ so reject H_0. There is sufficient evidence of negative correlation in the population.

c) Critical region $X < -0.7545$ or $X > 0.7545$. Actual value $0.7243 < 0.7545$ so accept H_0. There is in sufficient evidence of correlation in the population.

d) Critical region $X < -0.8745$ or $X > 0.8745$. Actual value $-0.8916 < -0.8745$ so reject H_0. There is sufficient evidence of correlation in the population.

Working with graphs

You are the examiner

B, C, G, E, A, D, F

Skill builder

1 B

2 D

3

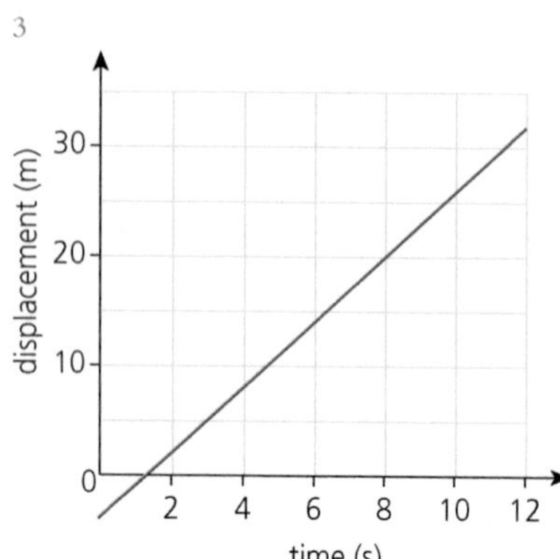

4

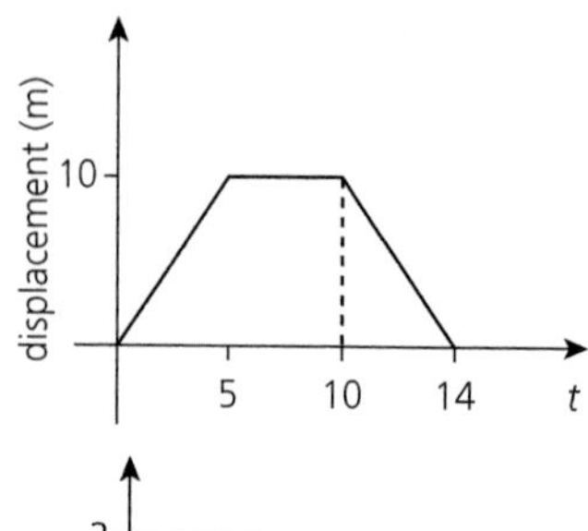

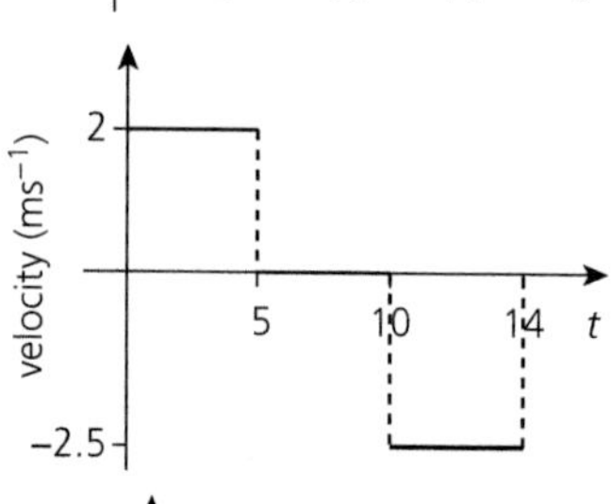

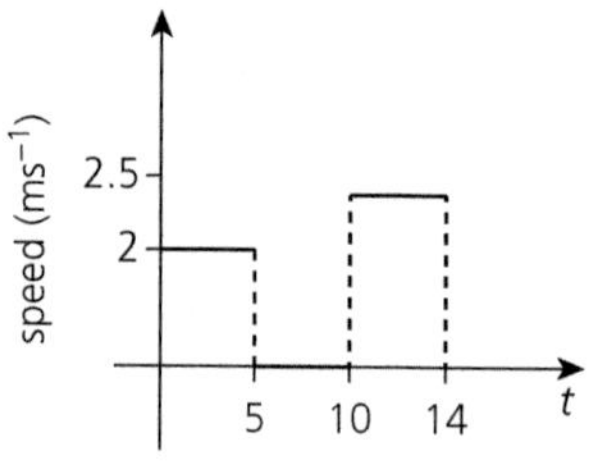

5 a)

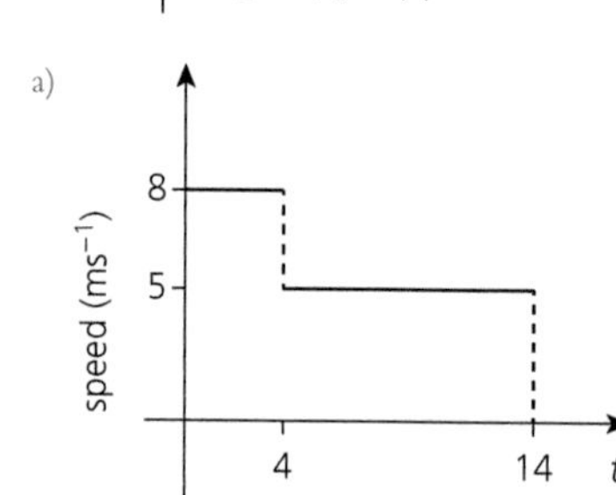

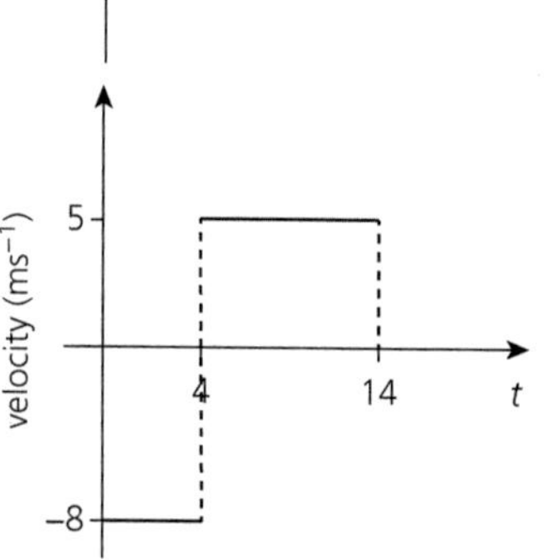

b)

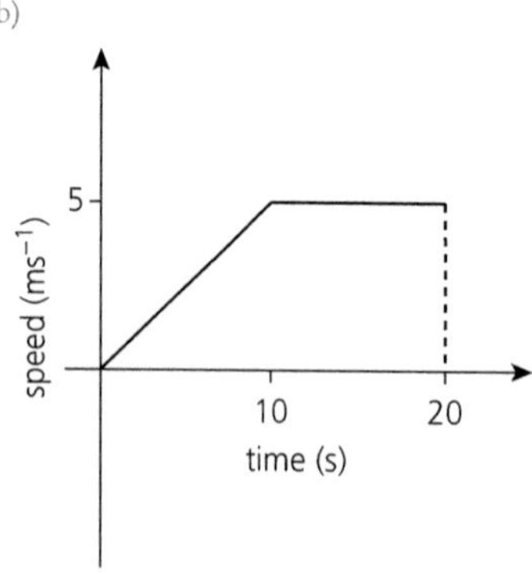

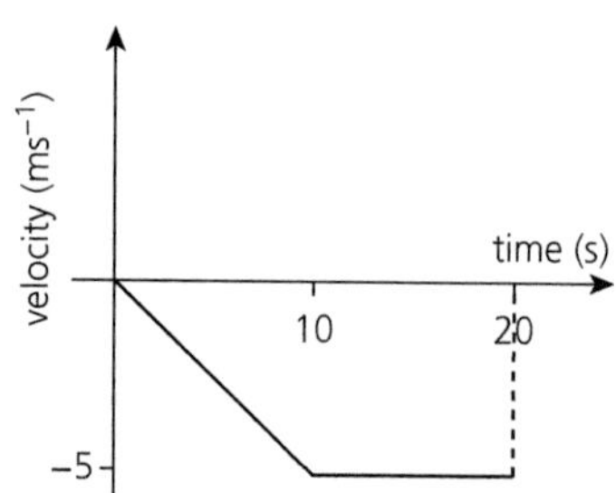

6 a) acceleration = $0\,\text{ms}^{-2}$ throughout, distance $= 1 \times 5 + 3 \times 5 = 20$ m, displacement $= (-1) \times 5 + 3 \times 5 = 10$ m

b) acceleration $= 1\,\text{ms}^{-2}$, then $0\,\text{ms}^{-2}$ during plateau, distance $= \frac{1}{2}(4 \times 4) + \frac{1}{2}(5 + 7) \times 2 = 20$ m, displacement $= -8 + 12 = 4$ m

c) acceleration across all 3 stages $= -1\,\text{ms}^{-2}, 0\,\text{ms}^{2}, 3\,\text{ms}^{-2}$, distance $= \frac{1}{2}(3 \times 3) + \frac{1}{2}(8 + 4) \times 3 = 22.5$ m, displacement $= 4.5 - 18 = -13.5$ m

suvat equations

You are the examiner

Mo is correct. Lilia has two sign errors. She has not used the initial height above the ground.(b) She has not used the upward part of the journey.

Skill builder

1 $v = 4 - 2t, \quad s = 4t - \frac{1}{2} \times 2t^2$

t	0	1	2	3	4	5
v	4	2	0	−2	−4	−6
s	0	3	4	3	0	−5

Mel starts at the origin, slows down and stops after 2 s 4 m beyond O. She changes direction and returns to the start after 4 s. She accelerates in the negative direction.

2 a) $u = 0, v = 7, a = 3.5$, Find t

b) $u = 20, v = 25, t = 40$, Find s

c) $u = 4, t = 5, a = -0.2$, Find s

d) $s = 200, u = 0, v = 18$, Find a

e) $s = 10, v = 0, a = -9.8, t = 3$ Find u

3 a) $v = u + at, \quad 7 = 0 + 3.5t \quad t = 2\,\text{s}$

b) $s = \frac{1}{2}(u + v)t = \frac{1}{2}(20 + 25) \times 40 = 900\,\text{m}$

c) $s = ut + \frac{1}{2}at^2$

$= 4 \times 5 - \frac{1}{2} \times 0.2 \times 5^2 = 17.5\,\text{m}$

d) $v^2 = u^2 + 2as$

$18^2 = 0^2 + 2a \times 200$

$a = 0.81\,\text{ms}^{-2}$

e) $v^2 = u^2 + 2as$

$0^2 = u^2 - 2 \times 9.8 \times 10$

$u = \sqrt{196} = 14\,\text{ms}^{-1}$

4 A

5 D

6 $2 = 8t - \frac{1}{2} \times 9.8t^2$ so $t = 0.3081, 1.3244$ so 1.02 s

7 AB $v = u + at = 1.5 \times 10 = 15\text{ms}^{-1}$
BC $v^2 = u^2 + 2as = 15^2 + 2 \times 0.5 \times 100$
so $v = \sqrt{325} = 18.0\ \text{ms}^{-1}$

Variable acceleration

You are the examiner

Nasreen is correct. Lila used *suvat* when the acceleration is not constant. She also simplified incorrectly.

Skill builder

1 B

2 A

3 A

4 a) $-3t^2 + 30t - 63 = 0$ So $t = 3, 7$ s. The insect changes direction.

b) $\int_2^3 (30t - 3t^2 - 63)\,dT = -7$ m

c) $\int_3^4 (30t - 3t^2 - 63)\,dT = 5$

d) total distance $7 + 5 = 12$ m

5 a) $v = \frac{ds}{dt} = 1.2t^2 - 1.4t$

b) $a = \frac{dv}{dt} = 2.4t - 1.4$ which is not constant

6 a) $v = \int a\,dt = 1.5t^2 - 0.25t^3 + c$ and $c = 0$

b) When $t = 4$, $v = 8\,\text{m s}^{-1}$

c) $s = \int v\,dt = 0.5t^3 - 0.0625t^4 + c$ and $c = 0$

d) When $t = 4$, $s = 16$ m

e) Distance left 84 m at $8\,\text{m s}^{-1}$ so 10.5 s. Total time 14.5 s.

Understanding forces

You are the examiner

Mo is correct. Lilia has no normal reaction or resistance, Peter has no normal reaction and Nasreen uses the wrong direction for weight, unlabelled force.

Skill builder

1 D

2 D

3 a)

N P F W

b)

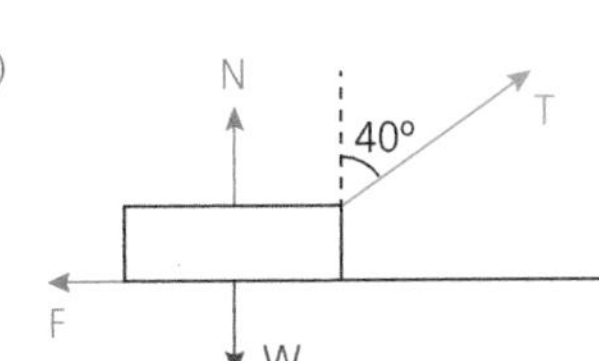

c)

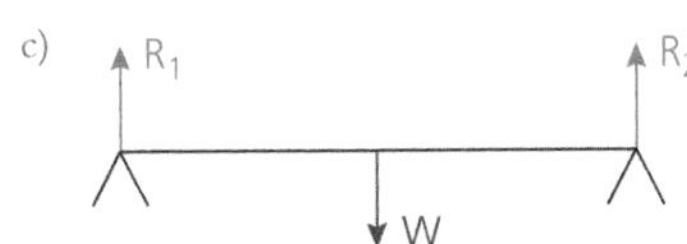

d)

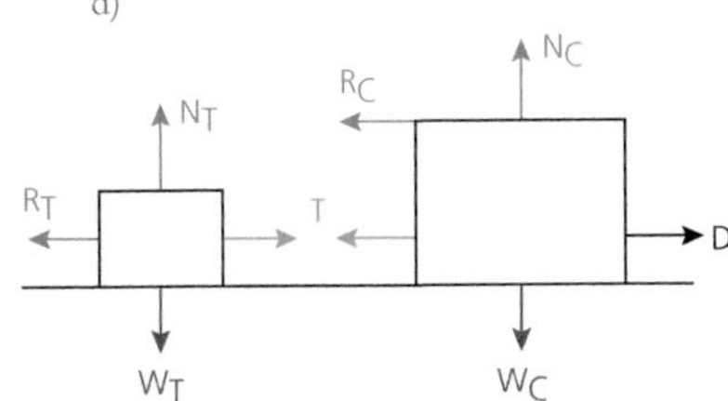

4

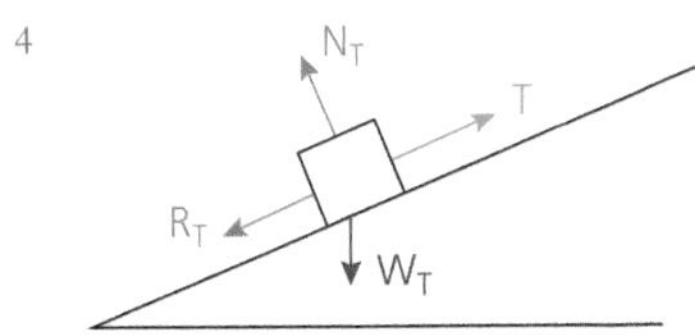

5 a)

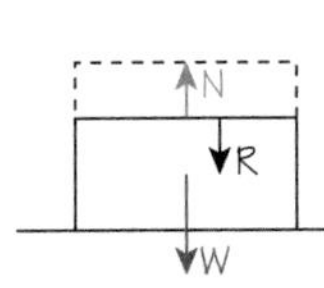

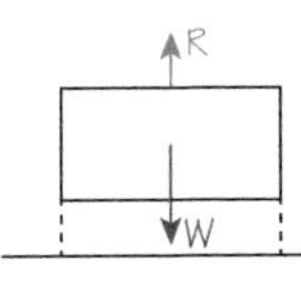

b) He is not right. The contact force between the upper and lower bricks has the same size and direction as the weight of the upper brick, but the weight of the upper brick acts on the upper brick.

Forces in a line

You are the examiner

Mo is correct. Lilia has incorrectly put the tension equal to the weight of the sphere. Peter has also done that, but has included the vertical weight in the horizontal motion of the box.

Skill builder

1 a) $700 - 200 = 800a$ so $a = 0.625\,\text{m s}^{-2}$

b) $720 - 150 = 800a$ so $a = 0.7125\,\text{m s}^{-2}$

c) $200 - 250 = 800a$ so $a = -0.0625\,\text{m s}^{-2}$

2 a) $D = ma = 4.2 \times 1.2 = 5.04$ N

b) $D - 8 = ma = 4.2 \times 2.5$ so $D = 18.5$ N

c) $D - 4.5 = ma = 4.2 \times (-0.5) = -2.1$ so $D = 2.4$ N

3 B

4 B

5 D

Using vectors

You are the examiner

Peter is correct. Mo has put the weight in upwards. Lilia has forgotten the weight altogether. Nasreen has found the magnitude of the forces which do not add to the magnitude of the resultant force.

Skill builder

1 a) $\mathrm{r} = \frac{1}{2}(\mathrm{u} + \mathrm{v})t$

$= \frac{1}{2}((\mathrm{i} - \mathrm{j}) + (-\mathrm{i} + 3\mathrm{j})) \times 2 = 2\mathrm{j}$

$\mathrm{v} = -\mathrm{i} + 3\mathrm{j} = \mathrm{u} + at = (\mathrm{i} - \mathrm{j}) + 2\mathrm{a}$

so $\mathrm{a} = -\mathrm{i} + 2\mathrm{j}$

b) $r = \frac{1}{2}(\mathrm{u} + \mathrm{v})t$

$= \frac{1}{2}((-\mathrm{i} + 3\mathrm{j}) + (-\mathrm{i} + 7\mathrm{j})) \times 4$

$= -4\mathrm{i} + 20\mathrm{j}$

$\mathrm{v} = -\mathrm{i} + 7\mathrm{j} = \mathrm{u} + at = (-\mathrm{i} + 3\mathrm{j}) + 4\mathrm{a}$

so $\mathrm{a} = \mathrm{j}$

c) $\mathbf{r} = \frac{1}{2}(\mathbf{u} + \mathbf{v})t$

$(t \neq 0) \Rightarrow \frac{2}{t}\mathbf{r} = \mathbf{u} + \mathbf{v}$

$\Rightarrow \mathbf{v} = \frac{2}{t}\mathbf{r} - \mathbf{u}$

We have $\mathbf{v} = \frac{2}{5}\begin{pmatrix}5\\0\end{pmatrix} - \begin{pmatrix}-1\\-1\end{pmatrix}$

$\Rightarrow \mathbf{v} = \begin{pmatrix}3\\1\end{pmatrix}$

$\mathbf{v} = \mathbf{u} + \mathbf{a}t$

$(t \neq 0) \Rightarrow \mathbf{a} = \frac{\mathbf{v} - \mathbf{u}}{t}$

We have $\mathbf{a} = \frac{1}{5}\left\{\begin{pmatrix}3\\1\end{pmatrix} - \begin{pmatrix}-1\\-1\end{pmatrix}\right\} = \frac{4}{5}\mathbf{i} + \frac{2}{5}\mathbf{j}$

d) $\mathrm{v} = \mathrm{i} - 3\mathrm{j} = \mathrm{u} + at$

$= (3\mathrm{i} - 7\mathrm{j}) + (-\mathrm{i} + 2\mathrm{j}) \times t$

$-2\mathrm{i} + 4\mathrm{j} = (-\mathrm{i} + 2\mathrm{j}) \times t$ so $t = 2$

$\mathrm{r} = \frac{1}{2}(\mathrm{u} + \mathrm{v})t$

$= \frac{1}{2}((3\mathrm{i} - 7\mathrm{j}) + (\mathrm{i} - 3\mathrm{j})) \times 2$

$= 4\mathrm{i} - 10\mathrm{j}$

2 a) $\mathrm{v} = \mathrm{u} + at = (2\mathrm{i} - 4\mathrm{j}) + (3\mathrm{i} + 5\mathrm{j}) \times 5$

$= 17\mathrm{i} + 21\mathrm{j}$

b) $\mathrm{r} = \mathrm{u}t + \frac{1}{2}at^2$

$= (2\mathrm{i} - 4\mathrm{j}) \times 6 + \frac{1}{2}(3\mathrm{i} + 5\mathrm{j}) \times 6^2$

$= 66\mathrm{i} + 66\mathrm{j}$

3 a) Resultant $\mathrm{F}_1 + \mathrm{F}_2 + \mathrm{F}_3 = -\mathrm{i} + 3\mathrm{j}$

b) $-\mathrm{i} + 3\mathrm{j} = m\mathrm{a} = 4\mathrm{a}$

so $\mathrm{a} = -0.25\mathrm{i} + 0.75\mathrm{j}$

4 $\mathrm{F}_1 + \mathrm{F}_2 = (25\mathrm{i} + 60\mathrm{j}) + \mathrm{F}_2 = m\mathrm{a}$

$= 7(4\mathrm{i} + 8\mathrm{j})$

$\mathrm{F}_2 = 7(4\mathrm{i} + 8\mathrm{j}) - (25\mathrm{i} + 60\mathrm{j}) = 3\mathrm{i} - 4\mathrm{j}$

5 a) $\mathrm{a} = \frac{d\mathrm{v}}{dt} = (3 - 6t)\mathrm{i} - 5\mathrm{j}$

b) $\mathrm{r} = \int \mathrm{v}\,dt$

$= (1.5t^2 - t^3)\mathrm{i} + (2t - 2.5t^2)\mathrm{j} + \mathrm{c}$

When $t = 0$, $\mathrm{r} = 2\mathrm{i}$ so $\mathrm{c} = 2\mathrm{i}$

So $\mathrm{r} = (1.5t^2 - t^3 + 2)\mathrm{i} + (2t - 2.5t^2)\mathrm{j}$

6 D

Resolving forces

You are the examiner

Nasreen is correct. In a) Lilia has not included the vertical component of the 80 N force. In b) She has put weight into her equation for horizontal motion.

Skill builder

1 B

2 B

3 a) Resolve up slope

$F\cos 35° - 5g\sin 35° = 0$

so $F = 34.3$ N

b) Resolve in the direction of R

$R - F\sin 35° - 5g\cos 35 = 59.8$ N

4 C

5 a) Resolve in the direction of motion

$270\cos 15° + 400\cos 10° - 150 = 600a$

Do $a = \frac{504.72}{600} = 0.841\,\text{m s}^{-2}$

b) Resolve at right angles:

$270\sin 15° - 400\sin 10° = 0.422$ N

which is very small.

Projectiles

You are the examiner

Peter is correct. Mo has used $y = 5$ instead of $y = -5$ when the ground is below the point of projection. The stone does not land twice. Mo has also mixed up the angle and the speed.

Skill builder

1 a) $x = 10$ $\quad y =$

$u_x = 16$ $\quad u_y = 0$

$v_x = 16$ $\quad v_y =$

$a = 0$ $\quad a = -g = -9.8$

$t =$ $\quad t =$

b) $x =$ $y =$

$u_x = 16\cos 45°$ $u_y = 16\sin 45°$

$v_x = 16\cos 45°$ $v_y = 0$

$a = 0$ $a = -g = -9.8$

$t =$ $t =$

c) $x =$ $y = 0$

$u_x = 3\cos 65°$ $u_y = 3\sin 65°$

$v_x = 3\cos 65°$ $v_y =$

$a = 0$ $a = -g = -9.8$

$t =$ $t =$

d) $x =$ $y = -28$

$u_x = 35\cos 25°$ $u_y = 35\sin 25°$

$v_x = 35\cos 25°$ $v_y =$

$a = 0$ $a = -g = -9.8$

$t =$ $t =$

2 a) x-direction $10 = 16t$ so $t = 0.625$ s

y-direction

$y = 0 - \frac{1}{2} \times 9.8 \times 0.625^2 = -1.91$m

b) y-direction $v = u + at$

$0 = 16\sin 45° - 9.8t$ so $t = 1.15$ s

y-direction

$y = 16\sin 45° \times 1.15 - 4.9 \times 1.15^2$

$= 6.53$m

c) y-direction $s = ut + \frac{1}{2}at$

$0 = 3\sin 65° t - 4.9t^2$ so $t = 0, 0.555$ s

x-direction

$x = 3\cos 65° \times 0.555 = 0.704$ m

d) y-direction $s = ut + \frac{1}{2}at^2$

$-28 = -35\sin 25° t - 4.9t^2$

so $t = 1.32, -4.34$

reject negative root so $t = 1.32\,s$

x-direction

$x = 1.32 \times 35\cos 25° = 41.8$ m

3 C

4 B

5 D

6 a) $x = 30\cos 60° t = 15t$

b) $y = 30\sin 60° t - 4.9t^2$

$= 15\sqrt{3}t - 4.9t^2$

c) $t = \frac{x}{15}$

d) $y = 15\sqrt{3}\left(\frac{x}{15}\right) - 4.9\left(\frac{x}{15}\right)^2$

$y = x\sqrt{3} - \frac{49x^2}{2250}$

Friction

You are the examiner

Peter is correct. Mo has not included the component of the 50 N force in his normal reaction. Mo has also assumed that friction takes its maximum value – the box will not move to the left if he pushes to the right.

Skill builder

1 a) $N = 5g$, $F_{max} = 0.4 \times 5g = 19.6$ N

b) $N = 5g\cos 5° = 48.8$ N,

$F_{max} = 0.4 \times 5g\cos 15° = 19.5$ N

c) $N = (5g - 10\sin 25°) = 44.8$ N,

$F_{max} = 0.4 \times 44.8 = 17.9$ N

d) $N = (5g + 10\sin 25°) = 53.2$ N,

$F_{max} = 0.4 \times 53.2 = 21.3$ N

2 B

3 a)

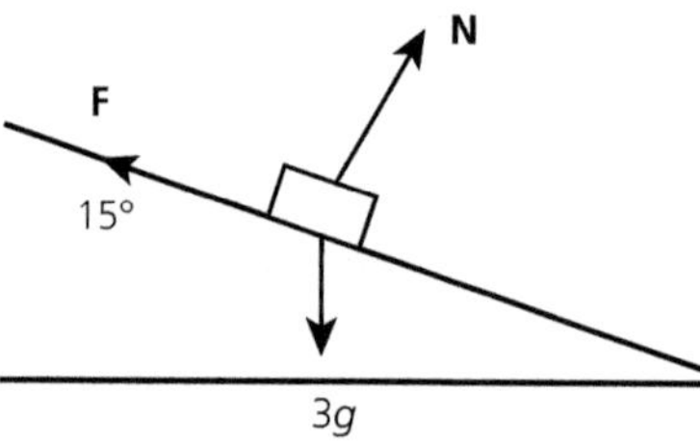

$N = 3g\cos 15° = 28.4$ so

$F \leq 0.3 \times 28.4 = 8.5$ N

Not to slide

$F = 3g\sin 15° = 7.61 < 8.5$ so the box does not move.

b) $N = 3g\cos 25° = 26.6$ so

$F = 0.3 \times 26.6 = 7.99$ N

Resolve down the slope

$3g\sin 25° - F = 3a$ so $a = 1.48$ ms^{-2}

4 C

5 A

6 a)

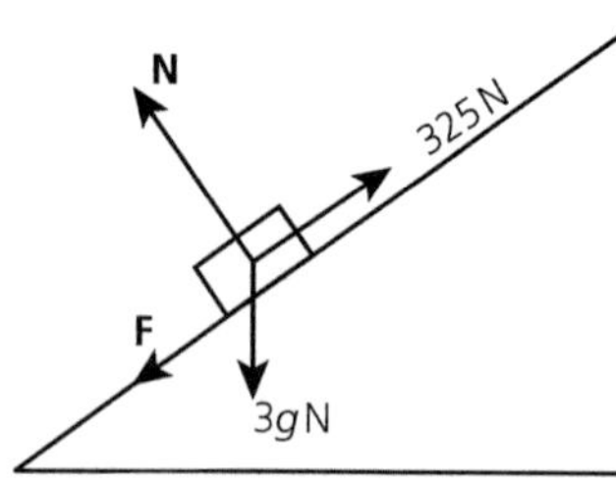

$N = 40g\cos 30° = 339$ N

b) $F = 0.2 \times 339.48 = 67.9$ N down the slope (opposing motion)

c) N2L up the slope

$325 - 67.9 - 40g\sin 30° = 40a$

so $a = 1.60$ ms^{-2}

Moments

You are the examiner

Lilia is right. Nasreen has the sign of the moment of Billy's weight incorrect.

Skill builder

1 D

2 R 5 cm 10 cm 7 cm R

0.05gN 0.035gN 0.05gN

x cm

$2R = 0.05g + 0.035g + 0.05g$

So $R = 0.6615$ N

$0.05g \times 5 + 0.035g \times 15 + 0.05g \times 22$

$= 0.6615x$

$x = 27.8$ cm

3 A

4 B

5

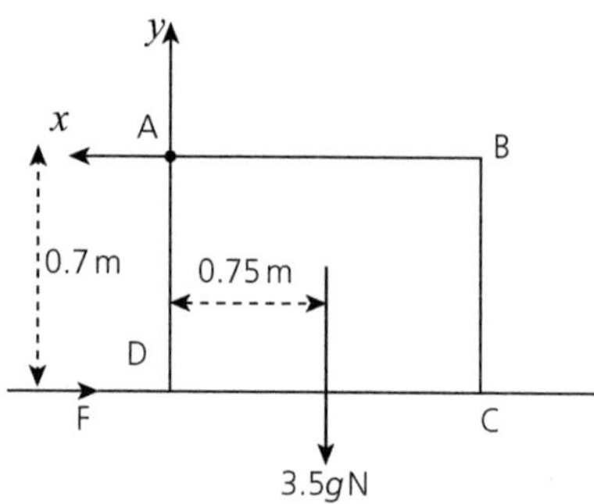

a) Moments about A

$F \times 0.7 - 3.5g \times 0.75 = 0$

so $F = 36.8$ N

b) $X = F = 36.8$ N and

$Y = 3.5g = 34.3$ N

c) Magnitude

$\sqrt{36.75^2 + 34.3^2} = 50.3$ N